Springer Optimization and Its Applications

Volume 196

Aims and Scope

Optimization has continued to expand in all directions at an astonishing rate. New algorithmic and theoretical techniques are continually developing and the diffusion into other disciplines is proceeding at a rapid pace, with a spot light on machine learning, artificial intelligence, and quantum computing. Our knowledge of all aspects of the field has grown even more profound. At the same time, one of the most striking trends in optimization is the constantly increasing emphasis on the interdisciplinary nature of the field. Optimization has been a basic tool in areas not limited to applied mathematics, engineering, medicine, economics, computer science, operations research, and other sciences.

The series **Springer Optimization and Its Applications (SOIA)** aims to publish state-of-the-art expository works (monographs, contributed volumes, textbooks, handbooks) that focus on theory, methods, and applications of optimization. Topics covered include, but are not limited to, nonlinear optimization, combinatorial optimization, continuous optimization, stochastic optimization, Bayesian optimization, optimal control, discrete optimization, multi-objective optimization, and more. New to the series portfolio include Works at the intersection of optimization and machine learning, artificial intelligence, and quantum computing.

Volumes from this series are indexed by Web of Science, zbMATH, Mathematical Reviews, and SCOPUS.

Ding-Zhu Du • Panos M. Pardalos • Xiaodong Hu •
Weili Wu

Introduction to Combinatorial Optimization

Ding-Zhu Du
Department of Computer Science
University of Texas, Dallas
Richardson, TX, USA

Panos M. Pardalos (iD)
Department of Industrial & Systems
Engineering
University of Florida
Gainesville, FL, USA

Xiaodong Hu
University of Chinese Academy of Sciences
Academy of Math and System Science
Chinese Academy of Sciences
Beijing, China

Weili Wu
Department of Computer Science
University of Texas at Dallas
Richardson, TX, USA

ISSN 1931-6828   ISSN 1931-6836   (electronic)
Springer Optimization and Its Applications
ISBN 978-3-031-11684-1   ISBN 978-3-031-10596-8   (eBook)
https://doi.org/10.1007/978-3-031-10596-8

This Springer imprint is published by the registered company Springer Nature Switzerland AG
The registered company address is: Gewerbestrasse 11, 6330 Cham, Switzerland

Since the fabric of the world is the most perfect and was established by the wisest Creator, nothing happens in this world in which some reason of maximum or minimum would not come to light.

—Euler

When you say it, it's marketing. When they say it, it's social proof.

—Andy Crestodina

God used beautiful mathematics in creating the world.

—Paul Dirac

Preface

The motivation for writing this book came from our previous teaching experience in the undergraduate, junior graduate, and senior graduate levels.

The first observation is about organization of courses on combinatorial optimization. Many textbooks are problem oriented. But our experience indicates that a methodology-oriented organization is preferred by students.

The second observation is about contents. At present, technological developments, such as wireless communication, cloud computing, social networks, and machine learning, involve many applications of combinatorial optimization and provide a platform for their new issues, new techniques, and new subareas to grow. This makes us update teaching materials.

This book is methodology oriented and organized along a line leading the reader step by step from the very beginning toward frontier of this field. Actually, all materials are selected from lecture notes from three courses, which are taught at undergraduate level, junior graduate (MS) level, and senior graduate (PhD) level, respectively.

The first part is selected from a course on computer algorithm design and analysis. This course does not clearly state that it is about combinatorial optimization. However, its contents have a very large portion overlapping with combinatorial optimization.

The second part comes from a course on the design and analysis of approximation algorithms. The third part comes from a course on nonlinear combinatorial optimization. These two parts are overlapping at a few chapters. Therefore, we combined and simplified them.

While all three parts have been used for teaching at the University of Texas at Dallas and the University of Florida for many years, the second and the third parts are also utilized for teaching in short summer courses at the University of Chinese Academy of Sciences, Beijing University, Tsinghua University, Beijing Jiaotong University, Xi'an Jiaotong University, Ocean University of China, Beijing University of Technology, Lanzhou University, Zhejiang Normal University, Shandong University, Harbin Institute of Technology, CityU of Hong Kong, and PolyU of Hong Kong. Therefore, we wish to thank Professors Andy Yao, Francis

Yao, Jianzhong Li, Hong Gao, Xiaohua Jia, Jiannong Cao, Qizhi Fang, Jianliang Wu, Naihua Xiu, Lingchen Kong, Dachuan Xu, Suixiang Gao, Wenguo Yang, Zhao Zhang, Xujin Chen, Xianyue Li, and Hejiao Huang for their support.

Finally, we would like to acknowledge the support in part by NSF of USA under grants 1747818, 1822985, and 1907472.

Richardson, TX, USA	Ding-Zhu Du
Gainesville, FL, USA	Panos M. Pardalos
Beijing, China	Xiaodong Hu
Richardson, TX, USA	Weili Wu
January 2022	

Contents

Chapter 1
Introduction

True optimization is the revolutionary contribution of modern research to decision processes.

—George Dantzig

Let us start this textbook from a fundamental question and tell you what will constitute this book.

1.1 What Is Combinatorial Optimization?

The aim of combinatorial optimization is to find an optimal object from a finite set of objects. Those candidate objects are called *feasible solutions*, while the optimal one is called an *optimal solution*. For example, consider the following problem.

Problem 1.1.1 (Minimum Spanning Tree) *Given a connected graph $G = (V, E)$ with nonnegative edge weight $c : E \to R_+$, find a spanning tree with minimum total weight, where "spanning" means that all nodes are included in the tree and hence a spanning tree interconnects all nodes in V. An example is as shown in Fig. 1.1.*

Clearly, the set of all spanning trees is finite, and the aim of this problem is to find one with minimum total weight from this set. Each spanning tree is a feasible solution, and the optimal solution is the spanning tree with minimum total weight, which is also called the *minimum spanning tree*. Therefore, this is a combinatorial optimization problem.

A combinatorial optimization problem may have more than one optimal solution. For example, in Fig. 1.1, there are two spanning trees with minimum total length. (The second one can be obtained by using edge (e, f) to replace edge (d, f).) Therefore, by the optimal solution as mentioned in the above, it means a general member in the class of optimal solutions.

The combinatorial optimization is a proper subfield of discrete optimization. In fact, there exist some problems in discrete optimization, which do not belong

© The Author(s), under exclusive license to Springer Nature Switzerland AG 2022
D.-Z. Du et al., *Introduction to Combinatorial Optimization*, Springer Optimization and Its Applications 196, https://doi.org/10.1007/978-3-031-10596-8_1

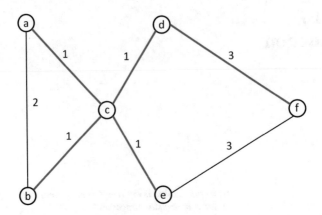

Fig. 1.1 A example for the minimum spanning tree

to combinatorial optimization. For example, consider the integer programming. It always belongs to discrete optimization. However, when feasible domain is infinite, it does not belong to combinatorial optimization. But such a difference is not recognized very well in the literature. Actually, if a paper on lattice-point optimization is submitted to *Journal of Combinatorial Optimization*, then usually, it will not be rejected due to out of scope.

In view of methodologies, combinatorial optimization and discrete optimization have very close relationship. For example, to prove NP-hardness of integer programming, we need to cut its infinitely large feasible domain into a finite subset containing optimal solution (see Chap. 8 for details), i.e., transform it into a combinatorial optimization problem.

Geometric optimization is another example. Consider the following problem.

Problem 1.1.2 (Minimum Length Guillotine Partition) *Given a rectangle with point-holes inside, partition it into smaller rectangles without hole inside by a sequence of guillotine cuts to minimize the total length of cuts. Here, the guillotine cut is a vertical or horizontal straight line segment which partitions a rectangle into two smaller rectangles. An example is as shown in Fig. 1.2.*

There exist infinitely many numbers of partitions. Therefore, it is not a combinatorial optimization problem. However, we can prove that optimal partition can be found from a finite set of partitions of a special type (for detail, see Chap. 3). Therefore, to solve the problem, we need only to study partitions of this special type, i.e., a combinatorial optimization problem.

Due to above, we do not make a clear cut to exclude other parts of discrete optimization. Actually, this book is methodology oriented. Problems are selected to illustrate methodology. Especially, for each method, we may select a typical problem as companion to explore the method, such as its requirements and applications. For example, we use the shortest path problem to explain dynamic

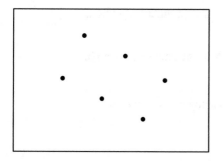 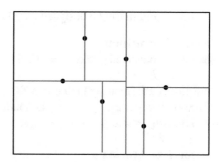

Fig. 1.2 Rectangular guillotine partition

programming, employ the minimum spanning tree problem to illustrate greedy algorithms, etc.

1.2 Optimal and Approximation Solutions

Let us show an optimality condition for the minimum spanning tree.

Theorem 1.2.1 (Path Optimality) *A spanning tree T^* is a minimum spanning tree if and only if it satisfies the following condition:*

Path Optimality Condition *For every edge (u, v) not in T^*, there exists a path p in T^*, connecting u and v, and moreover, $c(u, v) \geq c(x, y)$ for every edge (x, y) in path p.*

Proof Suppose, for contradiction, that $c(u, v) < c(x, y)$ for some edge (x, y) in the path p. Then $T' = (T^* \setminus (x, y)) \cup (u, v)$ is a spanning tree with cost less than $c(T^*)$, contradicting the minimality of T^*.

Conversely, suppose that T^* satisfies the path optimality condition. Let T' be a minimum spanning tree such that among all minimum spanning tree, T' is the one with the most edges in common with T^*. Suppose, for contradiction, that $T' \neq T^*$. We claim that there exists an edge $(u, v) \in T^*$ such that the path in T' between u and v contains an edge (x, y) with length $c(x, y) \geq c(u, v)$. If this claim is true, then $(T' \setminus (x, y)) \cup (u, v)$ is still a minimum spanning tree, contradicting the definition of T'.

Now, we show the claim by contradiction. Suppose the claim is not true. Consider an edge $(u_1, v_1) \in T^* \setminus T'$. the path in T' connecting u_1 and v_1 must contain an edge (x_1, y_1) not in T^*. Since the claim is not true, we have $c(u_1, v_1) < c(x_1, y_1)$. Next, consider the path in T^* connecting x_1 and y_1, which must contain an edge $(u_2, v_2) \notin T'$. Since T^* satisfies the path optimality condition, we have $c(x_1, y_1) \leq c(u_2, v_2)$. Hence, $c(u_1, v_1) < c(u_2, v_2)$. As this argument continues, we will find a sequence of edges in T^* such that $c(u_1, v_2) < c(u_2, v_2) < c(u_3, v_3) < \cdots$, contradicting the finiteness of T^*. $\qquad\square$

An algorithm can be designed based on path optimality condition.

Kruskal Algorithm
input: A connected graph $G = (V, E)$ with nonnegative edge weight
$\quad c : E \to R_+$.
output: A minimum spanning tree T.
$\quad$ Sort all edges $e_1, e_2, \ldots, e_m$ in nondecreasing order of weight,
$\quad$ i.e., $c(e_1) \le c(e_2) \le \cdots \le c(e_m)$;
$\quad T \leftarrow \emptyset$;
$\quad$ **for** $i \leftarrow 1$ **to** m **do**
$\quad\quad$ **if** $T \cup e_i$ does not contain a cycle
$\quad\quad\quad$ **then** $T \leftarrow T \cup e_i$;
$\quad$ **return** T.

From this algorithm, we see that it is not hard to find the optimal solution for the minimum spanning tree problem. If every combinatorial optimization problem likes the minimum spanning tree, then we would be very happy to find optimal solution for it. Unfortunately, there exist a large number of problems that it is unlikely to be able to compute their optimal solution efficiently. For example, consider the following problem.

Problem 1.2.2 (Minimum Length Rectangular Partition) *Given a rectangle with point-holes inside, partition it into smaller rectangles without hole to minimize the total length of cuts.*

Problems 1.1.2 and 1.2.2 are quite different. Problem 1.2.2 is intractable, while there exists an efficient algorithm to compute an optimal solution for Problem 1.1.2. Actually, in theory of combinatorial optimization, we need to study not only how to design and analyze algorithms for finding optimal solutions but also how to design and analyze algorithms for computing approximation solutions. When do we put our efforts on optimal solution? When should we pay attention to approximation solutions? Ability for making such a judgement has to be growth from study computational complexity.

The book consists of three building blocks, design and analysis of computer algorithm for exact optimal solution, design and analysis of approximation algorithms, and nonlinear combinatorial optimization.

The first block contains Chaps. 2–7, which can be divided into two parts (Fig. 1.3). The first part is on algorithms with self-reducibility, including the divide-and-conquer, the dynamic program, the greedy algorithm, the local search, the local ratio, etc., which are organized into Chaps. 2–4. The second part is on incremental method, including the primal algorithm, the dual algorithm, and the primal-dual algorithm, which are organized also into Chaps. 5–7. There is an intersection between algorithms with self-reducibility and primal-dual algorithms. In fact, in computation process of the former, an optimal feasible solution is built up step by step based on certain techniques, and the latter also has a process to build up an optimal primal solution by using information from dual side. Therefore, some

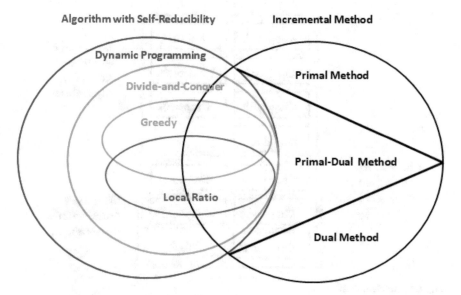

Fig. 1.3 Design and analysis of computer algorithms

algorithm can be illustrated as an algorithm with self-reducibility, and meanwhile it can also be explained as a prima-dual algorithm.

The second block contains Chaps. 8–11, covering the fundamental knowledge on computational complexity, including theory on NP-hardness and inapproximability, and basic techniques for design of approximation, including the restriction, the greedy approximation, and the relaxation with rounding.

The third block contains Chaps. 10, 11, and 12. Since Chaps. 10–11 serve both the second and the third blocks, selected examples are mainly coming from the submodular optimization. Then, Chaps. 12 is contributed to the nonsubmodular optimization. Nonsubmodular optimization is an active research area currently. There are a lot of recent publications in the literature. Probably, Chap. 12 can be seen as an introduction to this area. For a complete coverage, we may need a new book.

Now, we put above structure of this book into Fig. 1.4 for a clear overview.

1.3 Preprocessing

In Kruskal algorithm, the first line is to sort all edges into a nondecreasing order of cost. This requires a preprocessing procedure for solving the sorting problem as follows.

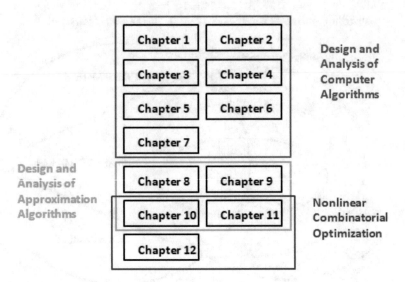

Fig. 1.4 Structure of this book

Problem 1.3.1 (Sorting) *Given a sequence of positive integers, sort them into nondecreasing order.*

The following is a simple algorithm to do sorting job.

Insertion Sort
input: An array A with a sequence of positive integers.
output: An array A with a sequence of positive integers in
           nondecreasing order.
    for $j \leftarrow 2$ **to** length[A]
        do $key \leftarrow A[j]$
          $i \leftarrow j - 1$
          while $i > 0$ and $A[i] > key$
             do $A[i + 1] \leftarrow A[i]$
               $i \leftarrow i - 1$
        end-while
        $A[i + 1] \leftarrow key$
  end-for
  return A.

An example for using insertion sort is as shown in Fig. 1.5.

Although insertion sort is simpler, it runs a little slow. Since sorting appears very often in algorithm design for combinatorial optimization problems, we have to spend some space in Chap. 2 to introduce faster algorithms.

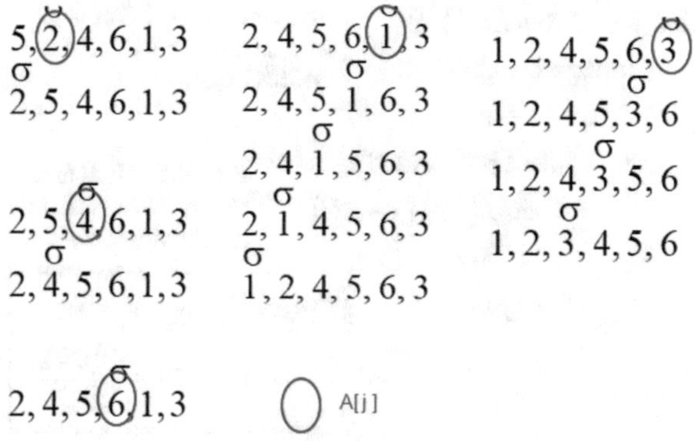

$$5,\textcircled{2}4,6,1,3 \quad 2,4,5,6,\textcircled{1}3 \quad 1,2,4,5,6,\textcircled{3}$$
$$\sigma \qquad\qquad\qquad \sigma \qquad\qquad\qquad \sigma$$
$$2,5,4,6,1,3 \quad 2,4,5,1,6,3 \quad 1,2,4,5,3,6$$
$$\qquad\qquad\qquad\quad \sigma \qquad\qquad\qquad \sigma$$
$$\qquad\qquad\qquad 2,4,1,5,6,3 \quad 1,2,4,3,5,6$$
$$\qquad\qquad\qquad\quad \sigma \qquad\qquad\qquad \sigma$$
$$2,5,\textcircled{4}6,1,3 \quad 2,1,4,5,6,3 \quad 1,2,3,4,5,6$$
$$\sigma \qquad\qquad\quad \sigma$$
$$2,4,5,6,1,3 \quad 1,2,4,5,6,3$$

$$2,4,5,\textcircled{6}1,3 \qquad \bigcirc \; A[j]$$

Fig. 1.5 An example for insertion sort. σ is the *key* lying outside of array A

1.4   Running Time

The most important measure of quality for algorithms is the running time. However, for the same algorithm, it may take different times when we run it in different computers. To give a uniform standard, we have to get an agreement that runs algorithms in a theoretical computer model. This model is the multi-tape Turing machine which has been accepted by a very large population. Based on the Turing machine, the theory of computational complexity has been built up. We will touch this part of theory in Chap. 8.

But, we will use RAM model to evaluate the running time for algorithms throughout this book except Chap. 8. In RAM model, assume that each line of pseudocode requires a constant time. For example, the running time of insertion sort is calculated in Fig. 1.6.

Actually, RAM model and Turing machine model are closely related. The running time estimated based on these two models is considered to be close enough. However, they are sometimes different in estimation of running time. For example, the following is a piece of pseudocode.

for $i = 1$ to n
    do assign First$(i) \leftarrow i$
end-for

According to RAM model, the running time of this piece is $O(n)$. However, based on the Turing machine, the running time of this piece is $O(n \log n)$ because the assigned value has to be represented by a string with $O(\log n)$ symbols.

Theoretically, a constant factor is often ignored. For example, we usually say that the running time of insertion sort is $O(n^2)$ instead of giving the specific quadratic function with respect to n. Here $f(n) = O(g(n))$ means that there exist constants $c > 0$ and $n_0 > 0$ such that

$$f(n) \leq c \cdot g(n) \text{ for } n \geq n_0$$

for $j \leftarrow 2$ to $length[A]$ This loop runs n-1 times
   do $key \leftarrow A[j]$ and each time runs at most
                         4+3(j-1) lines.
   $i \leftarrow j-1$
   while $i > 0$ and $A[i] > key$ This loop runs at most
     do $A[i+1] \leftarrow A[i]$ j-1 times and each time
                        runs at most 3 lines.
     $i \leftarrow i-1$
   $A[i+1] \leftarrow key$

$$T(n) \le \sum_{j=2}^{n} (4 + 3(j-1))$$

$$= n - 1 + 3\frac{(n-1)(n+2)}{2}$$

Fig. 1.6 Running time calculation

There are two more notations which appear very often in representation of running time. $f(n) = \Omega(g(n))$ means that there exist constant $c > 0$ and $n_0 > 0$ such that

$$0 \le c \cdot g(n) \le f(n) \text{ for } n \ge n_0.$$

$f(n) = \Theta(g(n))$ means that there exist constants $c_1 > 0$, $c_2 > 0$ and $n_0 > 0$ such that

$$c_1 \cdot g(n) \le f(n) \le c_2 \cdot g(n) \text{ for } n \ge n_0.$$

Finally, let us make a remark on input numbers.

In the minimum spanning tree problem, every edge has a nonnegative weight which is an input number. In the problem definition, we assumed that this is a real number. However, a real number cannot be exactly input in a computer. Actually, the computation of a real number is sometimes a very hard problem in the theory of computational complexity [260]. Therefore, we have to treat each real number with an oracle which can provide a rational number with expected accuracy, without computation cost. In other words, we ignore the computation trouble of the real number.

However, when our analysis of running time has to consider the size of input numbers, e.g., in analysis of weakly polynomial-time algorithms, we have to change the setting from the real number to the integer.

1.5   Data Structure

A data structure is a data storage format which is organized and managed to have efficient access and modification. Each data structure has several standard operations. They are building bricks to construct algorithms. The data structure plays an important role in improving efficiency of algorithms. For example, we may introduce a data structure "Disjoint Sets" to improve Kruskal algorithm.

Consider a collection of disjoint sets. For each set S, let First(S) denote the first node in set S. For each element x in set S, denote First(x) = First(S). Define three operations as follows:

Make-Set(x) creates a new set containing only x.
Union(x, y) unions sets S_x and S_y containing x and y, respectively, into $S_x \cup S_y$,
   Moreover, set

$$\text{First}(S_x \cup S_y) = \begin{cases} \text{First}(S_x) & \text{if } |S_x| \geq |S_y|, \\ \text{First}(S_y) & \text{otherwise.} \end{cases}$$

Find-Set(x) returns First(S_x) where S_x is the set containing element x.

With this data structure, Kruskal algorithm can be modified as follows.

Kruskal Algorithm
input: A connected graph $G = (V, E)$ with nonnegative edge weight $c : E \to R_+$.
output: A minimum spanning tree T.
   Sort all edges $e_1, e_2, \ldots, e_m$ in nondecreasing order of weight,
   i.e., $c(e_1) \leq c(e_2) \leq \cdots \leq c(e_m)$;
   $T \leftarrow \emptyset$;
   for each node $v \in V$ **do**
         Make-Set(v);
   end-for
   for $i \leftarrow 1$ **to** m **do**
         if Find-Set(x) $\neq$ Find-Set(y) where $e_i = (x, y)$
               then $T \leftarrow T \cup e_i$
                     and Union(x, y);
   end-for
   return T.

An example for running this algorithm is as shown in Fig. 1.7.
Denote $m = |E|$ and $n = |V|$. Let us estimate the running time of Kruskal algorithm.

- Sorting on all edges takes $O(m \log n)$ time.
- Assigning First(v) for all $v \in V$ takes $O(n)$ time.
- For each node v, the value of First(v) can be changed at most $O(\log n)$ time. This is because the value of First(v) is changed only if v is included in Union operation, and after the operation, the set containing v has size doubled.
- Thus, the second "for-loop" takes $O(n \log n)$ time.
- Put total together, the running time is $O(m \log n) = O(m \log n + n \log n)$.

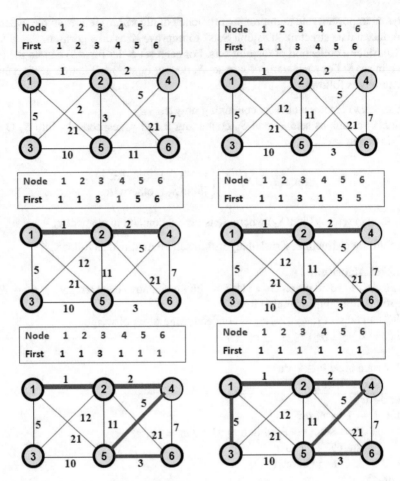

Fig. 1.7 An example for the Kruskal algorithm

Exercises

1. In a city there are N houses, each of which is in need of a water supply. It costs W_i dollars to build a well at house i, and it costs C_{ij} to build a pipe in between houses i and j. A house can receive water if either there is a well built there or there is some path of pipes to a house with a well. Give an algorithm to find the minimum amount of money needed to supply every house with water.
2. Consider a connected graph G with all distinct edge weights. Show that the minimum spanning tree of G is unique.
3. Consider a connected graph $G = (V, E)$ with nonnegative edge weight $c :$ $E \to R_+$. Suppose $e_1^*, e_2^*, \ldots, e_k^*$ are edges generated by Kruskal algorithm, and $e_1, e_2, \ldots, e_k$ are edges of a spanning tree in ordering $c(e_1) \le c(e_2) \le \cdots \le c(e_k)$. Show that $c(e_i^*) \le c(e_i)$ for all $1 \le i \le k$.

4. Let V be a fixed set of n vertices. Consider a sequence of m undirected edges $e_1, e_2, \ldots, e_m$. For $1 \le i \le m$, let G_i denote the graph with vertex set V and edge set $E_i = \{e_1, \ldots, e_i\}$. Let c_i denote the number of connected components of G_i. Design an algorithm to compute c_i for all i. Your algorithm should be asymptotically as fast as possible. What is the running time of your algorithm?

5. There are n points lying in the Euclidean plane. Show that there exists a minimum spanning tree on these n points such that every node (i.e., point) has degree at most five.

6. Can you modify Kruskal algorithm to compute a maximum weight spanning tree?

7. Consider a connected graph $G = (V, E)$ with edge weight $c : E \to R$, i.e., the weight is possibly negative. Does Kruskal algorithm work for computing a minimum weight spanning tree?

8. Consider a connected graph $G = (V, E)$ with nonnegative edge weight $c : E \to R_+$. Suppose edge e is unique longest edge in a cycle. Show that e cannot be included in any minimum spanning tree.

9. Consider a connected graph $G = (V, E)$ with nonnegative edge weight $c : E \to R_+$. While a cycle exists, delete the longest edge from the cycle. Show that this computation ends at a minimum spanning tree.

10. Consider a game with two players on an undirected given graph G. Two players take turn alternatively. The first player removes an edge of his/her choice. Then the second player adds an edge of her/his choice. The rule is that no removed edge can be added back, and no added edge can be removed. The first player wins if he/she manages to disconnect the graph. The second player wins if she/he manages to have a spanning tree in the graph. Prove that the second player has a winning strategy if and only if G contains two edge-disjoint spanning trees.

Historical Notes

There are many books which have been written for combinatorial optimization [72, 105, 264, 272, 275, 335, 358, 367, 368]. There are also many books published in design and analysis of computer algorithms [73, 280], which cover a large portion on combinatorial optimization problems. However, those books mainly on computing exact optimal solutions and possibly a small part on approximation solutions. For approximation solutions, a large part of materials are usually covered in separated books [100, 387, 408]. For issues on computational complexity, the reader may refer to [99, 260].

In recent developments of technology, combinatorial optimization gets a lot of new applications and new research directions [337, 338, 422, 425]. In this book, we try to meet requests from various areas for teaching, research, and reference, to put together three components, the classic part of combinatorial optimization, approximation theory developed in recent years, and newly appeared nonlinear combinatorial optimization.

Chapter 2
Divide-and-Conquer

> *Defeat Them in Detail: The Divide and Conquer Strategy. Look at the parts and determine how to control the individual parts, create dissension and leverage it.*
>
> —Robert Greene

The divide-and-conquer is an important technique for design of algorithms. In this chapter, we will employ several examples to introduce this technique, including the rectilinear minimum spanning tree, the Fibonacci search method, and the sorting problem. Sorting is not a combinatorial optimization problem. However, it appears in algorithms very often as a procedure, especially in algorithms for solving combinatorial optimization problems. Therefore, we would like to make more discussion in this chapter.

2.1 Algorithms with Self-Reducibility

There exist a large number of algorithms in which the problem is reduced to several subproblems, each of which is the same problem on a smaller-size input. Such a problem is said to have the self-reducibility, and the algorithm is said to be with self-reducibility.

For example, consider sorting problem again. Suppose input contains n numbers. We may divide these n numbers into two subproblems. One subproblem is the sorting problem on $\lfloor n/2 \rfloor$ numbers, and the other subproblem is the sorting problem on $\lceil n/2 \rceil$ numbers. After completely sorting each subproblem, combine two sorted sequences into one. This idea will result in a sorting algorithm, called the *merge sort*. The pseudocode of this algorithm is shown in Algorithm 1.

The main body calls a procedure. This procedure contains two self-calls, which means that the merge sort is a recursive algorithm, that is, the divide will continue until each subproblem has input of single number. Then this procedure employs another procedure (Merge) to combine solutions of subproblems with smaller

© The Author(s), under exclusive license to Springer Nature Switzerland AG 2022
D.-Z. Du et al., *Introduction to Combinatorial Optimization*, Springer Optimization and Its Applications 196, https://doi.org/10.1007/978-3-031-10596-8_2

Algorithm 1 Merge sort

Input: n numbers $a_1, a_2, \ldots, a_n$ in array $A[1 \ldots n]$.
Output: n numbers $a_{i_1} \le a_{i_2} \le \cdots \le a_{i_n}$ in array A.

1: Sort$(A, 1, n)$
2: **return** $A[1 \ldots n]$

Procedure Sort(A, p, r).
% Sort $r - p + 1$ numbers $a_p, a_{p+1}, \ldots, a_r$ in array $A[p \ldots r]$. %

1: **if** $p < r$ **then**
2:     $q \leftarrow \lfloor (p + r)/2 \rfloor$
3:     Sort(A, p, q)
4:     Sort$(A, q + 1, r)$
5:     Merge(A, p, q, r)
6: **end if**
7: **return** $A[p \ldots r]$

Procedure Merge(A, p, q, r).
% Merge sorted two arrays $A[p \ldots q]$ and $A[p + 1 \ldots r]$ into one. %

1: **for** $i \leftarrow 1$ to $q - p + 1$ **do**
2:     $B[i] \leftarrow A[p + i - 1]$
3: **end for**
4: $i \leftarrow 1$
5: $j \leftarrow p + 1$
6: $B[q - p + 2] \leftarrow +\infty$
7: $A[r + 1] \leftarrow +\infty$
8: **for** $k \leftarrow p$ to r **do**
9:     **if** $B[i] \le A[j]$ **then**
10:         $A[k] \leftarrow B[i]$
11:         $i \leftarrow i + 1$
12:     **else**
13:         $A[k] \leftarrow A[j]$
14:         $j \leftarrow j + 1$
15:     **end if**
16: **end for**
17: **return** $A[p \ldots r]$

---

inputs into subproblems with larger inputs. This computation process on input $\{5, 2, 7, 4, 6, 8, 1, 3\}$ is shown in Fig. 2.1.

Note that the running time of procedure Merge at each level is $O(n)$. Let $t(n)$ be the running time of merge sort on input of size n. By the recursive structure, we can obtain that $t(1) = 0$ and

$$t(n) = t(\lfloor n/2 \rfloor) + t(\lceil n/2 \rceil) + O(n).$$

Suppose

$$t(n) \le 2 \cdot t(\lceil n/2 \rceil) + c \cdot n$$

for some positive constant c. Define $T(1) = 0$ and

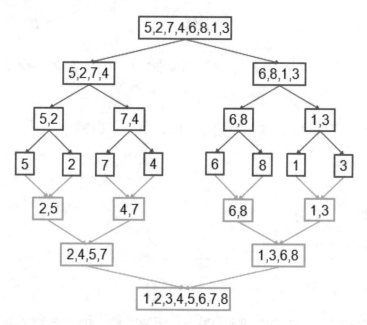

Fig. 2.1 Computation process of merge sort

$$T(n) = 2 \cdot T(\lceil n/2 \rceil) + c \cdot n.$$

By induction, we can prove that

$$t(n) \leq T(n) \text{ for all } n \geq 1.$$

For base step, $t(1) = 0 = T(1)$. For induction step,

$$
\begin{aligned}
t(n) &\leq 2 \cdot t(\lceil n/2 \rceil) + c \cdot n \\
&\leq 2 \cdot T(\lceil n/2 \rceil) + c \cdot n \quad \text{(by induction hypothesis)} \\
&= T(n).
\end{aligned}
$$

Now, let us discuss how to solve recursive equation about $T(n)$. Usually, we use two stages. In the first stage, we consider special numbers $n = 2^k$ and employ the recursive tree to find $T(2^k)$ (Fig. 2.2), that is,

$$
\begin{aligned}
T(2^k) &= 2 \cdot T(2^{k-1}) + c \cdot 2^k \\
&= 2 \cdot (2 \cdot T(2^{k-2}) + c \cdot 2^{k-1}) + c \cdot 2^k \\
&= \ldots \\
&= 2^k T(1) + kc \cdot 2^k
\end{aligned}
$$

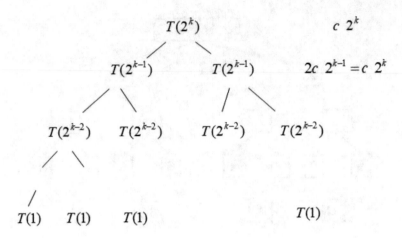

Fig. 2.2 Recursive tree

$$= c \cdot k2^k.$$

In general, we may guess that $T(n) \leq c' \cdot n \log n$ for some constant $c' > 0$. Let us show it by mathematical induction.

First, we choose c' to satisfy $T(n) \leq c'$ for $n \leq n_0$ where n_0 will be determined later. This choice will make $T(n) \leq c' n \log n$ for $n \leq n_0$, which meets the requirement for the basic step of mathematical induction.

For induction step, consider $n \geq n_0 + 1$. Then we have

$$
\begin{aligned}
T(n) &= 2 \cdot T(\lceil n/2 \rceil) + c \cdot n \\
&\leq 2 \cdot c' \lceil n/2 \rceil \log \lceil n/2 \rceil + c \cdot n \\
&\leq 2 \cdot c'((n+1)/2)(\log(n+1) - 1) + c \cdot n \\
&= c' \cdot (n+1) \log(n+1) - c'(n+1) + c \cdot n \\
&\leq c'(n+1)(\log n + 1/n) - (c' - c)n - c' \\
&= c' n \log n + c' \log n - (c' - c)n + c'/n.
\end{aligned}
$$

Now, we choose n_0 sufficiently large such that $n/2 > \log n + 1/n$ and $c' > \max(2c, T(1), \ldots, T(n_0))$. Then the above mathematical induction proof will be passed. Therefore, we obtained the following.

Theorem 2.1.1 *Merge sort runs in* $O(n \log n)$ *time.*

By the mathematical induction, we can also prove the following result.

Theorem 2.1.2 *Let* $T(n) = aT(n/b) + f(n)$ *where constants* $a > 1$, $b > 1$, *and* n/b *mean* $\lceil n/b \rceil$ *or* $\lfloor n/b \rfloor$. *Then we have the following:*

1. If $f(n) = O(n^{\log_b a - \varepsilon})$ for some positive constant ε, then $T(n) = \Theta(n^{\log_b a})$.
2. If $f(n) = \Theta(n^{\log_b a})$, then $T(n) = \Theta(n^{\log_b a} \log n)$.
3. If $f(n) = \Omega(n^{\log_b a + \varepsilon})$ for some positive constant ε and moreover, $af(n/b) \leq cf(n)$ for sufficiently large n and some constant $c < 1$, then $T(n) = \Theta(f(n))$.

In Fig. 2.1, we see a tree structure between problem and subproblems. In general, for any algorithm with self-reducibility, its computational process will produce a set of subproblems on which we can also construct a graph to describe relationship between them by adding an edge from subproblem A to subproblem B if at an iteration, subproblem A is reduced to several problems, including subproblem B. This graph is called the *self-reducibility structure* of the algorithm.

All algorithms with tree self-reducibility structure form a class, called *divide-and-conquer*, that is, **an algorithm is in class of divide-and-conquer if and only if its self-reducibility structure is a tree.** Thus, the merge sort is a divide-and-conquer algorithm.

In a divide-and-conquer algorithm, it is not necessary to divide a problem evenly or almost evenly. For example, we consider another sorting algorithm, called *Quick Sort*. The idea is as follows.

In merge sort, the procedure Merge takes $O(n)$ time, which is the main consumption of time. However, if $A[i] \leq A[q]$ for $p \leq i < q$ and $A[q] \leq A[j]$ for $q < j \leq r$, then this procedure can be skipped, and after sort $A[p \ldots q - 1]$ and $A[q + 1 \ldots r]$, we can simply put them together to obtain sorted $A[p \ldots r]$.

In order to have above property satisfied, Quick Sort uses $A[r]$ to select all elements $A[p \ldots r - 1]$ into two subsequences such that one contains elements less than $A[r]$ and the other one contains elements at least $A[r]$. A pseudocode of quick sort is as shown in Algorithm 2.

The division is not balanced in Quick Sort. In the worst case, one part contains nothing, and the other contains $r - p$ elements. This will result in running time $O(n^2)$. However, Quick Sort has expected running time $O(n \log n)$. To see it, let $T(n)$ denote the running time for n numbers. Note that the procedure Partition runs in linear time. Then, we have

$$
\begin{aligned}
E[T(n)] \leq \frac{1}{n}(& \quad E[T(n-1)] + c_1 n) \\
& + \frac{1}{n}(E[T(1)] + E[T(n-2)] + c_1 n) \\
& + \cdots \\
& + \frac{1}{n}(E[T(n-1)] \quad\quad + c_1 n) \\
= c_1 n & + \frac{2}{n} \sum_{i=1}^{n-1} E[T(i)].
\end{aligned}
$$

Algorithm 2 Quick sort

Input: n numbers $a_1, a_2, \ldots, a_n$ in array $A[1 \ldots n]$.
Output: sorted numbers $a_{i_1} \leq a_{i_2} \leq \cdots \leq a_{i_n}$ in array A.
 1: Quicksort$(A, 1, n)$
 2: **return** $A[1 \ldots n]$
Procedure Quicksort(A, p, r).
% Sort $r - p + 1$ numbers $a_p, a_{p+1}, \ldots, a_r$ in array $A[p \ldots r]$. %
 1: **if** $p < r$ **then**
 2:     $q \leftarrow$ Partition(A, p, r)
 3:     Quicksort$(A, p, q - 1)$
 4:     Quicksort$(A, q + 1, r)$
 5: **end if**
 6: **return** $A[p \ldots r]$
Procedure Partition(A, p, r).
% Find q such that there are $q - p + 1$ elements less than $A[r]$ and others bigger than or equal to $A[r]$ %
 1: $x \leftarrow A[r]$
 2: $i \leftarrow p - 1$
 3: **for** $j \leftarrow p - 1$ to $r - 1$ **do**
 4:     **if** $A[j] < x$ **then**
 5:         $i \leftarrow i + 1$ and exchange $A[i] \leftrightarrow A[j]$
 6:     **end if**
 7:     exchange $A[i + 1] \leftrightarrow A[r]$
 8: **end for**
 9: **return** $i + 1$

Now, we prove by induction on n that

$$E[T(n)] \leq cn \log n$$

for some constant c. For $n = 1$, it is trivial. Next, consider $n \geq 2$. By induction hypothesis,

$$E[T(n)] \leq c_1 n + \frac{2c}{n} \sum_{i=1}^{n-1} i \log i$$

$$= c_1 n + c(n - 1) \log \left(\Pi_{i=1}^{n-1} i^i \right)^{2/(n(n-1))}$$

$$\leq c_1 n + c(n - 1) \log \frac{1^2 + 2^2 + \cdots + (n - 1)^2}{n(n - 1)/2}$$

$$= c_1 n + c(n - 1) \log \frac{2n - 1}{3}$$

$$\leq c_1 n + cn \log \frac{2n}{3}$$

$$= cn \log n + \left(c_1 - c \log \frac{3}{2} \right) n.$$

Choose $c \geq c_1 / \log \frac{3}{2}$. We obtain $E[T(n)] \leq cn \log n$.

Theorem 2.1.3 *Expected running time of Quick Sort is $O(n \log n)$.*

2.2 Rectilinear Minimum Spanning Tree

Consider two points $A = (x_1, y_1)$ and $B = (x_2, y_2)$ in the plane. The *rectilinear distance* of A and B is defined by

$$d(A, B) = |x_1 - x_2| + |y_1 - y_2|.$$

The *rectilinear plane* is the plane with the rectilinear distance, denoted by L_1-plane. In this section, we study the following problem.

Problem 2.2.1 (Rectilinear Minimum Spanning Tree) *Given n points in the rectilinear plane, compute the minimum spanning tree on those n given points.*

In Chap. 1, we already present Kruskal algorithm which can compute a minimum spanning tree within $O(m \log n)$ time. In this section, we will improve this result by showing that the rectilinear minimum spanning tree can be computed in $O(n \log n)$ time. To do so, we first study an interesting problem as follows.

Problem 2.2.2 (All Northeast Nearest Neighbors) *Consider a set P of n points in the rectilinear plane. For each $A = (x_A, y_A) \in P$, another point $B = (x_B, y_B) \in P$ is said to lie in northeast (NE) area of A if $x_A \leq x_B$ and $y_A \leq y_B$, but $A \neq B$. Furthermore, B is the NE nearest neighbor of A if B has the shortest distance from A among all points lying in the NE area of A. This problem is required to compute the NE nearest neighbor for every point in P. (The NE nearest neighbor of a point A is "none" if no given point lies in the northeast area of A.)*

Let us design a divide-and-conquer algorithm to solve this problem. For simplicity of description, assume all n points have distinct x-coordinates and distinct y-coordinates. Now, we bisect n points by a vertical line L. Let P_l be the set of points lying on the left side of L and P_r the set of points lying on the right side of L. Suppose we already solve the all NE nearest neighbors problem on input point sets P_l and P_r, respectively. Let us discuss how to combine solutions for two subproblems into a solution for all NE nearest neighbors on P.

For point A in P_r, the NE nearest neighbor in P_r is also the NE nearest neighbor in P. However, for point A in P_l, the NE nearest neighbor in P_l may not be the NE nearest neighbor in P. Actually, let B_1 denote the NE nearest neighbor of A in P_l and B_r the NE nearest neighbor of A for B_2 in P_r. Then, if $d(A, B_1) \leq d(A, B_2)$, then the NE nearest neighbor of A in P is B_1; otherwise, it is B_2. Therefore, to complete the combination task, it is sufficient to compute the NE nearest neighbors in P_r for all points in P_l. We will show that this computation takes $O(n)$ time. To do so, let us first show a lemma.

Lemma 2.2.3 *Consider four points A, B, C, and D in the rectilinear plane. Suppose C and D lie in the northeast area of A and the northeast area of B. Then $d(A, C) \leq d(A, D)$ if and only if $d(B, C) \leq d(B, D)$.*

Proof Clearly, we have

$$d(A, C) = (x_C - x_A) + (y_C - y_A) \leq d(A, D) = (x_D - x_A) + (y_D - y_A)$$

$$\Leftrightarrow x_C + y_C \leq x_D + y_D$$

$$\Leftrightarrow d(B, C) = (x_C - x_B) + (y_C - y_B) \leq d(B, D) = (x_D - x_B) + (y_D - y_B).$$

$\square$

By this lemma, we can compute the NE nearest neighbors in P_r for all points in P_l as follows.

- For P_l, put all points in decreasing ordering of y-coordinates. For P_r, also, put all points in deceasing ordering of y-coordinates. Put *none* in P_r as the first element. Assume that *none* has y-coordinate $+\infty$ and for any point $A \in P_l$, $d(A, none) = +\infty$.
- Employ three pointers *left*, *right*, and *min*. *left* will be located in P_l. *right* and *min* work in P_r and *none*.
- Initially, assign *left* with the first point in P_l, and assign *right* and *min* with the first element P_r.
- If *right* has y-coordinate higher than *left* and $d(left, right) \geq d(left, min)$, then move *right* to next point in P_r.
- If *right* has y-coordinate higher than *left* and $d(left, right) < d(left, min)$, then set *min = right*, and move *right* to next one in P_r.
- If *right* has y-coordinate lower than *left*, then *min* is the NE nearest neighbor of *left*. Put this fact in record, and move *left* to next one in P_l.

Since *left*, *right*, and *min* always move down and never move up, above procedure runs in $O(n)$ time. Let $T(n)$ be the running time for computing all NE nearest neighbors for n points. Then we obtain $T(n) = 2T(\lceil n/2 \rceil) + O(n)$. Therefore, $T(n) = O(n \log n)$.

We make a remark on the case that P contains points with the same x-coordinate or y-coordinates. If P has some points with the same x-coordinate, then in order to partition P into two even parts, we may also consider their y-coordinates. If P has some points with same y-coordinate, then we may need to give a little adjustment for combination procedure.

Theorem 2.2.4 *Computing all NE nearest neighbors for n points can be done in $O(n \log n)$ time.*

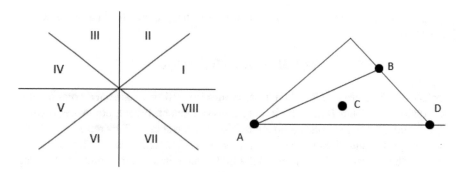

Fig. 2.3 Eight octants of point *A*

Now, we move back our attention to the rectilinear minimum spanning tree.

Consider any point *A*. As shown in Fig. 2.3, divide the area surrounding *A* into eight octants. To make them disjoint, we assume that each octant contains only one boundary, which can be reached from an interior ray to turn in counterclockwise direction.

Lemma 2.2.5 *Suppose* (A, B) *is an edge in a rectilinear minimum spanning tree. Then B must be the nearest neighbor of A in an octant.*

Proof Without loss of generality, assume that *B* lies in octant I of point *A*. For contradiction, suppose that *B* is not the nearest neighbor of *A* in octant I. Let *C* be the nearest neighbor of *A* in octant I, i.e., *C* lies in octant I and $d(A, C) < d(A, B)$. Note that

$$\{D \in \text{octant I} \mid d(A, D) \leq d(A, B)\}$$

is a triangle without boundary (A, B), which has a property that every two points in this triangle has rectilinear distance less than $d(A, B)$ (Fig. 2.3). Therefore, $d(C, B) < d(A, B)$.

Remove edge (A, B) from the rectilinear minimum spanning tree, which will partition the tree into two connected components, containing points *A* and *B*, respectively. If *A* and *C* lie in the same component, then add edge (C, B); otherwise, *C* and *B* must lie in the same component, and add edge (A, C). Therefore, we will obtain a shorter rectilinear minimum spanning tree, a contradiction.   □

Construct a graph *G* in the following way: For each point *A*, if an octant of *A* contains another given point, then find a nearest neighbor *B* for *A* in this octant, and add edge (A, B) to *G*.

Lemma 2.2.6 *G contains a rectilinear minimum spanning tree.*

Proof Consider a rectilinear minimum spanning tree *T*. For each point *A*, *T* must contain an edge (A, B). By Lemma 2.2.5, *B* is the nearest neighbor of *A* in an octant. Suppose (A, B) is not an edge of *G*. Then *G* must contain an edge (A, C)

lying in the same octant, and C is another nearest neighbor of A in the same octant. Note that

$$d(A, B) = d(A, C) > d(B, C).$$

Delete (A, B) from tree T. Then T is partitioned into two connected components. We claim that C and B must lie in the same component. In fact, otherwise, assume that A and C lie in the same component. Then we can shorten T by replacing (A, B) with (B, C), contradicting the minimality of T. Therefore, our claim is tree. It follows that replacing (A, B) by (A, C) in T results in another minimum spanning tree T'. Continue above operations; we will find a rectilinear minimum spanning tree contained in G. □

Lemma 2.2.7 *G can be constructed in $O(n \log n)$ time.*

Proof For each point A, its octant represents a direction. Fix an octant, i.e., fix a direction. We claim that computing all octant nearest neighbors for all given points in P takes $O(n \log n)$ time. Without loss of generality, consider octant I. Define a mapping

$$\phi : (x, y) \to ((x - y)/2, y).$$

Then ϕ has the following properties:

- Point B lies in octant I of point A if and only if $\phi(B)$ lies in the first quadrant of $\phi(A)$.
- $d(A, B) \le d(A, C)$ if and only $d(\phi(A), \phi(B)) \le d(\phi(A), \phi(C))$.

It follows from those properties that B is the octant nearest neighbor of A if and only if $\phi(B)$ is the NE nearest neighbor of $\phi(A)$. This means that computing all octant nearest neighbors can be reduced to computing all NE nearest neighbors. By Theorem 2.2.4, this computation can be done in $O(n \log n)$ time.

For each of eight octants, $O(n \log n)$ time is required. The total time is still bounded by $O(n \log n)$. □

Theorem 2.2.8 *The rectilinear minimum spanning tree can be computed in $O(n \log n)$ time where n is the number of given points.*

Proof By Lemmas 2.2.6 and 2.2.7, it is sufficient to compute the minimum spanning tree of graph G with Kruskal algorithm, since the number of edges in G is bounded by $O(n)$. Kruskal algorithm will spend $O(n \log n)$ time on G. □

2.3 Fibonacci Search

Consider a sequence of n distinct integers which are stored in an array $A[1..n]$. An element $A[i]$ is a *local maximum* if $A[i-1] < A[i]$ and $A[i] > A[i+1]$ for $1 < i < n$, $A[i] > S[i+1]$ for $i = 1$, and $A[i-1] < A[i]$ for $i = n$. The sequence $A[1..n]$ is said to be *bitonic* if it contains exactly one local maximum, which is actually the global maximum one. Consider the following problem.

Problem 2.3.1 (Maximum Element in Bitonic Sequence) *Given a sequence $A[1..n]$ of n distinct integers, find the maximum element.*

The problem can be solved by the following lemma.

Lemma 2.3.2 *Assume $1 \leq i < j \leq n$. If $A[i] < A[j]$, then $A[i+1..n]$ must contain a local maximum. If $A[i] > A[j]$, then $A[1..j-1]$ must contain a local maximum.*

Proof First, assume $A[i] < A[j]$. Consider two cases.

Case 1. $A[j] < A[j+1]$. In this case, if none of $A[j+1], \ldots, A[n-1]$ is a local maximum, then $A[j+1] < A[j+2] < \cdots < A[n]$. Hence, $A[n]$ is a local maximum.

Case 2. $A[j] > A[j+1]$. In this case, if none of $A[j], A[j-1], \ldots, A[i-1]$ is a local maximum, then $A[j] < A[j-1] < \cdots < A[i]$, contradicting to $A[i] < A[j]$.

Similarly, we can show the second statement.   □

For $n \geq 4$, we can choose i and j such that $1 \leq i < j \leq n$, $i \geq n/3$, and $n - j + 1 \geq n/3$. With such i and j, for each comparison, the sequence can be cut off at least one third. Therefore, the maximum element can be found within $O(\log n)$ comparisons.

Next, we consider a situation that $A[i] = f(i)$, that is, $A[i]$ has to be obtained through evaluation of a function $f(i)$. Therefore, we want to find the maximum element with the minimum number of evaluations. In this situation, i and j will be selected based on a rule with Fibonacci number F_i defined as follows:

$$F_0 = F_1 = 1, F_i = F_{i-2} + F_{i-1} \text{ for } i \geq 2.$$

Associated Fibonacci search method is as follows.

Step 0.   Select the maximum m such that $F_m \leq n$. Set $k \leftarrow 0$.
Step 1.   Set $i \leftarrow k + F_{m-1}$ and $j \leftarrow k + F_m$.
Step 2.   If $A[i] < A[j]$, then set $k \leftarrow i$.
Step 3.   Set $m \leftarrow m - 1$.
Step 4.   If $m \geq 2$, then go back to Step 1. Else, return $A[i+1]$, i.e., $A[k+1]$ is the maximum element.

About this method, we have the following result.

Theorem 2.3.3 *Let m be the maximum integer such that $F_m \leq n$. Then, it is sufficient to make m evaluations to search the maximum element from a bitonic sequence of n distinct integers. Moreover, in the worst case, it is necessary to make m evaluations.*

Proof Sufficiency can be seen from the Fibonacci search algorithm. We prove the necessity by induction on m. For $m = 1$, we have $n = 1$, and evaluation for $A[1]$ is required. For $m = 2$, we have $n = 2$, and evaluations for $A[1]$ and $A[2]$ are necessary. For $m \geq 3$, suppose we compare $A[i]$ and $A[j]$ for $1 \leq i < j \leq n$. Consider two cases.

Case 1. $i \leq F_{m-2}$. In this case, $n - i \geq F_{m-1}$. If $A[i] < A[j]$, then the maximum element is in $A[i+1..n]$. By induction hypothesis, in the worst case, $m - 1$ evaluations are required to find the maximum element. Add $A[i]$. Total number of evaluations is m.

Case 2. $i > F_{m-2}$. If $A[i] > A[j]$, then the maximum element is in $A[1..j - 1]$. In the next step, we need to select a number $k \in \{1, \ldots, i - 1\} \cup \{i + 1, \ldots, j - 1\}$ and compare $A[i]$ and $A[k]$. It does not matter if $k \in \{1, \ldots, i - 1\}$ or $k \in \{i + 1, \ldots, j - 1\}$, we can have a comparison result to keep $A[1..i - 1]$ left. Since $i - 1 \geq F_{m-2}$. In the worst case, we need at least $m - 2$ evaluations to find the maximum element from the subsequence containing $A[1..i - 1]$. Add evaluations on $A[i]$ and $A[j]$. Total number of required evaluations is at least m.   □

2.4 Heap

Heap is a quite useful data structure. Let us introduce it here and, by the way, give another sorting algorithm, Heap Sort.

A heap is a nearly complete binary tree, stored in an array (Fig. 2.4). What is *nearly complete binary tree*? It is a binary tree satisfying the following conditions:

• Every level other than bottom is complete.
• On the bottom, nodes are placed as left as possible.

For example, binary trees in Fig. 2.5 are not nearly complete. An advantage of nearly complete binary tree is to operate on it easily. For example, for node i (i.e., a node with address i), its parent, left-child, and right-child can be easily figured out as follows:

Parent(i)
   return $\lfloor i/2 \rfloor$.

Left(i)
   return $2i$.

Fig. 2.4 A heap

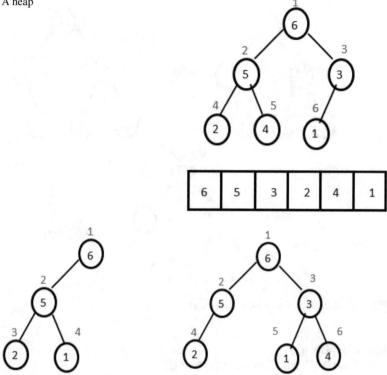

Fig. 2.5 They are not nearly complete

Right(i)

    return $2i + 1$.

There are two types of heaps with special properties, respectively.

Max-heap: For every node i other than root, $A[\text{Parent}(i)] \geq A[i]$.

Min-heap: For every node i other than root, $A[\text{Parent}(i)] \leq A[i]$.

Max-heap has two special operations: Max-Heapify and Build-Max-Heap. We describe them as follows.

When operation Max-Heapify(A, i) is called, two subtrees rooted, respectively, at Left(i) and Right(i) are max-heaps, but $A[i]$ may not satisfy the max-heap property. Max-Heapify(A, i) makes the subtree rooted at $A[i]$ become a max-heap by moving $A[i]$ downside. An example is as shown in Fig. 2.6.

The following is an algorithmic description for this operation.

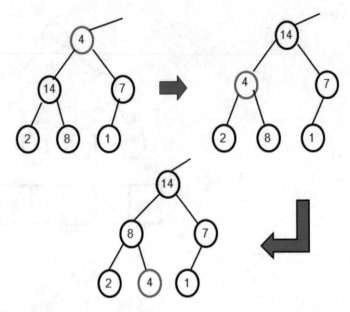

Fig. 2.6 An example for Max-Heapify(A, i)

Max-Heapify(A, i)
   if Left(i) $\geq$Right(i) and Left(i) $>$ $A[i]$
      then Exchange $A[i]$ and Left(i)
         Max-Heapify(A, Left(i))
   if Left(i) $<$Right(i) and Right(i) $>$ $A[i]$
      then Exchange $A[i]$ and Right(i)
         Max-Heapify(A, Right(i));

Operation Build-Max-Heap applies to a heap and makes it become a max-heap, which can be described as follows. (Note that Parent($size[A]$) $= \lfloor size[A]/2 \rfloor$.)

Build-Max-Heap(A)
   for $i \leftarrow \lfloor size[A]/2 \rfloor$ down to 1
      do Max-Heapify(A, i);

An example is as shown in Fig. 2.7.

It is interesting to estimate the running time of this operation. Let h be the height of heap A. Then $h = \lfloor \log_2 n \rfloor$. At level i, A has 2^i nodes, at each of which Max-Heapify spends at most $h - i$ steps to float down. Therefore, the running time of Build-Max-Heap(A) is

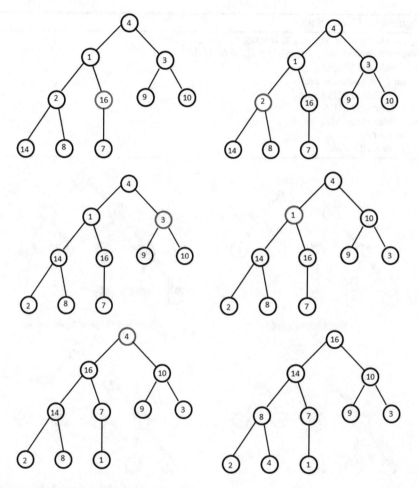

Fig. 2.7 An example for Build-Max-Heap(A)

$$O\left(\sum_{i=0}^{h} 2^i (h-i)\right) = O\left(2^h \sum_{i=0}^{h} \frac{h-i}{2^{h-i}}\right)$$

$$= O\left(2^h \sum_{i=0}^{h} \frac{i}{2^i}\right)$$

$$= O(n).$$

Now, as shown in Algorithm 3, a sorting algorithm can be designed with max-heap. Initially, build a max-heap A. In each subsequent step, the algorithm first exchanges $A[1]$ and $A[heap - size(A)]$ and then reduces $heap - size(A)$ by 1, meanwhile with Max-Heapify($A, 1$) to recover the max-heap. An example is as shown in Fig. 2.8.

Algorithm 3 Heap Sort

Input: n numbers $a_1, a_2, \ldots, a_n$ in array $A[1 \ldots n]$.
Output: n numbers $a_{i_1} \leq a_{i_2} \leq \cdots \leq a_{i_n}$ in array A.

1: Build-Max-Heap(A)
2: **for** $i \leftarrow n$ down to 2 **do**
3:     exchange $A[1] \leftrightarrow A[i]$
4:     heap-size $[A] \leftarrow i - 1$
5:     Max-Heapify($A, 1$)
6: **end for**
7: **return** $A[1 \ldots n]$

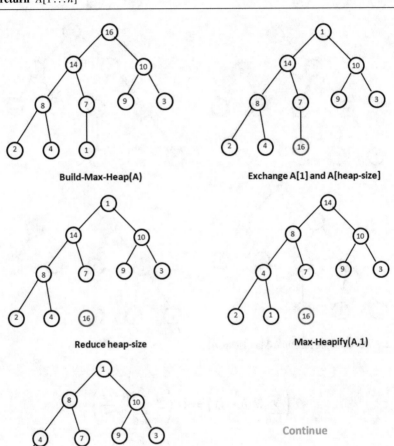

Fig. 2.8 An example for Heap Sort

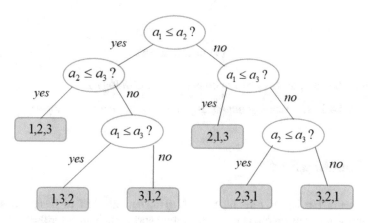

Fig. 2.9 Decision tree

Since the number of steps is $O(n)$ and Max-Heapify$(A, 1)$ takes $O(\log n)$ time, the running time of Heap Sort is $O(n \log n)$.

Theorem 2.4.1 *Heap Sort runs in $O(n \log n)$ time.*

We already have two sorting algorithms with $O(n \log n)$ running time and one sorting algorithm with expected $O(n \log n)$ running time. But, there is no sorting algorithm with running time faster than $O(n \log n)$. Is $O(n \log n)$ a barrier of running time for sorting algorithm? In some sense, the answer is yes. All sorting algorithms presented previously belong to a class, called *comparison sort*.

In comparison sort, order information about input sequence can be obtained only by comparison between elements in the input sequence. Suppose input sequence contains n positive integers. Then there are $n!$ possible permutations. The aim of sorting algorithm is to determine a permutation which gives a nondecreasing order. Each comparison divides the set of possible permutations into two subsets. The comparison result tells which subset contains a nondecreasing order. Therefore, every comparison sort algorithm can be represented by a binary decision tree (Fig. 2.9). The (worst case) running time of the algorithm is the height (or depth) of the decision tree.

Since the binary decision tree has $n!$ leaves, its height $T(n)$ satisfies

$$1 + 2 + \cdots + 2^{T(n)} \geq n!$$

that is,

$$2^{T(n)+1} - 1 \geq \sqrt{2\pi n} \left(\frac{n}{e}\right)^n.$$

Thus,

$$T(n) = \Omega(n \log n).$$

Therefore, no comparison sort can do better than $O(n \log n)$.

Theorem 2.4.2 *The running time of any comparison sort is* $\Omega(n \log n)$.

2.5   Counting Sort

To break the barrier of running time $O(n \log n)$, one has to design a sorting algorithm without using comparison. Counting sort is such an algorithm.

Let us use an example to illustrate Counting Sort as shown in Algorithm 4. This algorithm contains three arrays, A, B, and C. Array A contains input sequence of positive integers. Suppose $A = \{4, 6, 5, 1, 4, 5, 2, 5\}$. Let k be the largest integer in input sequence. Initially, the algorithm makes preprocessing on array C in three stages:

1. Clean up array C.
2. For $1 \leq i \leq k$, assign $C[i]$ with the number of i's appearing in array A. (In the example, $C = \{1, 1, 0, 2, 3, 1\}$ at this stage.)
3. Update $C[i]$ such that $C[i]$ is equal to the number of integers with value at most i appearing in A. (In the example, $C = \{1, 2, 2, 4, 7, 8\}$ at this stage.)

With the help of array C, the algorithm moves element $A[j]$ to array B for $j = n$ down to 1, by

$$B[C[A[j]]] \leftarrow A[j]$$

Algorithm 4 Counting Sort

Input: n numbers $a_1, a_2, \ldots, a_n$ in array $A[1 \ldots n]$.
Output: n numbers $a_{i_1} \leq a_{i_2} \leq \cdots \leq a_{i_n}$ in array B.
 1: **for** $i \leftarrow 1$ to k **do**
 2:     $C[i] \leftarrow 0$
 3: **end for**
 4: **for** $j \leftarrow 1$ to n **do**
 5:     $C[A[j]] \leftarrow C[A[j]] + 1$
 6: **end for**
 7: **for** $i \leftarrow 2$ to k **do**
 8:     $C[i] \leftarrow C[i] + C[i-1]$
 9: **end for**
10: **for** $j \leftarrow n$ down to 1 **do**
11:     $B[C[A[j]]] \leftarrow A[j]$
12:     $C[A[j]] \leftarrow C[A[j]] - 1$
13: **end for**
14: **return** $B[1 \ldots n]$

---

and then updates array C by

$$C[A[j]] \leftarrow C[A[j]] - 1.$$

This part of computation about the example is as follows.

$$
\begin{array}{c|c}
C & 1\ 2\ 2\ 4\ 7\ 8 \\
\hline
A & 4\ 6\ 5\ 1\ 4\ 5\ 2\ \hat{5} \\
B & \qquad\qquad 5
\end{array}
$$

$$
\begin{array}{c|c}
C & 1\ 2\ 2\ 4\ 6\ 8 \\
\hline
A & 4\ 6\ 5\ 1\ 4\ 5\ \hat{2}\ 5 \\
B & 2\qquad\quad 5
\end{array}
$$

$$
\begin{array}{c|c}
C & 1\ 1\ 2\ 4\ 6\ 8 \\
\hline
A & 4\ 6\ 5\ 1\ 4\ \hat{5}\ 2\ 5 \\
B & 2\qquad\quad 5\ 5
\end{array}
$$

$$
\begin{array}{c|c}
C & 1\ 1\ 2\ 4\ 5\ 8 \\
\hline
A & 4\ 6\ 5\ 1\ \hat{4}\ 5\ 2\ 5 \\
B & 2\quad\ 4\ \ 5\ 5
\end{array}
$$

$$
\begin{array}{c|c}
C & 1\ 1\ 2\ 3\ 5\ 8 \\
\hline
A & 4\ 6\ 5\ \hat{1}\ 4\ 5\ 2\ 5 \\
B & 1\ 2\quad\ 4\ \ \ 5\ 5
\end{array}
$$

$$
\begin{array}{c|c}
C & 0\ 1\ 2\ 3\ 5\ 8 \\
\hline
A & 4\ 6\ \hat{5}\ 1\ 4\ 5\ 2\ 5 \\
B & 1\ 2\quad\ 4\ 5\ 5\ 5
\end{array}
$$

$$
\begin{array}{c|c}
C & 0\ 1\ 2\ 3\ 4\ 8 \\
\hline
A & 4\ \hat{6}\ 5\ 1\ 4\ 5\ 2\ 5 \\
B & 1\ 2\quad\ 4\ 5\ 5\ 5\ 6
\end{array}
$$

$$
\begin{array}{c|c}
C & 0\ 1\ 2\ 3\ 4\ 7 \\
\hline
A & \hat{4}\ 6\ 5\ 1\ 4\ 5\ 2\ 5 \\
B & 1\ 2\ 4\ 4\ 5\ 5\ 5\ 6
\end{array}
$$

Now, let us estimate the running time of Counting Sort.

Theorem 2.5.1 *Counting Sort runs in $O(n + k)$ time.*

Proof The loop at line 1 takes $O(k)$ time. The loop at line 4 takes $O(n)$ time. The loop at line 7 takes $O(k)$ time. The loop at line 10 takes $O(n)$ time. Putting all together, the running time is $O(n + k)$. $\square$

A student found a simple way to improve Counting Sort. Let consider the same example. At the second stage, $C = \{1, 1, 0, 2, 3, 1\}$ where $C[i]$ is equal to the number of i's appearing in array A. The student found that with this array C, array B can be put in integers immediately without array A.

C	1 1 0 2 3 1
B	1
B	1 2
B	1 2 4 4
B	1 2 4 4 5 5 5
B	1 2 4 4 5 5 5 6

Is this method acceptable? The answer is no. Why not? Let us explain.

First, we should note that those numbers in input sequence may come from labels of objects. The same numbers may come from different objects. For example, consider a sequence of objects $\{329, 457, 657, 839, 436, 720, 355\}$. If we use their first digits from left as labels, then we will obtain a sequence $\{9, 7, 7, 9, 6, 0, 5\}$. When apply Counting Sort on this sequence, we will obtain a sequence $\{720, 355, 436, 457, 657, 329, 839\}$. This is because a label gets moved together with its object in Counting Sort.

Moreover, consider two objects 329 and 839 with the same label 9. In input sequence, 329 lies on the left side of 839. After Counting Sort, 329 lies still on the left side of 839.

A sorting algorithm is *stable* if for different objects with the same label, after labels are sorted, the ordering of objects in output sequence is the same as their ordering in input sequence. The following can be proved easily.

Lemma 2.5.2 *Counting Sort is stable.*

The student's method cannot keep stable property.

With stable property, we can use Counting Sort in the following way. Remember, after sorting the leftmost digit, we obtain sequence

$$\{720, 355, 436, 457, 657, 329, 839\}.$$

Now, we continue to sort this sequence based on the second leftmost digit. Then we will obtain sequence

$$\{720, 329, 436, 839, 355, 457, 657\}.$$

Continue to sort based on the rightmost digit, we will obtain sequence

$$\{329, 355, 436, 457, 657, 720, 839\}.$$

Now, let us use this technique to solve a problem.

Example 2.5.3 There are n integers between 0 and $n^2 - 1$. Design an algorithm to sort them. The algorithm is required to run in $O(n)$ time.

Each integer between 0 and $n^2 - 1$ can be represented as

$$an + b \text{ for } 0 \le a \le n - 1, 0 \le b \le n - 1.$$

Apply Counting Sort first to b and then to a. Each application takes $O(n) = O(n + k)$ time since $k = n$. Therefore, total time is still $O(n)$.

In general, suppose there are n integers, each of which can be represented in the form

$$a_d k^d + a_{d-1} 1 k^{d-1} + \cdots + a_0$$

where $0 \le a_i \le k - 1$ for $0 \le i \le d$. Then we can sort these n integers by using Counting Sort first on a_0, second on a_1, ..., finally on a_d. This method is called *Radix Sort*.

Theorem 2.5.4 *Radix Sort takes $O(d(n + k))$ time.*

2.6 More Examples

Let us study more examples with divide-and-conquer technique and sorting algorithms.

Example 2.6.1 (Maximum Consecutive Subsequence Sum) Given a sequence of n integers, find a consecutive subsequence with maximum sum.

Divide input sequence S into two subsequence S_1 and S_2 such that $|S_1| = \lfloor n/2 \rfloor$ and $|S_2| = \lceil n/2 \rceil$. Let $MaxSub(S)$ denote the consecutive subsequence of S with maximum sum. Then there are two cases.

Case 1. $MaxSub(S)$ is contained in either S_1 or S_2. In this case, $MaxSub(s) = MaxSub(S_1)$ or $MaxSub(s) = MaxSub(S_2)$.

Case 2. $MaxSub(S) \cap S_1 \ne \emptyset$ and $MaxSub(S) \cap S_2 \ne \emptyset$. In this case, $MaxSub(S) \cap S_1$ is the tail subsequence with maximum sum. That is, suppose $S_1 = \{a_1, a_2, \ldots, a_p\}$. Then among subsequences $\{a_p\}$, $\{a_{p-1}, a_p\}, \ldots, \{a_1, \ldots, a_p\}$, $MaxSub(S) \cap S_1$ is the one with maximum sum. Therefore, it can be found in $O(n)$ time. Similarly, $MaxSub(S) \cap S_2$ is the head subsequence with maximum sum, which can be computed in $O(n)$ time.

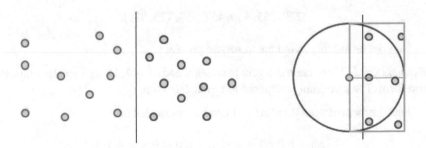

Fig. 2.10 Closest pair of points

Suppose $MaxSub(S)$ can be computed in $T(n)$ time. Summarized from the above two cases, we obtain

$$T(n) = 2T(\lceil n/2 \rceil) + O(n).$$

Therefore, $T(n) = O(n \log n)$.

Next, we present another algorithm running in $O(n)$ time.

Let S_j be the maximum sum of a consecutive subsequence ending at the jth integer a_j. Then, we have

$$S_1 = a_1$$

$$S_{j+1} = \begin{cases} S_j + a_{j+1} & \text{if } S_j > 0, \\ a_{j+1} & \text{if } S_j \leq 0. \end{cases}$$

This recursive formula gives a linear time algorithm to compute S_j for all $1 \leq j \leq n$. From them, find the maximum one, which is the solution for the maximum consecutive subsequence sum problem.

Example 2.6.2 (Closest Pair of Points) Given n points in the Euclidean plane, find a pair of points to minimize the distance between them.

Initially, we may assume that all n points have distinct x-coordinates since, if not, we may rotate the coordinate system a little.

Now, divide all points into two half parts based on x-coordinates. Find the closest pair of points in each part. Suppose δ_1 and δ_2 are distances of closest pairs in two parts, respectively. Let $\delta = \min(\delta_1, \delta_2)$. We next study if there is a pair of points lying in both parts, respectively, and with distance less than δ (Fig. 2.10).

For each point $u = (x_u, y_u)$ in the left part (Fig. 2.10), consider the rectangle $R_u = \{(x, y) \mid x_u \leq x \leq x_u + \delta, \ y_u - \delta \leq y \leq y_u + \delta\}$. It has the following properties:

- Every point in the right part and within distance δ from u lies in this rectangle.
- This rectangle contains at most six points in the right part because every two points have distance at least δ.

Fig. 2.11 x is selected
through first three steps

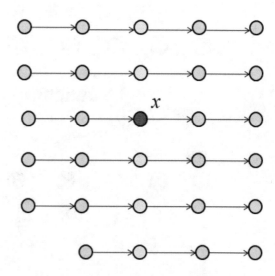

For each u in the left part, check every point v lying in R_u, if distance $d(u, v) < \delta$. If
yes, then we keep the record and choose the closest pair of points from them, which
should be the solution. If not, then the solution should either be the closest pair of
points in the left part or the closest pair of points in the right part.

Let $T(n)$ be the time for finding the closest pair of points from n points. Above
method gives a recursive relation

$$T(n) = 2T(\lceil n/2 \rceil) + O(n).$$

Therefore, $T(n) = O(n \log n)$.

Example 2.6.3 (The ith Smallest Number) Given a sequence of n distinct numbers
and a positive integer i, find ith smallest number in $O(n)$ time.

This algorithm consists of five steps. Let us name this algorithm as $A(n, i)$ for
convenience of recursive call.

Step 1. Divide n numbers into $\lceil n/5 \rceil$ groups of five elements, possibly except the
last one of less than five elements (Fig. 2.11).

Step 2. Find the median of each group by merge sort. Possibly, for the last group,
there are two median; in such a case, take the smaller one (Fig. 2.11).

Step 3. Make a recursive call $A(\lceil n/5 \rceil, \lceil \lceil n/5 \rceil/2 \rceil)$. This call will find the median
x of $\lceil n/5 \rceil$ group median and, moreover, will select the smaller one in case that two
candidates of x exist (Fig. 2.11).

Step 4. Exchange x with the last element in input array, and partition all numbers
into two parts by using Partition procedure in Quick Sort. One part (on the left)
contains numbers less than x, and the other part (on the right) contains numbers
larger than x (Fig. 2.12).

Step 5. Let k be the number of elements in the left part (Fig. 2.12). If $k = i - 1$,
then x is the ith smallest number. If $k \geq i$, then the ith smallest number lies on the

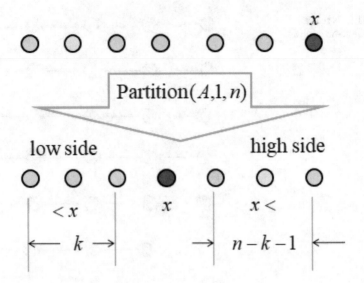

Fig. 2.12 x is selected through the first three steps

left of x and hence makes a recursive call $A(k, i)$. If $k \le i - 2$, then the ith smallest number lies in the right of x and hence makes a recursive call $A(n - k - 1, i - k - 1)$.

Now, let us analyze this algorithm. Let $T(n)$ be the running time of $A(n, i)$.

- Steps 1 and 2 take $O(n)$ time.
- Step 3 takes $T(\lceil n/5 \rceil)$ time.
- Step 4 takes $O(n)$ time.
- Step 5 takes $T(\max(k, n - k - 1))$ time.

Therefore,

$$T(n) = T(\lceil n/5 \rceil) + T(\max(k, n - k - 1)) + O(n).$$

We claim that

$$\max(k, n - k - 1) \le n - \left(3 \left\lceil \frac{1}{2} \left\lceil \frac{n}{5} \right\rceil \right\rceil - 2 \right).$$

In fact, as shown in Fig. 2.13,

$$k + 1 = 3 \left\lceil \frac{1}{2} \left\lceil \frac{n}{5} \right\rceil \right\rceil$$

and

$$n - k \ge 3 \left\lceil \frac{1}{2} \left\lceil \frac{n}{5} \right\rceil \right\rceil - 2.$$

Fig. 2.13 Estimation of
$k + 1$ and $n - k$

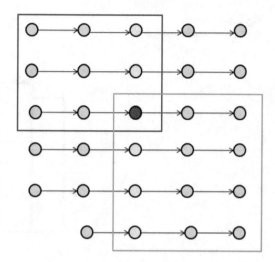

Therefore,

$$n - k - 1 \leq n - 3 \left\lceil \frac{1}{2} \left\lceil \frac{n}{5} \right\rceil \right\rceil$$

and

$$k \leq n - \left(3 \left\lceil \frac{1}{2} \left\lceil \frac{n}{5} \right\rceil \right\rceil - 2 \right).$$

Note that

$$n - \left(3 \left\lceil \frac{1}{2} \left\lceil \frac{n}{5} \right\rceil \right\rceil - 2 \right) \leq n - \left(\frac{3n}{10} - 2 \right) \leq \frac{7n}{10} + 2.$$

By the claim,

$$T(n) \leq T(\lceil n/5 \rceil) + T\left(\frac{7n}{10} + 2 \right) + c'n$$

for some constant $c' > 0$. Next, we show that

$$T(n) \leq cn \tag{2.1}$$

for some constant $c > 0$. Choose

$$c = \max(20c', T(1), T(2)/2, \ldots, T(59)/59).$$

Fig. 2.14 Largest
rectangular area in histogram

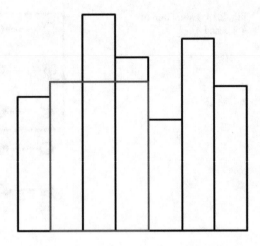

Fig. 2.14 Largest
rectangular area in histogram

Therefore, (2.1) holds for $n \leq 59$. Next, consider $n \geq 60$. By induction hypothesis, we have

$$T(n) \leq c(n/5 + 1) + c(7n/10 + 2) + c'n$$
$$\leq cn - (cn/10 - 3c - c'n)$$
$$\leq cn$$

since

$$c(n/10 - 3) \geq n/20 \geq c'n.$$

The first inequality is due to $n \geq 60$, and the second one is due to $c \geq 20c'$. This ends the proof of $T(n) = O(n)$.

Example 2.6.4 (Largest Rectangular Area in Histogram) Consider a histogram as shown in Fig. 2.14. Assume every bar has unit width and heights are $h_1, h_2, \ldots, h_n$, respectively. Find the largest rectangular area.

Let $h_k = \min(h_i, h_2, \ldots, h_j)$. Denote by $m(i, j)$ the largest rectangular area in histogram with bars between i and j. Then, we can obtain the following recursive formula.

$$m(i, j) = \max(m(i, k - 1), m(k + 1, j), h_k(j - i + 1)).$$

It is similar to Quicksort that expected running time can be proved to be $O(n \log n)$.

Exercises

1. Use a recursion tree to estimate a good upper bound on the recurrence $T(n) = 3T(\lfloor n/2 \rfloor) + n$ and $T(1) = 0$. Use the mathematical induction to prove correctness of your estimation.

2. Draw the recursion tree for $T(n) = 3T(\lfloor n/2 \rfloor) + cn$, where c is a positive constant, and guess an asymptotic upper bound on its solution. Prove your bound by mathematical induction.

3. Show that for input sequence in decreasing order the running time of Quick Sort is $\Theta(n^2)$.

4. Show that Counting Sort is stable.

5. Find an algorithm to sort n integers in the range 0 to $n^3 - 1$ in $O(n)$ time.

6. Let $A[1 : n]$ be an array of n distinct integers sorted in increasing order. (Assume, for simplicity, that n is a power of 2.) Give an $O(\log n)$-time algorithm to decide if there is an integer i, $1 \le i \le n$, such that $A[i] = i$.

7. Given an array A of integers, please return an array B such that $B[i] = |\{A[k] \mid k > i \text{ and } A[k] < A[i]\}|$.

8. Given a string S and an integer $k > 0$, find the longest substring of s such that each symbol appears at least k times if it appears in the substring.

9. Given an integer array A, please compute the number of pairs $\{i, j\}$ with $A[i] > 2 \cdot A[j]$.

10. Given a sorted sequence of distinct nonnegative integers, find the smallest missing number.

11. Given two sorted sequences with m, n elements, respectively, design and analyze an efficient divide-and-conquer algorithm to find the kth element in the merge of the two sequences. The best algorithm runs in time $O(\log(\max(m, n)))$.

12. Design a divide-and-conquer algorithm for the following longest ascending subsequence problem: Given an array $A[1..n]$ of natural numbers, find the length of the longest ascending subsequence. (A subsequence is a list $A[i_1]$, $A[i_2], \ldots, A[i_m]$ where m is the length.)

13. Show that in a max-heap of length n, the number of nodes rooted at which the subtree has height h is at most $\lceil \frac{n}{2^{h+1}} \rceil$.

14. Let A be an $n \times n$ matrix of integers such that each row is strictly increasing from left to right and each column is strictly increasing from top to bottom. Give an $O(n)$-time algorithm for finding whether a given number x is an element of A, i.e., whether $x = A(i, j)$ for some i, j.

15. Let S be a set of n points, $p_i = (x_i, y_i)$, $1 \le i \le n$, in the plane. A point $p_j \in S$ is a *maximal point of* S if there is no other point $p_k \in S$ such that $x_k \ge x_j$ and $y_k \ge y_j$. In Fig. 2.15, it illustrates the maximal points of a point-set S. Note that the maximal points form a "staircase" which descends rightward. Give an efficient divide-and-conquer algorithm to determine the maximal points of S.

16. Let $A[1..n]$ be an array of n distinct integers where $n \ge 2$. An element $A[i]$ is a local maximum if $A[i - 1] < A[i]$ and $A[i] > A[i + 1]$ for $1 < i < n$,

Fig. 2.15 Maximal points and non-maximal points

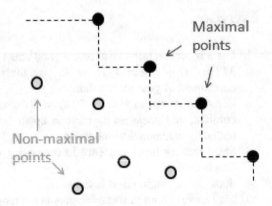

$A[i] > S[i + 1]$ for $i = 1$, and $A[i - 1] < A[i]$ for $i = n$. Please design an algorithm to find a local maximum in $O(\log n)$ time.

17. The maximum subsequence sum problem is defined as follows: Given an array $A[1..n]$ of integer numbers, find values of i and j with $1 \leq i \leq j \leq n$ such that $\sum_{k=i}^{j} A[i]$ is maximized. Design a divide-and-conquer algorithm for solving the maximum subsequence sum problem in time $O(n \log n)$.

18. In the plane, there are n distinct points $p_1, p_2, \ldots, p_n$ lying on line $y = 0$ and also n distinct points $q_1, q_2, \ldots, q_n$ lying on line $y = 0$. Consider n segments $[p_1, q_1], [p_2, q_2], \ldots, [p_n, q_n]$. Design an algorithm to count how many cross pairs in these n segments. Your algorithm should run in $O(n \log n)$ time.

19. Design a divide-and-conquer algorithm for multiplying n complex numbers using only $3(n - 1)$ real multiplications.

20. Consider a 0-1 matrix of order $(2^n - 1) \times n$. All rows have distinct 0-1 sequences of length n, that is, no two rows are identical. Design a $O(n)$ time algorithm to find the missing sequence.

21. Given a sequence of n distinct integers and a positive integer i, finding the ith smallest one in the sequence can be done in $O(n)$ time (see Example 2.6.3). Now, consider the problem of finding the ith smallest one for every $i = 1, 2, \ldots, k$. Can you do it in $O(n \log k)$ time?

22. An inversion in an array $A[1..n]$ is a pair of indices i and j such that $i < j$ and $A[i] > A[j]$. Design an algorithm to count the number of inversions in an n-element array in $O(n \log n)$ time.

23. In Example 2.6.3, a linear time algorithm is given for finding the ith smallest number in a unsorted list of n distinct integers. Now, let us modify the first two steps as follows: Initially, suppose all n integers are given in array A. Partition all input integers into groups of three elements. Then sort each group, and place its median into another array B. Repeat the same process for B, that is, partition elements in B into groups of three elements, and then place the median of each group into array C. Now, make a recursive call to find the median x of C. The remaining part is the same as later steps in the linear time algorithm. Please analyze the running time of this modified algorithm.

24. Design an $O(n^{\log_2 3})$ step algorithms for multiplication of two n-digit numbers. A single step only allows the multiplication/division or addition/subtraction of single digit numbers. Could you improve your algorithm with running $O(n^{\log_3 5})$ steps?

Historical Notes

Divide-and-conquer is a popular technique for algorithm design. It has a special case, decrease-and-conquer. In decrease-and-conquer, the problem is reduced to a single subproblem. Both divide-and-conquer and decrease-and-conquer have a long history. Their stamps can be found in many earlier works, such as Gauss's work on Fourier transform in 1850 [209], John von Neumann's work on merge sort in 1945 [258], and John Mauchly's work in 1946 [258]. Quick Sort was developed by Tony Hoare in 1959 [210] (published in 1962 [211]). Counting Sort and its applications to Radix Sort were found by Harold H. Seward in 1954 [73, 258, 362].

The closest-point problem and its variations, such as the problem of all nearest neighbors, have many applications. Construction of rectilinear minimum spanning tree in $O(n \log n)$ time [192] is one of them. There is another way to obtain $O(n \log n)$-time algorithm for the rectilinear minimum spanning tree [222], in which the Voronoi diagram in L_1 is constructed in $O(n \log n)$ time [274, 363] and then compute the rectilinear minimum spanning tree in the Voronoi diagram in $O(n)$ time. The idea was initiated by Yao [433] to consider closest neighbors in different directions in construction of minimum spanning tree in a plane. In a Euclidean plane, the minimum spanning tree can also be computed in $O(n \log n)$ time [273, 274, 363]. For planar graphs, the minimum spanning tree can be computed in $O(n)$ time [63]. Several algorithms exist for a long time for computing the minimum spanning tree with arbitrary distance [36, 65, 266, 339].

Fibonacci search [140] is motivated from Golden section search [244] to find the maximum value of a unimodal function since $F_k/F_{k+1} \leftarrow (\sqrt{5} - 1)/2$, which is called the Golden ratio. The Golden section search has received a great deal of applications [140, 218, 258, 333].

Chapter 3
Dynamic Programming and Shortest Path

The art of programming is the art of organizing complexity.

—Edsger Dijkstra

A divide-and-conquer algorithm consists of many iterations. Usually, each iteration contains three steps. In the first step (called the divide step), divide the problem into smaller subproblems. In the second step (called conquer step), solve those subproblems. In the third step (called the combination step), combine solutions for subproblems into a solution for the original problem. **Is it true that every algorithm with each iteration consisting of the above three steps belongs to the class of divide-and-conquer?** The answer is No. In this chapter, we would like to introduce a class of algorithms, called *dynamic programming*. Every algorithm in this class consists of discrete iterations, each of which contains the divide step, the conquer step, and the combination step. However, they may not be the divide-and-conquer algorithms. Actually, their self-reducibility structure may not be a tree.

3.1 Dynamic Programming

Let us first study several examples and start from a simpler one.

Example 3.1.1 (Fibonacci Number) Fibonacci number F_i for $i = 0, 1, \ldots$ is defined by

$$F_0 = 0, \ F_1 = 1, \ \text{and} \ F_i = F_{i-1} + F_{i-2}.$$

The computational process can be considered as a dynamic programming with self-reducibility structure as shown in Fig. 3.1.

Example 3.1.2 (Labeled Tree) Let $a_1, a_2, \ldots, a_n$ be a sequence of n positive integers. A *labeled tree* for this sequence is a binary tree T of n leaves named $v_1, v_2, \ldots, v_n$ from left to right. We label v_i by a_i for all i, $1 \leq i \leq n$. Let D_i

© The Author(s), under exclusive license to Springer Nature Switzerland AG 2022
D.-Z. Du et al., *Introduction to Combinatorial Optimization*, Springer Optimization and Its Applications 196, https://doi.org/10.1007/978-3-031-10596-8_3

Fig. 3.1 Fibonacci numbers

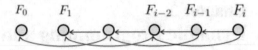

$$F_0 = 0, F_1 = 1, \text{ and } F_i = F_{i-1} + F_{i-2}$$

Fig. 3.2 The table of subproblems $T(i, j)$

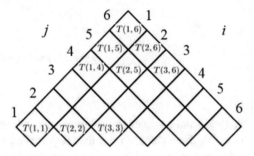

be the length of the path from v_i to the root of T. The cost of T is defined by

$$cost(T) = \sum_{i=1}^{n} a_i D_i.$$

The problem is to construct a labeled tree T to minimize the cost $cost(T)$ for a given sequence of positive integers $a_1, a_2, \ldots, a_n$.

Let $T(i, j)$ be the optimal labeled tree for subsequence $\{a_i, a_{i+1}, \ldots, a_j\}$ and $sum(i, j) = a_i + a_{i+1} + \cdots + a_j$. Then

$$cost(T(i, j)) = \min_{i \le k < j} \{cost(T(i, k)) + cost(T(k + 1, j))\} + sum(i, j)$$

where

$$sum(i, j) = \begin{cases} a_i & \text{if } i = j \\ a_i + sum(i + 1, j) & \text{if } i < j. \end{cases}$$

As shown in Fig. 3.2, there are $1 + 2 + \cdots + n = \frac{n(n+1)}{2}$ subproblems $T(i, j)$ in the table. From recursive formula, it can be seen that solution of each subproblem $T(i, j)$ can be computed in $O(n)$ time. Therefore, this dynamic programming runs totally in $O(n^3)$ time.

Actually, the running time of a dynamic programming is often estimated by the following formula:

$$running\ time = (number\ of\ subproblems) \times (computing\ time\ of\ recursion).$$

Algorithm 5 Algorithm for labeled tree

Input: A sequence of positive integers $a_1, a_2, \ldots, a_n$.
Output: Minimum cost of a labeled tree.
 1: **return** $cost(T(1, n))$.

function $cost(T(i, j))$ $(i \le j)$

 1: **if** $i = j$ **then**
 2:     $temp \leftarrow a_i$
 3: **else**
 4:     $temp \leftarrow +\infty$
 5:     **for** $k = i$ to $j - 1$ **do**
 6:         $temp \leftarrow \min(temp, cost(T(i, k)) + cost(T(k + 1, j)) + sum(i, j))$
 7:     **end for**
 8: **end if**
 9: **return** $cost(T(i, j)) \leftarrow temp$;

function $sum(i, j)$ $(i \le j)$

 1: **if** $i = j$ **then**
 2:     **return** $sum(i, j) \leftarrow a_i$
 3: **else**
 4:     **return** $sum(i, j) \leftarrow a_i + sum(i + 1, j)$
 5: **end if**

Algorithm 6 Algorithm for labeled tree

Input: A sequence of positive integers $a_1, a_2, \ldots, a_n$.
Output: Minimum cost of a labeled tree.
 1: **for** $i = 1$ to n **do**
 2:     $cost(T(i, i)) \leftarrow a_i; sum(i, i) \leftarrow a_i$
 3: **end for**
 4: **for** $l = 2$ to n **do**
 5:     **for** $i = 1$ to $n - l + 1$ **do**
 6:         $j \leftarrow i + l - 1$
 7:         $cost(T(i, j)) \leftarrow +\infty; sum(i, j) \leftarrow sum(i, j - 1) + a_j$
 8:         **for** $k = i$ to $j - 1$ **do**
 9:             $q \leftarrow cost(T(i, k)) + cost(T(k + 1, j)) + sum(i, j)$
10:             $cost(T(i, j)) \leftarrow \min(cost(T(i, j)), q)$
11:         **end for**
12:     **end for**
13: **end for**
14: **return** $cost(T(1, n))$

There are two remarks on this formula: (1) There are some exceptional cases. We will see one in the next section. (2) The divide-and-conquer can be considered as a special case of the dynamic programming. Therefore, its running time can also be estimated with this formula. However, the outcome is usually too rough.

It is similar to the divide-and-conquer that there are two ways to write software codes for the dynamic programming. The first way is to employ recursive call as shown in Algorithm 5. The second way is as shown in Algorithm 6 which saves the recursive calls, and hence in practice, it runs faster with smaller space requirement.

Fig. 3.3 A rectangle with point-holes inside

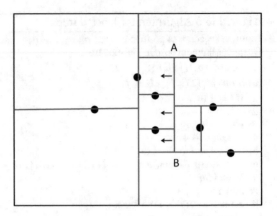

Before we study the next example, let us first introduce a concept, guillotine cut. Consider a rectangle P, a cut on P is called a *guillotine cut* if it cuts P into two parts. A guillotine partition is a sequence of guillotine cuts.

Example 3.1.3 (Minimum Length Guillotine Partition) Given a rectangle with point-holes inside, partition it into smaller rectangles without a hole inside by a sequence of guillotine cuts to minimize the total length of cuts. Here, the guillotine cut is a vertical or horizontal straight line segment which partitions a rectangle into two smaller rectangles. An example is shown in Fig. 1.2.

Example 3.1.3 is a geometric optimization problem. It has infinitely many feasible solutions. Therefore, strictly speaking, it is not a combinatorial optimization problem. However, it can be reduced to a combinatorial optimization problem.

Lemma 3.1.4 (Canonical Partition) *There exists a minimum length guillotine partition such that every guillotine cut passes through a point-hole.*

Proof Suppose there exists a guillotine cut AB not passing through any point-hole (Fig. 3.3). Without loss of generality, assume that AB is a vertical cut. Let n_1 be the number of guillotine cuts touching AB on the left and n_2 the number of guillotine cuts touching AB on the right. Without loss of generality, assume $n_1 \geq n_2$. Then we can move AB to the left without increasing the total length of rectangular guillotine partition, until a point-hole is met. If this moving cannot meet a point-hole, then AB can be moved to meet another vertical cut or vertical boundary, and in either case, AB can be deleted, contradicting the optimality of the partition. $\square$

By Lemma 3.1.4, we may consider only canonical guillotine partitions. During the canonical guillotine partition, each subproblem can be determined by a rectangle in which each boundary edge is obtained by a guillotine cut or a boundary edge of a given rectangle, and hence there are $O(n)$ possibility. This implies that the number of subproblems is $O(n^4)$.

To find an optimal one, let us study a guillotine cut on a rectangle P. Let n be the number of point-holes. Since the guillotine cut passes a point-hole, there are at most $2n$ possible positions. Suppose P_1 and P_2 are two rectangles obtained from P by the guillotine cut. Let $opt(P)$ denote the minimum total length of guillotine partition on P. Then we have

$$opt(P) = \min_{\text{candidate cuts}} [opt(P_1) + opt(P_2) + (\text{cut length})],$$

The computation time for this recurrence is $O(n)$. Therefore, the optimal rectangular guillotine partition can be computed by a dynamic programming in $O(n^5)$ time.

One of the important techniques for design of dynamic programming for a given problem is to replace the original problem by a proper one which can be easily found to have a self-reducibility. The following is such an example.

Example 3.1.5 Consider a horizontal strip. There are n target points lying inside and m unit disks with centers lying outside of the strip where each unit disk d_i has radius one and a positive weight $w(d_i)$. Each target point is covered by at least one unit disk. The problem is to find a subset of unit disks, with minimum total weight, to cover all target points.

First, without loss of generality, assume all target points have distinct x-coordinates; otherwise, we may rotate the strip together with coordinate system a little to reach such a property. Line up all target points $p_1, p_2, \ldots, p_n$ in the increasing ordering of x-coordinate. Let $\mathcal{D}_a$ be the set of all unit disks with centers lying above the strip and $\mathcal{D}_b$ the set of all unit disks with centers lying below the strip. Let $\ell_1, \ell_2, \ldots, \ell_n$ be vertical lines passing through $p_1, p_2, \ldots, p_n$, respectively. For any two disks $d, d' \in \mathcal{D}_a$, define $d \prec_i d'$ if the lowest intersection between the boundary of disk d and ℓ_i is not lower than the lowest intersection between the boundary of disk d' and ℓ_i. Similarly, for any two sensors $d, d' \in \mathcal{D}_b$, define $d \prec_i d'$ if the highest intersection between the boundary of disk d and ℓ_i is not higher than the highest intersection between the boundary of disk d' and ℓ_i.

For any two disks $d_a \in \mathcal{D}_a$ and $d_b \in \mathcal{D}_b$ with p_i covered by d_a or d_b, let $D_i(d_a, d_b)$ be an optimal solution of the following problem.

$$\min \ w(D) = \sum_{d \in D} w(d) \tag{3.1}$$

$$\text{subject to } d_a, d_b \in D,$$

$$\forall d \in D \cap \mathcal{D}_a : d \prec_i d_a,$$

$$\forall d \in D \cap \mathcal{D}_b : d \prec_i d_b,$$

$$D \text{ covers target points } p_1, p_2, \ldots, p_i.$$

Then, we have the following recursive formula.

Lemma 3.1.6

$$w(D_i(d_a, d_b)) = \min\{w(S_{i-1}(d_a', d_b')) + [d_a \neq d_a']w(d_a) + [d_b \neq d_b']w(d_b)$$

$$\mid d_a' \prec_i d_a, \; d_b' \prec_i d_b, \; \text{and } p_{i-1} \text{ is covered by } d_a' \text{ or } d_b'\}$$

where

$$[d \neq d'] = \begin{cases} 1 & \text{if } d \neq d', \\ 0 & \text{otherwise}. \end{cases}$$

Proof Let d_a' be the disk in $D_i(d_a, d_b) \cap \mathcal{D}_a$ whose boundary has the lowest intersection with ℓ_{i-1} and d_b' the disk in $D_i(d_a, d_b) \cap \mathcal{D}_b$ whose boundary has the highest intersection with ℓ_{i-1}. We claim that

$$w(D_i(d_a, d_b)) = w(D_{i-1}(d_a', d_b')) + [d_a \neq d_a']w(d_a) + [d_b \neq d_b']w(d_b). \qquad (3.2)$$

To prove it, we first show that if $d_a \neq d_a'$, then $d_a \notin D_{i-1}(d_a', d_b')$ for $w(d_a) > 0$. In fact, for otherwise, there exists $i' < i - 1$ such that $p_{i'}$ is covered by d_a, but not covered by d_a'.

This is impossible (Fig. 3.4). To see this, let A be the lowest intersection between the boundary of disk d_a' and $\ell_{i'}$ and B the lowest intersection between the boundary of disk d_a' and ℓ_i. Then A and B lie inside the disk d_a. Let C and D be intersection points between line AB and the boundary of disk d_a. Let E be the lowest intersection between the boundary of disk d_a and ℓ_{i-1} and F the lowest intersection between

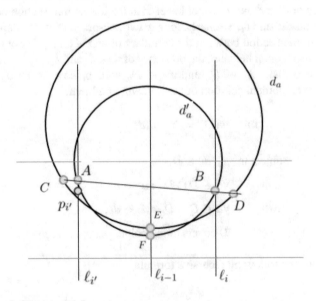

Fig. 3.4 Proof of Lemma 3.1.6

the boundary of disk d_a' and ℓ_{i-1}. Note that d_a and d_a' lie above the strip. We have $\angle CED > \angle AFB > \pi/2$ and hence $\sin \angle CED < \sin \angle AFB$. Moreover, we have $|AB| < |CD|$. Thus,

$$\text{radius } (d_a) = \frac{|CD|}{2 \sin \angle CED} > \frac{|AB|}{2 \sin \angle AFB} = \text{radius}(d_a'),$$

contradicting the homogeneous assumption of disks. Therefore, our claim is true. Similarly, if $d_b \neq d_b'$, then $d_b \notin S_{i-1}(d_a', d_b')$ for $w(s_b) > 0$. Therefore, (3.2) holds. This means that for equation in Lemma 3.1.6, the left-side $\geq$ the right-side.

To see the left-side $\leq$ the right-side for the equation in Lemma 3.1.6, we note that in the right side, $S_{i-1}(d_a', d_b') \cup \{d_a, d_b\}$ is always a feasible solution of the problem (3.1). □

Let us employ the recursive formula in Lemma 3.1.6 to compute all $S_i(d_a, d_b)$. There are totally $O(m^2 n)$ problems. With the recursive formula, each $S_i(d_a, d_b, k)$ can be computed in time $O(m^2)$. Therefore, all $S_i(d_a, d_b, k)$ can be computed by dynamic programming in time $O(m^4 n)$. The solution of Example 3.1.5 can be computed by

$$S = \text{argmin}_{S_n(d_a, d_b)} w(S_n(d_a, d_b))$$

where $d_a \in \mathcal{D}_a$, $d_b \in \mathcal{D}_b$, and p_n is covered by d_a or d_b. This requires additional computation within time $O(m^2)$. Therefore, putting all computations together, the time is $O(m^4 n)$.

Example 3.1.7 Consider a directed graph $G = (V, E)$. A node v is said to be influenced by another node u if there exists a path from u to v. Given a positive integer k, the influence maximization problem is to find a subset of at most k nodes, called *seeds*, to influence the maximum number of nodes. Suppose G is an in-arborescence, i.e., a directed tree with all arc directed to the root. A dynamic programming algorithm can give a polynomial-time solution to the influence maximization problem.

First, note that for an arc (u, v), selecting u as a seed is better than selecting v. It follows that all seeds should be selected from leaves.

Second, note that we may introduce some virtual nodes to transform the input in-arborescence into a binary in-arborescence that every internal node has in-degree at most two (Fig. 3.5). Then, assign every original node with weight one and every virtual node with weight zero. Then, the number of virtual nodes is at most $|E|$ ($< |V|$ for any arborescence).

Let $f(v, k)$ denote the maximum total weight of influenced nodes when k seeds are placed in the sub-arborescence rooted at v. If v is a leaf, then v must be an original node and hence, $f(v, k) = 1$ for every $k \geq 1$. If v is not a leaf, then v have either one coming neighbor u or two coming neighbors u_1 and u_2. In the former case,

Fig. 3.5 Transform an
arborescence to a binary
arborescence by adding
virtual nodes

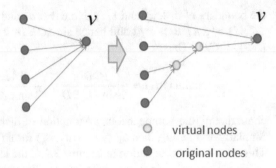

○ virtual nodes
● original nodes

$$f(v, k) = f(u, k) + w_v$$

where w_v is the weight of node v. In the latter case,

$$f(v, k) = w_v + \max_{k_1+k_2=k, k_1, k_2 \geq 0} [f(u_1, k_1) + f(u_2, k_2)].$$

This recurrence suggests a dynamic programming algorithm with running time $O(nk^2)$ where $n = |V|$.

3.2 Shortest Path

Often, the running time of a dynamic programming algorithm can be estimated by the product of the table size (the number of subproblems) and the computation time of the recursive formula (i.e., the time for recursively computing the solution of each subproblem). **Does this estimation hold for every dynamic programming algorithm?** The answer is No. In this section, we would like to provide a counterexample, the shortest path problem. For this problem, we must consider something else in order to estimate the running time of a dynamic programming algorithm.

Problem 3.2.1 (Shortest Path) *Given a directed graph $G = (V, E)$ with arc cost $c : E \to R$, a source node s, and a sink node t in V, where R is the set of real numbers, find a path from s to t with minimum total arc cost.*

In the study of shortest path, arcs coming to s and arc going out from t are useless. Therefore, we assume that those arcs do not exist in G, which may simplify some statements later.

For any node $u \in V$, let $d^*(s, u)$ denote the total cost of the shortest path from node s to node u and $N^-(u)$, the in-neighbor set of u, i.e., the set of nodes each with an arc coming to u. Then, we can obtain the following recursive formula (Fig. 3.6).

$$d^*(s) = 0,$$

$$d^*(u) = \min_{v \in N^-(v)} \{d^*(v) + c(v, u)\}.$$

Fig. 3.6 Recursive relation
of $d^*(u)$

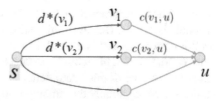

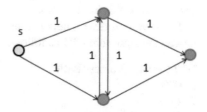

Fig. 3.7 When $S = \{s\}$,
there is no node u such that
$N^-(u) \subseteq S$

Based on this recursive formula, we may write down an algorithm as follows:

DP1 for the Shortest Path
$S \leftarrow \{s\}$;
$T \leftarrow V - S$;
while $T \neq \emptyset$ **do begin**
        find $u \in T$ such that $N^-(u) \subseteq S$;
        compute $d^*(u) = \min_{v \in N^-(u)}\{d^*(v) + c(v, u)\}$;
        $S \leftarrow S \cup \{u\}$;
        $T \leftarrow T - \{u\}$;
end-while
output $d^*(t)$.

This is a dynamic programming algorithm which works correctly for all acyclic
digraphs due to the following.

Theorem 3.2.2 *Consider an acyclic network $G = (V, E)$ with a source node s and
a sink node t. Assume that for any $v \in V - \{s\}$, $N^-(v) \neq \emptyset$. Let (S, T) be a partition
of V such that $s \in S$ and $t \in T$. Then there exists $u \in T$ such that $N^-(u) \subseteq S$.*

Proof Note that for any $u \in T$, $N^-(u) \neq \emptyset$. If $N^-(u) \not\subseteq S$, then there exists
$v \in N^-(u)$ such that $v \in T$. If $N^-(v) \not\subseteq S$, then there exists $w \in N^-(v)$ such
that $w \in T$. This process cannot go forever. Finally, we would find $z \in T$ such that
$N^-(z) \subseteq S$.   □

In this theorem, the acyclic condition cannot be dropped. In Fig. 3.7, a counterexample is shown that a simple cycle may make no node u in T satisfy $N^-(u) \subseteq S$.

To estimate the running time of algorithm DP1, we note that $d^*(u)$ needs to be
computed for u over all nodes, that is, the size of table for holding all subproblems is
$O(n)$ where n is the number of nodes. In the recursive formula for computing each
$d^*(u)$, the "min" operation is over all nodes in $N^-(u)$ which may contain $O(n)$

nodes. Thus, the product of the table size and the computation time of recursive formula is $O(n^2)$. However, this estimation for the running time of algorithm DP1 is not correct. In fact, we need also to consider the time for finding $u \in T$ such that $N^-(u) \subseteq S$. This requires to check if a set is a subset of another set. What is the running time of this computation? Roughly speaking, this may take $O(n \log n)$ time, and hence, totally the running time of algorithm DP1 is $O(n(n + n \log n)) = O(n^2 \log n)$.

Can we improve this running time by a smarter implementation? The answer is yes. Let us do this in two steps.

First, we introduce a new number $d(u) = \min_{v \in N^-(u) \cap S}(d^*(v) + c(v, u))$ and rewrite the algorithm DP1 as follows.

DP2 for the Shortest Path
$S \leftarrow \emptyset$;
$T \leftarrow V$;
while $T \neq \emptyset$ **do begin**
    find $u \in T$ such that $N^-(u) \subseteq S$;
    $S \leftarrow S \cup \{u\}$;
    $T \leftarrow T - \{u\}$;
    $d^*(u) = d(u)$;
    for every $w \in N^+(u)$ update $d(w) \leftarrow \min(d(w), d^*(u) + c(u, w))$;
end-while
output $d^*(t)$.

In this algorithm, updating value of $d(u)$ would be performed on all edges, and for each edge, update once. Therefore, the total time is $O(m)$ where m is the number of edges, i.e., $m = |E|$.

Secondly, we introduce the topological sort. The *topological sort* of nodes in a digraph $G = (V, E)$ is an ordering such that for any arc $(u, v) \in E$, node u has position before node v. Please note that the topological sort exists only for directed acyclic graphs, which are exactly those networks where the dynamic programming can work for the shortest path problem by Theorem 3.2.2.

There is an algorithm with running time $O(m)$ for topological sort as shown in Algorithm 7. Actually, in Algorithm 7, line 3 takes $O(n)$ time, and line 7 takes $O(m)$ time. Hence, it runs totally in $O(m + n)$ time. However, for the shortest path problem, input directed graph is connected if ignoring the arc direction, and hence $n = O(m)$. Therefore, $O(m + n) = O(m)$.

An example for topological sort is shown in Fig. 3.8. In each iteration, yellow node is the one selected from S to initiate the iteration. During the iteration, the yellow node will be moved from S to end of L, and all arcs from the yellow node will be deleted; meanwhile, new nodes will be added to S.

Algorithm 7 Topological sort

Input: A directed graph $G = (V, E)$.
Output: A topologically sorted sequence of nodes.
1: $L \leftarrow \emptyset$
2: $S \leftarrow \{s\}$
3: **while** $S \neq \emptyset$ **do**
4:     remove a node u from S
5:     put u at tail of L
6:     **for** each node $v \in N^+(u)$ **do**
7:         remove arc (u, v) from graph G
8:         **if** v has no other incoming arc **then**
9:             insert v into S
10:         **end if**
11:     **end for**
12: **end while**
13: **if** graph G has an arc **then**
14:     **return** error (G contains at least one cycle)
15: **else**
16:     **return** L
17: **end if**

Algorithm 8 Dynamic programming for shortest path

Input: A directed graph $G = (V, E)$ with arc weight $c : E \rightarrow Z$, and two nodes s and t in V.
Output: The length of shortest path from s to t.
1: $S \leftarrow \emptyset$
2: $T \leftarrow V$
3: do topological sort on T
4: $d(s) \leftarrow 0$
5: **for** every $u \in V \setminus \{s\}$ **do**
6:     $d(u) \leftarrow \infty$
7: **end for**
8: **while** $T \neq \emptyset$ **do**
9:     remove the first node u from T
10:     $S \leftarrow S \cup \{u\}$
11:     $d^*(u) \leftarrow d(u)$
12:     **for** every $(u, v) \in E$ **do**
13:         $d(v) \leftarrow \min(d(v), d^*(u) + c(u, v))$
14:     **end for**
15: **end while**
16: **return** $d^*(t)$

Now, we can first do topological sort and then carry out dynamic programming, which will result in a dynamic programming (Algorithm 8 for the shortest path problem, running in $O(m)$ time).

An example is shown in Fig. 3.9. At the beginning, the topological sort is done in the previous example as shown in Fig. 3.8. In Fig. 3.9, the yellow node represents the one removed from the front of T to initiate an iteration. During the iteration, all red

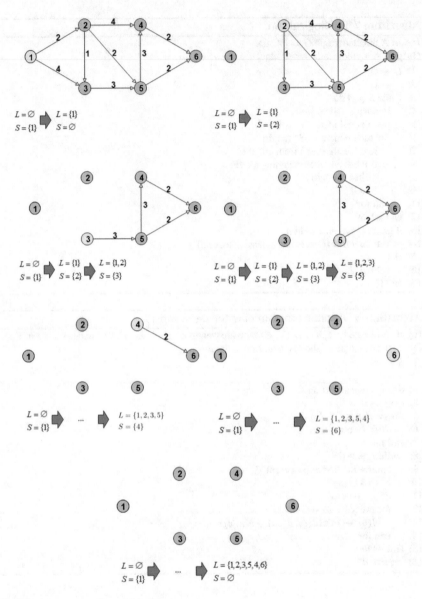

Fig. 3.8 An example of topological sort

arcs from the yellow node are used for updating the value of $d(\cdot)$, and meanwhile, the yellow node is added to S whose $d^*(\cdot)$'s value equals to $d(\cdot)$'s value.

It may be worth mentioning that Algorithm 8 works for acyclic directed graph without restriction on arc weight, i.e., arc weight can be negative. This implies that the longest path problem can be solved in $O(m)$ time if input graph is acyclic. For definition of the longest path problem, please find it in Chap. 8. The longest path

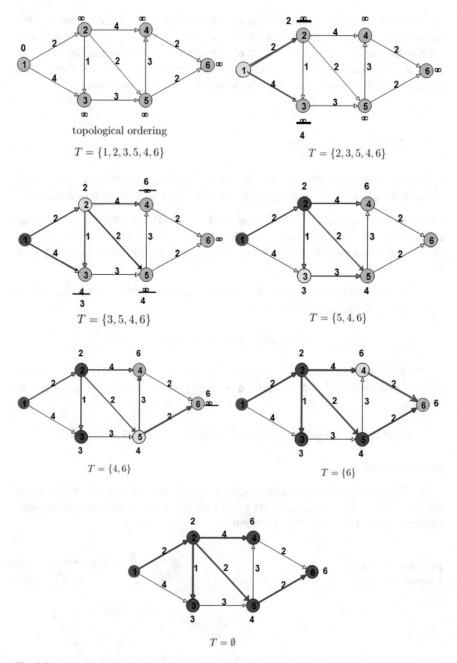

Fig. 3.9 An example of dynamic programming for shortest path

problem is NP-hard and hence unlikely to have a polynomial-time solution. This means that for the shortest path problem, if input directed graph is not acyclic and arc weights can be negative, then solution may not be polynomial-time computable. What about the case when input directed graph is not acyclic and all arc weights are nonnegative? In the next section, we present a polynomial-time solution.

3.3 Dijkstra Algorithm

Dijkstra algorithm is able to find the shortest path in any directed graph with nonnegative arc weights. Its design is based on the following important discovery.

Theorem 3.3.1 *Consider a directed network $G = (V, E)$ with a source node s and a sink node t and every arc (u, v) has a nonnegative weight $c(u, v)$. Suppose (S, T) is a partition of V such that $s \in S$ and $t \in T$. If $d(u) = \min_{v \in T} d(v)$, then $d^*(u) = d(u)$.*

Proof For contradiction, suppose $d(u) = \min_{v \in T} d(v) > d^*(u)$. Then there exists a path p (Fig. 3.10) from s to u such that

$$length(p) = d^*(u) < d(u).$$

Let w be the first node in T on path p. Then $d(w) = length(p(s, w))$ where $p(s, w)$ is the piece of path p from s to w. Since all arc weights are nonnegative, we have

$$length(p) \geq length(p(s, w)) = d(w) \geq d(u) > d^*(u) = length(p),$$

a contradiction.   □

By Theorem 3.3.1, in dynamic programming for shortest path, we may replace $N^-(u) \subseteq S$ by $d(u) = \min_{v \in T} d(v)$ when all arc weights are nonnegative. This replacement results in Dijkstra algorithm.

Fig. 3.10 In proof of
Theorem 3.3.1

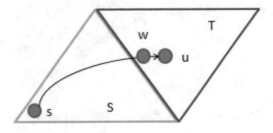

Dijkstra Algorithm
$S \leftarrow \emptyset$;
$T \leftarrow V$;
while $T \neq \emptyset$ **do begin**
    find $u \leftarrow \operatorname{argmin}_{v \in T} d(v)$;
    $S \leftarrow S \cup \{u\}$;
    $T \leftarrow T - \{u\}$;
    $d^*(u) = d(u)$;
    for every $w \in N^+(u)$, update $d(w) \leftarrow \min(d(w), d^*(u) + c(u, w))$;
end-while
output $d^*(t)$.

With different data structures, Dijkstra algorithm can be implemented with different running times.

With min-priority queue, Dijkstra algorithm can be implemented in time $O((m + n) \log n)$.

With Fibonacci heap, Dijkstra algorithm can be implemented in time $O(m + n \log n)$.

With Radix heap, Dijkstra algorithm can be implemented in time $O(m + n \log c)$ where c is the maximum arc weight.

We will pick up one of them to introduce in the next section. Before doing it, let us first implement Dijkstra algorithm with simple buckets (also known as Dial algorithm). This implementation requires all arc weights as integers. It can achieve running time $O(m + nc)$ where c is the maximum arc weight, of course, also an integer. When c is small, e.g., $c = 1$, it could be a good choice. (This case occurs in the study of Edmonds-Karp algorithm for maximum flow in Chap. 5.)

In this implementation, $(n - 1)c + 2$ buckets are prepared with labels $0, 1, \ldots, (n - 1)c, \infty$. They are used for storing nodes in T such that every node u is stored in bucket $d(u)$. Therefore, initially, s is in bucket 0, and other nodes are in bucket ∞. As $d(u)$'s value is updated, node u will be moved from a bucket to another bucket with smaller label. Note that if $d(u) < \infty$, then there must exist a simple path from s to u such that $d(u)$ is equal to the total weight of this path. Therefore, $d(u) \leq c(n-1)$, i.e., buckets set up as above are enough for our purpose. In Fig. 3.11, an example is computed by Dijkstra algorithm with simple buckets.

Now, let us estimate the running time of Dijkstra algorithm with simple buckets.

- Time to create buckets is $O(nc)$.
- Time for finding u to satisfy $d(u) = \min_{v \in T} d(v)$ is $O(nc)$. In fact, u can be chosen arbitrarily from the nonempty bucket with smallest label. Such a bucket in Fig. 3.11 is pointed by a red arrow, which is traveling from left to right without going backward. This is because, after update $d(w)$ for $w \in T$, we always have $d^*(u) \leq d(w)$ for $w \in T$.
- Time to update $d(w)$ for $w \in T$ and update buckets is $O(m)$.
- Therefore, total time is $O(m + nc)$.

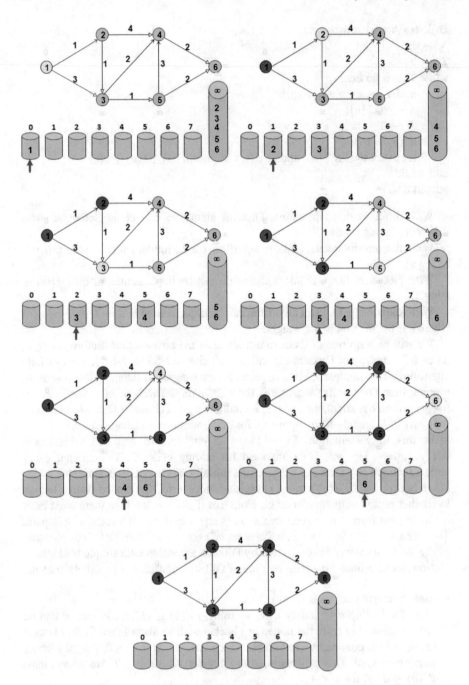

Fig. 3.11 An example for Dijkstra algorithm with simple buckets

3.4 Priority Queue

Although Dijkstra algorithm with simple buckets runs faster for small c, it cannot be counted as a polynomial-time solution. In fact, the input size of c is $\log c$. Therefore, we would like to select a data structure which implements Dijkstra algorithm in polynomial-time. This data structure is priority queue.

A priority queue is a data structure for maintaining a set S of elements, each with an associated value, called a key. All keys are stored in an array A such that an element belongs to set S if and only if its key is in array A. There are two types of priority queues, the min-priority queue and the max-priority queue. Since they are similar, we introduce one of them, the min-priority queue.

A min-priority queue supports the following operations: Minimum(S), Extract-Min(S), Increase-Key(S, x, k), and Insert(S, x).

The min-heap can be employed in implementation of those operations.

Minimum(S) returns the element of S with the smallest key, which can be implemented as follows.

Heap-Minimum(A)
   return $A[1]$.

Extract-Min(S) removes and returns the element of S with the smallest key, which can be implemented by using min-heap as follows.

Heap-Extract-Min(A)
   min $\leftarrow A[1]$;
   $A[1] \leftarrow A[\text{heap-size}[A]]$;
   heap-size$[A] \leftarrow$ heap-size$[A]$-1;
   Min-Heapify($A, 1$);
   return min.

Decrease-Key(S, x, k) decreases the value of element x's key to the new value k, which is assumed to be no more than x's current key value. Suppose that $A[i]$ contains x's key. Then, Decrease-Key(S, x, k) can be implemented as an operation of min-heap as follows.

Heap-Decrease-Key(A, i, key)
   if $key > A[i]$
      then error "new key is larger than current key";
   $A[i] \leftarrow key$;
   while $i > 1$ and $A[\text{Parent}(i)] > A[i]$
      do exchange $A[i] \leftrightarrow A[\text{Parent}(i)]$
        and $i \leftarrow \text{Parent}(i)$.

Insert($S, x.key$) inserts the element x into S, which is implemented in the following.

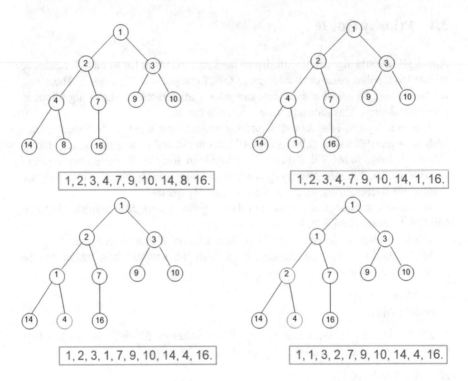

1, 2, 3, 4, 7, 9, 10, 14, 8, 16.

1, 2, 3, 4, 7, 9, 10, 14, 1, 16.

1, 2, 3, 1, 7, 9, 10, 14, 4, 16.

1, 1, 3, 2, 7, 9, 10, 14, 4, 16.

Fig. 3.12 Heap-Decrease-Key$(A, 9, 1)$

Insert(A, key)
    array-size$[A] \leftarrow$ array-size$[A] + 1$;
    $A[$array-size$[A]] \leftarrow +\infty$;
    Decrease-Key$(A,$ array-size$[A], key)$.

Now, we analyze these four operations. Minimum(S) runs clearly in $O(1)$ time. Each of the other three operations runs in $O(\log n)$ time. Actually, since Min-Heapify$(A, 1)$ runs in $O(\log n)$ time, so does Extract-Min(S). For Decrease-Key(S, x, k), as shown in Fig. 3.12, computation is along a path from a node approaching to the root of the heap and hence runs in $O(\log n)$ time. This also implies that Insert$(S, x.key)$ can be implemented in $O(\log n)$ time.

In Algorithm 9, Dijkstra algorithm is implemented with priority queue as follows.

- Use min-priority queue to keep set T, and for every node $u \in T$, use $d(u)$ for the key of u.
- Use operation Extract-Min(T) to obtain u satisfying $d(u) = \min_{v \in T} d(v)$. This operation at line 9 will be used for $O(n)$ times.
- Use operation Decrease-Key(T, v, key) on each edge (u, v) to update $d(v)$ and the min-heap. This operation on line 14 will be used for $O(m)$ times.
- Therefore, the total running time is $O((m + n) \log n)$.

Algorithm 9 Dijkstra algorithm with priority queue

Input: A directed graph $G = (V, E)$ with arc weight $c : E \to Z$ and source node s and sink node t in V.

Output: The length of shortest path from s to t.

```
1:  S ← ∅
2:  T ← V
3:  d(s) ← 0
4:  for every u ∈ V \ {s} do
5:      d(u) ← ∞
6:  end for
7:  build a min-priority queue on T with key d(u) for each node u ∈ T, i.e., use keys to build a
    min-heap.
8:  while T ≠ ∅ do
9:      u ← Extract-Min(T)
10:     S ← S ∪ {u}
11:     d*(u) ← d(u)
12:     for every (u, v) ∈ E do
13:         if d(v) > d*(u) + c(u, v) then
14:             Decrease-Key(T, v, d*(u) + c(u, v))
15:         end if
16:     end for
17: end while
18: return d*(t)
```

An example is as shown in Fig. 3.13.

3.5 Bellman-Ford Algorithm

Bellman-Ford algorithm allows negative arc cost; only restriction is no negative weight cycle. The disadvantage of this algorithm is that the running time is a little slow. The idea behind this algorithm is very simple.

Initially, assign $d(s) = 0$ and $d(u) = \infty$ for every $u \in V \setminus \{s\}$. Then, algorithm updates $d(u)$ such that in iteration i, $d(u)$ is equal to the shortest distance from s to u passing through at most i arcs. If there is no negative weight cycle, then the shortest path contains at most $n - 1$ arcs. Therefore, after $n - 1$ iterations, $d(t)$ is the minimum weight of the path from s to t. If in the nth iteration, $d(u)$ is still updated for some u, then it means that a negative weight cycle exists.

Bellman-Ford Algorithm

input: A directed graph $G = (V, E)$ with weight $c : E \to R_+$,
        a source node s and a sink node t.

output: The minimum weight of path from s to t,
        or a message "G contains a negative weight cycle".

$d(s) \leftarrow 0;$
for $u \in V \setminus \{s\}$ **do**

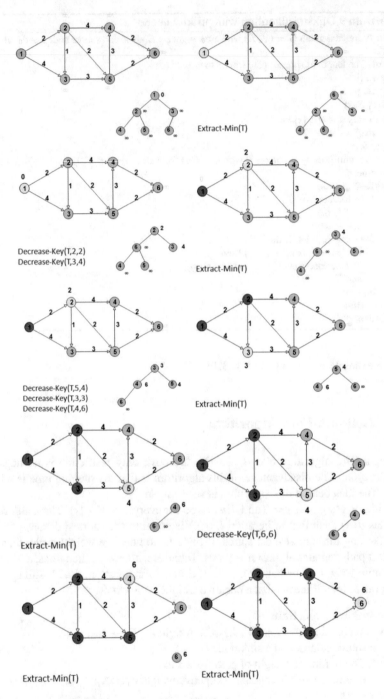

Fig. 3.13 An example for Dijkstra algorithm with priority queue

$$d(u) \leftarrow \infty;$$
for $i \leftarrow 1$ **to** $n - 1$ **do**
  for each arc $(u, v) \in E$ **do**
    if $d(u) + c(u, v) < d(v)$
      then $d(v) \leftarrow d(u) + c(u, v);$
for each arc $(u, v) \in E$ **do**
  if $d(u) + c(u, v) < d(v)$
    then return "G contains a negative weight cycle".
    else return $d(t).$

Its running time is easily estimated.

Theorem 3.5.1 *Bellman-Ford algorithm computes a shortest path from s to t within $O(mn)$ time where n is the number of nodes and m is the number of arcs in input directed graph.*

3.6 All Pairs Shortest Paths

In this section, we study the following problem.

Problem 3.6.1 (All-Pairs-Shortest-Paths) *Given a directed graph $G = (V, E)$, find the shortest path from s to t for all pairs $\{s, t\}$ of nodes.*

If we apply the Bellman-Ford algorithm for single pair of nodes for each of $O(n^2)$ pairs, then the total time for computing a solution of the all-pairs-shortest-paths problem is $O(n^3 m)$. In the following, we will present two faster algorithms, with running time $O(n^3 \log n)$ and $O(n^3)$, respectively, with only restriction that no negative weight cycle exists. Before doing so, let us consider an example on which we introduce an approach which can be used for the all-pairs-shortest-paths problem.

Example 3.6.2 (Path Counting) Given a directed graph $G = (V, E)$ and a positive integer k, count the number of paths with exactly k arcs from s to t for all pairs $\{s, t\}$ of nodes

Let $a_{st}^{(k)}$ denote the number of paths with exactly k arcs from s to t. Then, we have

$$a_{st}^{(1)} = \begin{cases} 1 \text{ if } (s, t) \in E, \\ 0 \text{ otherwise.} \end{cases}$$

This means that $(a_{st}^{(1)})$ is the adjacency matrix of graph G. Denote

$$A(G) = (a_{st}^{(1)}).$$

We claim that

$$A(G)^k = (a_{st}^{(k)}).$$

Let us prove this claim by induction on k. Suppose it is true for k. Consider a path from s to t with exactly $k + 1$ arcs. Decompose the path at a node h such that the subpath from s to h contains exactly k arcs and (h, t) is an arc. Then the subpath from s to h has $a_{sh}^{(k)}$ choices and (h, t) has $a_{ht}^{(1)}$ choices. Therefore,

$$a_{st}^{(k+1)} = \sum_{h \in V} a_{sh}^{(k)} a_{ht}^{(1)}.$$

It follows that

$$(a_{st}^{(k+1)}) = (a_{sh}^{(k)})(a_{ht}^{(1)}) = A(G)^k \cdot A(G) = A(G)^{k+1}.$$

Now, we come back to the all-pairs-shortest-paths problem. First, we assume that G has no loop. In fact, a loop with nonnegative weight does not play any role in a shortest path and a loop with negative weight means that the problem is meaningless.

Let $\ell_{st}^{(k)}$ denote the weight of the shortest path with at most k arcs from s to t. For $k = 1$, we have

$$\ell_{st}^{(1)} = \begin{cases} c(s, t) & \text{if } (s, t) \in E, \\ \infty & \text{if } (s, t) \notin E \text{ and } s \neq t, \\ 0 & \text{if } s = t. \end{cases}$$

Denote

$$L(G) = (\ell_{st}^{(1)}).$$

This is called the *weighted adjacency matrix*. For example, the graph in Fig. 3.14 has weighted adjacency matrix

$$\begin{pmatrix} 0 & 4 & \infty \\ \infty & 0 & 6 \\ 5 & \infty & 0 \end{pmatrix}.$$

Fig. 3.14 A weighted directed graph

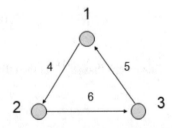

We next establish a recursive formula for $\ell_{st}^{(k)}$.

Lemma 3.6.3

$$\ell_{st}^{(k+1)} = \min_{h \in V}(\ell_{sh}^{(k)} + \ell_{st}^{(1)}).$$

Proof Since the shortest path from s to h with at most k arcs and the shortest path from h to t with at most one arc form a path from s to t with at most $k+1$ arcs, we have

$$\ell_{st}^{(k+1)} \leq \min_{h \in V}(\ell_{sh}^{(k)} + \ell_{ht}^{(1)}).$$

Next, we show

$$\ell_{st}^{(k+1)} \geq \min_{h \in V}(\ell_{sh}^{(k)} + \ell_{ht}^{(1)}).$$

To do so, consider two cases.

Case 1. There is a path with weight $\ell_{st}^{(k+1)}$ from s to t containing at most k arcs. In this case, we have

$$\ell_{st}^{(k+1)} = \ell_{st}^{(k)}$$
$$= \ell_{st}^{(k)} + \ell_{ht}^{(1)}$$
$$\geq \min_{h \in V}(\ell_{sh}^{(k)} + \ell_{ht}^{(1)}).$$

Case 2. Every path with weight $\ell_{st}^{(k+1)}$ from s to t contains exactly $k+1$ arcs. In this case, we can find a node h' on the path such that the piece from s to h' contains exactly k arcs and $(h', t) \in E$. Their weights should be $\ell_{sh'}^{(k)}$ and $\ell_{h't}^{(1)}$, respectively. Therefore,

$$\ell_{st}^{(k+1)} = \ell_{sh'}^{(k)} + \ell_{h't}^{(1)}$$
$$\geq \min_{h \in V}(\ell_{sh}^{(k)} + \ell_{ht}^{(1)}).$$

$\square$

If G does not contain negative weight cycle, then each shortest path does not need to contain a cycle. Therefore, we have

Theorem 3.6.4 *If G does not have a negative weight cycle, then $\ell_{st}^{(n-1)}$ is the weight of shortest path from s to t where $n = |V|$.*

This suggests a dynamic programming to solve the all-pairs-shortest-paths problem by using recursive formula in Lemma 3.6.3. Since each $\ell_{st}^{(k)}$ is computed

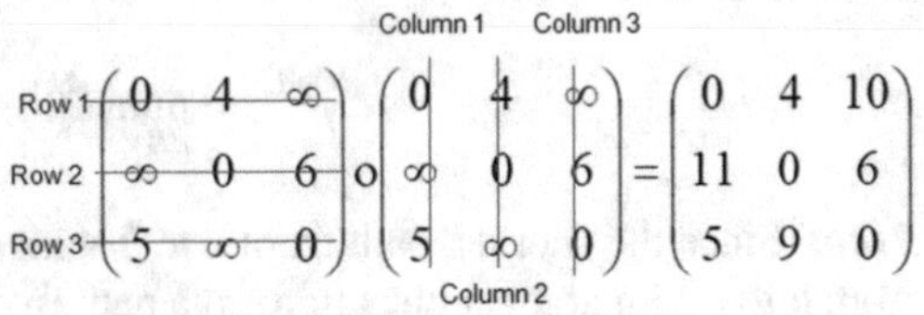

Fig. 3.15 A new matrix multiplication

$$(a_{ih})_{m \times n} \text{ o}(b_{hj})_{n \times p} = (\min_{1 \le h \le n}\{a_{ih} + b_{hj}\})_{m \times p}$$

in $O(n)$ time, this algorithm will run in $O(n^4)$ time to compute $\ell_{st}^{(n-1)}$ for all pairs $\{s, t\}$.

Next, we give a method to speed up this algorithm. To do so, let us define a new operation for matrixes. Consider two $n \times n$ square matrixes $A = (a_{ij})$ and $B = (b_{ij})$. Define

$$A \text{ o } B = \left(\min_{1 \le h \le n} (a_{ih} + b_{hj}) \right).$$

An example is as shown in Fig. 3.15.

This operation satisfies associative law.

Lemma 3.6.5

$$(A \text{ o } B) \text{ o } C = A \text{ o } (B \text{ o } C).$$

We leave proof of this lemma as an exercise.

By this lemma, the following is well-defined.

$$A^{(k)} = \underbrace{A \text{ o } \cdots \text{ o } A}_{k}.$$

Note that if G has no negative weight cycle, then for $m \ge n - 1$, $L(G)^{(m)} = L(G)^{(n-1)}$. This observation suggests the following algorithm to compute $L(G)^{(n-1)}$.

> $n \leftarrow |V|$;
> $m \leftarrow 1$;
> $L^{(1)} \leftarrow L(G)$;
> **while** $m < n - 1$
>   **do** $L^{2m} \leftarrow L^{(m)} \text{ o } L^{(m)}$ and
> $m \leftarrow 2m$;
> **return** $L^{(m)}$.

With this improvement, the dynamic programming with recursive formula in Lemma 3.6.3 is called the *faster-all-pairs-shortest-paths* algorithm, which runs in $O(n^3 \log n)$ time.

Fig. 3.16 Proof of
Lemma 3.6.7

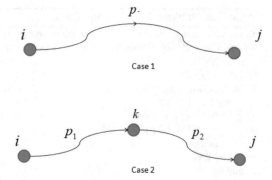

The above result is derived under assumption that G does not have a negative weight cycle. Suppose G is unknown to have a negative weight cycle or not. Can we modify the faster-all-pairs-shortest-paths algorithm to find a negative weight cycle if G has one? The answer is yes. However, we need to compute $L(G)^{(m)}$ for $m \geq n$.

Theorem 3.6.6 *G contains a negative weight cycle if and only if $L(G)^{(n)}$ contains a negative diagonal entry. Moreover, if $L(G)^{(n)}$ contains a negative diagonal entry, then such an entry keeps negative sign in every $L(G)^{(m)}$ for $m \geq n$.*

Proof It follows immediately from the fact that a simple cycle contains at most n arcs.   □

Next, let us study another algorithm for the all-pairs-shortest-paths problem. First, we show a lemma.

Lemma 3.6.7 *Assume $V = \{1, 2, \ldots, n\}$. Let $d_{ij}^{(k)}$ denote the weight of shortest path from i to j with internal nodes in $\{1, 2, \ldots, k\}$. Then for $i \neq j$,*

$$d_{ij}^{(k)} = \begin{cases} c(i, j) & \text{if } k = 0, \\ \min(d_{ij}^{(k-1)}, d_{ik}^{(k-1)} + d_{kj}^{(k-1)}) & \text{if } 1 \leq k \leq n, \end{cases}$$

and $d_{ij}^{(k)} = 0$ for $i = j$ and $k \geq 0$.

Proof We need only to consider $i \neq j$. Let p be the shortest path from i to j with internal nodes in $\{1, 2, \ldots, k\}$. For $k = 0$, p does not contain any internal node. Hence, its weight is $c(i, j)$. For $k \geq 1$, there are two cases (Fig. 3.16).

Case 1. p does not contain internal node k. In this case,

$$d_{ij}^{(k)} = d_{ij}^{(k-1)}.$$

Case 2. p contains an internal node k. Since p does not contain a cycle, node k appears exactly once. Suppose that node k decomposes path p into two pieces p_1 and p_2, from i to k and from k to j, respectively. Then the weight of p_1 should be $d_{ik}^{(k-1)}$, and the weight of p_2 should be $d_{ij}^{(k-1)}$. Therefore, in this case, we have

---

Algorithm 10 Floyd-Warshall algorithm

Input: A directed graph $G = (V, E)$ with arc weight $c : E \to Z$.
Output: The weight of shortest path from s to t for all pairs of nodes s and t.

1: $n \leftarrow |V|$
2: $D^{(0)} \leftarrow L(G)$
3: **for** $k \leftarrow 1$ to n **do**
4:   **for** $i \leftarrow 1$ to n **do**
5:     **for** $j \leftarrow 1$ to n **do**
6:       $d_{ij}^{(k)} \leftarrow \min(d_{ij}^{(k-1)}, d_{ik}^{(k-1)} + d_{kj}^{(k-1)})$
7:     **end for**
8:   **end for**
9: **end for**
10: **return** $D^{(n)}$

$$d_{ij}^{(k)} = d_{ik}^{(k-1)} + d_{kj}^{(k-1)}.$$

□

Denote $D^{(k)} = (d_{ij}^{(k)})$. Based on recursive formula in Lemma 3.6.7, we obtain a dynamic programming as shown in Algorithm 10, which is called *Floyd-Warshall algorithm*.

From algorithm description, we can see the following.

Theorem 3.6.8 *If G does not contain a negative weight cycle, then Floyd-Warshall algorithm computes all-pairs shortest paths in $O(n^3)$ time.*

If G contains a negative weight cycle, could Floyd-Warshall algorithm tell us this fact? The answer is yes. Actually, we also have

Theorem 3.6.9 *G contains a negative weight cycle if and only if $D^{(n)}$ contains a negative diagonal element.*

Exercises

1. Please construct a directed graph $G = (V, E)$ with arc weight and source vertex s such that for every arc $(u, v) \in E$, there is a shortest-paths tree rooted at s that contains (u, v), and there is another shortest-paths tree rooted at s that does not contain (u, v).
2. Show that the graph G contains a negative weight cycle if and only if $A(G)^{(n-1)} \neq A(G)^{(2n-1)}$.
3. Please design an $O(n^2)$-time algorithm to compute the longest monotonically increasing subsequence for a given sequence of n numbers.
4. How can we use the output of the Floyd-Warshall algorithm to detect the presence of a negative weight cycle?

Fig. 3.17 PA stair

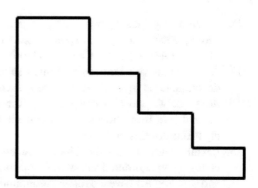

5. A stair is a rectilinear polygon as shown in Fig. 3.17. Show that the minimum length rectangular partition for a given stair can be computed by a dynamic programming in time $O(n^2 \log n)$.
6. Given a rectilinear polygon with hole free, design a dynamic programming to partition it into small rectangles with minimum total length of cuts.
7. Consider a horizontal line. There are n points lying below the line and m unit disks with centers above the line. Every one of the n points is covered by some unit disk. Each unit disk has a weight. Design a dynamic programming to find a subset of unit disks covering all n points, with the minimum total weight. The dynamic programming should run in polynomial time with respect to m and n.
8. Given a convex polygon in the Euclidean plane, partition it into triangles with minimum total length of cuts. Design a dynamic programming to solve this problem in time $O(n^3)$ where n is the number of vertices of input polygon.
9. Does Dijkstra's algorithm for shortest path work for input with negative weight and without negative weight cycle? If yes, please give a proof. If not, please give a counterexample and a way to modify the algorithm to work for input with negative weight and without negative weight cycle.
10. Given a directed graph $G = (V, E)$ and a positive integer k, count the number of paths with at most k arcs from s to t for all pairs of nodes s and t.
11. Given a graph $G = (V, E)$ and a positive integer k, count the number of paths with at most k edges from s to t for all pairs of nodes s and t.
12. Given a directed graph $G = (V, E)$ without loop, and a positive integer k, count the number of paths with at most k arcs from s to t for all pairs of nodes s and t.
13. Show Lemma 3.6.5, that is, $A \circ (B \circ C) = (A \circ B) \circ C$.
14. Does FASTER-ALL-PAIR-SHORTEST-PATH algorithm work for input with negative weight and without negative weight cycle? If yes, please give a proof. If not, please give a counterexample.
15. Does Floyd-Warshall algorithm work for input with negative weight and without negative weight cycle? If yes, please give a proof. If not, please give a counterexample.

16. Given a sequence $x_1, x_2, \ldots, x_n$ of (not necessary positive) integers, find a subsequence $x_i, x_{i+1}, \ldots, x_{i+j}$ of consecutive elements to maximize the sum $x_i + x_{i+1} + \cdots + x_{i+j}$. Can your algorithm run in linear time?

17. Assume that you have an unlimited supply of coins in each of the integer denominations $d_1, d_2, \ldots, d_n$, where each $d_i > 0$. Given an integer amount $m \geq 0$, we wish to make change for m using the minimum number of coins drawn from the above denominations. Design a dynamic programming algorithm for this problem.

18. Recall that $C(n, k)$—the binomial coefficient—is the number of ways of choosing an unordered subset of k objects from a set of n objects. (Here $n \geq 0$ and $0 \leq k \leq n$.) Give a dynamic programming algorithm to compute the value of $C(n, k)$ in $O(nk)$ time and $O(k)$ space.

19. Given a directed graph $G = (V, E)$, with nonnegative weight on its edges, and in addition, each edge is colored red or blue. A path from u to v in G is characterized by its total length and the number of times it switches colors. Let $\delta(u, k)$ be the length of a shortest path from a source node s to u that is allowed to change color at most k times. Design a dynamic program to compute $\delta(u, k)$ for all $u \in V$. Explain why your algorithm is correct, and analyze its running time.

20. Modify Dijkstra's algorithm in order to solve the bottleneck path problem: Given a directed graph $G = (V, E)$ with edge weight $c : E \rightarrow R$ and two nodes $s, t \in V$, find an s-t-path whose longest edge is shortest possible. Describe the whole algorithm, and show the correctness of your algorithm.

21. Given a set of n distinct points $x_1, x_2, \ldots, x_n$ and a set of m weighted closed intervals $I_1, I_2, \ldots, I_m$ on the real number line, find a subset of intervals to cover all points with the minimum total weight.

22. Consider a directed graph $G = (V, E)$ with a nonnegative arc weight $w : E \rightarrow R_+$. Let s and t be two distinct nodes in V. Suppose H is a subgraph obtained from G by deleting some arc. Design a $O(|E| \log |V|)$-time algorithm to find the shortest distance from s to t among all graphs from H by putting back one arc in $G \setminus H$.

23. Consider a directed graph $G = (V, E)$ with a nonnegative arc weight $w : E \rightarrow R_+$. Let s and t be two distinct nodes in V. Design a polynomial-time algorithm to find the second shortest path from s to t.

24. Design a dynamic programming algorithm to solve the knapsack problem as follows.

$$\max \ c_1 x_1 + c_2 x_2 + \cdots + c_n x_n$$
$$\text{subject to} \ a_1 x_1 + a_2 x_2 + \cdots + a_n x_n \leq S,$$
$$x_1, x_2, \ldots, x_n \in \{0, 1\},$$

where $c_1, c_2, \ldots, c_n, a_1, a_2, \ldots, a_n$ and S are given positive integers.

25. Given two sequences X and Y, design dynamic programming to compute solution of the following problems.

(a) Find the longest common subsequence.

(b) Find the longest consecutive subsequence.

(c) Find the shortest sequence which contains both X and Y as subsequences.

26. Consider two sequences $X = x_1 x_2 \cdots x_m$ and $Y = y_1 y_2 \cdots y_n$ and a set P of pairs (i, j) where $1 \leq i \leq m$ and $1 \leq j \leq n$. Suppose $Z = z_1 z_2 \cdots z_k$ is a common subsequence of X and Y such that $z_1 = x_{i_1} = y_{j_1}$, $z_2 = x_{i_2} = y_{j_2}$, $\ldots, z_k = x_{i_k} = y_{j_k}$. Then $(i_h, j_h) \in P$ is called a *loved pair*. Design a dynamic programming algorithm to find a longest common subsequence Z of X and Y such that among all longest common subsequence, Z contains the maximum number of loved pairs.

27. Consider a horizontal strip. There are n target points p_i for $1 \leq i \leq n$, lying inside of the strip and m unit disks d_j for $1 \leq j \leq m$ with centers lying outside of the strip, where each unit disk d_i has radius one. Assume that each target point is covered by at least one unit disk. Given a positive integer k, design an algorithm to find a subset of at most k unit disks, covering the maximum number of target points.

28. Consider a horizontal strip. There are n target points p_i for $1 \leq i \leq n$, lying inside of the strip and m unit disks d_j for $1 \leq j \leq m$ with centers lying outside of the strip, where each unit disk d_i has radius one. Assume that each target point is covered by at least one unit disk. Given a positive integer k, design an algorithm to find the smallest subset of unit disks to cover at least k target points.

29. Design a dynamic programming algorithm to solve the following problem: Given n activities each with a time period $[s_i, f_i)$ and a positive weight w_i, find a nonoverlapping subset of activities to maximize the total weight.

30. Consider the following recurrence:

$$c(i, i) = 0$$

$$c(i, j) = w(i, j) + \min_{i < k \leq j} (c(i, k - 1) + c(k, j)).$$

Suppose w satisfies the quadrangle inequality

$$w(i, k) + w(j, l) \leq w(j, k) + w(i, l),$$

and monotone,

$$w(j, k) \leq w(i, l),$$

for all $i \leq j \leq k \leq l$. Show that a dynamic programming algorithm can compute all $c(i, j)$ for $1 \leq i \leq j \leq n$ in $O(n^2)$ time.

31. Consider a set of n points $(x_1, y_1), (x_2, y_2), \ldots, (x_n, y_n)$ lying in the plane with all $x_i \geq 0$ and $y_i \geq 0$ for $1 \leq i \leq n$. Design a dynamic programming algorithm to compute the minimum length rectilinear arborescence rooted at

original $(0, 0)$, in which every arc can have its direction either going up or to the right.

32. Consider three strings X, Y, Z over alphabet Σ. Z is said to be a *shuffle* of X and Y if $Z = X_1 Y_1 X_2 Y_2 \cdots X_n Y_n$ where $X_1, X_2, \ldots, X_n$ are substring of X such that $X = X_1 X_2 \cdots X_n$ and $Y_1, Y_2, \ldots, Y_n$ are substrings of Y such that $Y = Y_1 Y_2 \cdots Y_n$. Design a dynamic programming algorithm to determine whether Z is a shuffle of X and Y for given X, Y, and Z.

33. Consider a tree $T = (V, E)$ with arbitrary integer weights $w : E \to Z$. Design an algorithm to compute the diameter of T, i.e., the maximum weight of a simple path in T.

34. Let $G = (V, E)$ be a planar graph lying in the Euclidean plane. The weight of any edge (u, v) is the Euclidean distance between nodes u and v, denoted by $L(u, v)$. For any two nodes x and y, denote by $d(x, y)$ the total weight of shortest path between x and y. If there is no path between x and y, then define $d(x, y) = \infty$. The *stretch factor* is defined to be the smallest upper bound for ratio $d(x, y)/L(x, y)$ for any two distinct nodes $x, y \in V$. Design an efficient algorithm to find the stretch factor for given graph G.

35. Given a set of n integers $a_1, a_2, \ldots, a_n$ and a target integer T, design a dynamic programming algorithm to determine whether there exists a subset of given integers such that their sum is equal to T. Your algorithm should run in $O(nT)$ time.

36. (Sensor Barrier Cover) Consider a rectangle R. Randomly deploy a set of sensors. Each sensor can monitor an area, called *sensing area*. Suppose that the sensing area of every sensor is a unit disk. A subset of sensors is called a *barrier cover* if their sensing areas cover a curve connecting two vertical edges of R. Given a set of sensors, find the minimum barrier cover. Please formulate this problem into a shortest path problem.

37. (Influence Maximization) A social network is a directed graph $G = (V, E)$ with an information diffusion model m. Suppose m is the linear threshold (LT) model as follows: Each arc (u, v) is associated with a weight $w_{uv} \in [0, 1]$ such that for any node v, the total weight of arcs coming to v is at most one. Each node has two possible states, active and inactive. Initially, all nodes are inactive. To start an information diffusion process, we may select a few nodes, called *seed*, to activate and select, for each node v, a threshold θ_v uniformly and randomly from $[0, 1]$. Then step by step, more nodes will be activated. In each step, every inactive node v gets evaluated for the total weight of coming arcs from active nodes. If it is less than its threshold θ_v, then v is kept in inactive state; otherwise, v is activated. This process ends at the step in which no new node can be activated. Given a positive integer k, select k seeds to maximize the expected number of active nodes (including themselves) at the end of the process. This problem is called the *influence maximization*. Suppose G is an in-arborescence. Design a dynamic programming algorithm for the influence maximization problem. Could your algorithm run in a polynomial-time with respect to $|V|$?

38. (Effector Detection) Consider a social network $G = (V, E)$ with the information diffusion model LT as stated in previous problem. The *effector detection* problem is as follows: given a set of active nodes, A, at the end of a process, find the set of seeds, S, called *effectors* to minimize

$$\sum_{u \in A} |1 - Prob(S, u)| + \sum_{u \in V \setminus A} |0 - Prob(S, u)|,$$

where $Prob(S, u)$ is the probability that node u becomes active when S is selected as the seed set. Suppose G is a tree with undirected edges. Design a dynamic programming algorithm to show that the effector detection problem can be solved in polynomial-time.

39. (Active Friending) Consider a social network $G = (V, E)$ with the information diffusion model LT as stated in previous problem. Given a positive integer k, a subset Q of nodes ($k \leq Q$), and a node $t \notin Q$, find a subset of k seeds in Q to maximize the probability that t is activated. This problem is called the *active friending*. Suppose that G is an in-arborescence rooted at t. Design a dynamic programming algorithm to solve this problem in a polynomial-time.

Historical Notes

Dynamic programming was proposed by Richard Bellman in 1953 [95] and later became a popular method in optimization and control theory. The basic idea is stemmed from self-reducibility. In the design of computer algorithms, it is a powerful and elegant technique to find an efficient solution for many optimization problems, which attracts a lot of researchers' efforts in the literature, especially in the direction of speed-up dynamic programming. For example, Yao [434] and Borchers and Gupta [35] speed up dynamic programming with the quadrangle inequality, including a construction for the rectilinear Steiner arborescence [345] from $O(n^3)$ time to $O(n^2)$ time.

The shortest path problem became a classical graph problem as early as in 1873 [407]. A. Schrijver [361] provides a quite detail historical note with a large list of references. There are many algorithms in the literature. Those closely related to dynamic programming algorithms can be found in Bellman [24], Dijkstra [83], Dial [82], and Fredman and Tarjan [149].

All-pair-shortest-paths problem was first studied by Alfonso Shimbel in 1953 [366], who gave a $O(n^4)$-time solution. Floyd [144] and Marshall [402] found a $O(n^3)$-time solution independently in the same year.

Examples and exercises about disk (or sensor) coverage can be found in [4, 160, 219, 296, 421], and those about social influence can be found in [393] for the influence maximization, [271] for the effector detection, and [427] for the active friending. Extended reading materials can be found in [188, 304, 379, 380, 435, 445].

Chapter 4
Greedy Algorithm and Spanning Tree

Greed, in the end, fails even the greedy.

—Cathryn Louis

Self-reducibility is the backbone of each greedy algorithm in which self-reducibility structure is a tree of special kind, i.e., its internal nodes lie on a path. In this chapter, we study algorithms with such a self-reducibility structure and related combinatorial theory supporting greedy algorithms.

4.1 Greedy Algorithms

A problem that the greedy algorithm works for computing optimal solutions often has the self-reducibility and a simple exchange property. Let us use two examples to explain this point.

Example 4.1.1 (Activity Selection) Consider n activities with starting times $s_1, s_2, \ldots, s_n$ and ending times $f_1, f_2, \ldots, f_n$, respectively. They may be represented by intervals $[s_1, f_1), [s_2, f_2), \ldots,$ and $[s_n, f_n)$. The problem is to find a maximum subset of nonoverlapping activities, i.e., nonoverlapping intervals.

This problem has the following exchange property.

Lemma 4.1.2 (Exchange Property) *Suppose $f_1 \leq f_2 \leq \cdots \leq f_n$. In a maximum solution without interval $[s_1, f_1)$, we can always exchange $[s_1, f_1)$ with the first activity in the maximum solution preserving the maximality.*

Proof Let $[s_i, f_i)$ be the first activity in the maximum solution mentioned in the lemma. Since $f_1 \leq f_i$, replacing $[s_i, f_i)$ by $[s_1, f_1)$ will not cost any overlapping. □

The following lemma states a self-reducibility.

© The Author(s), under exclusive license to Springer Nature Switzerland AG 2022
D.-Z. Du et al., *Introduction to Combinatorial Optimization*, Springer Optimization and Its Applications 196, https://doi.org/10.1007/978-3-031-10596-8_4

Lemma 4.1.3 (Self-Reducibility) *Suppose* $\{I_1^*, I_2^*, \ldots, I_k^*\}$ *is an optimal solution. Then,* $\{I_2^*, \ldots, I_k^*\}$ *is an optimal solution for the activity problem on input* $\{I_i \mid I_i \cap I_1^*\}$ *where* $I_i = [s_i, f_i)$.

Proof For contradiction, suppose that $\{I_2^*, \ldots, I_k^*\}$ is not an optimal solution for the activity problem on input $\{I_i \mid I_i \cap I_1^*\}$. Then, $\{I_i \mid I_i \cap I_1^*\}$ contains k nonoverlapping activities, which all are not overlapping with I_1^*. Putting I_1^* in these k activities, we will obtain a feasible solution containing $k + 1$ activities, contradicting the assumption that $\{I_1^*, I_2^*, \ldots, I_k^*\}$ is an optimal solution.   □

Based on Lemmas 4.1.2 and 4.1.3, we can design a greedy algorithm in Algorithm 11 and obtain the following result.

Algorithm 11 Greedy algorithm for activity selection

Input: A sequence of n activities $[s_1, f_1), [s_2, f_2), \ldots, [s_n, f_n)$.
Output: A maximum subset of nonoverlapping activities.
1: sort all activities into ordering $f_1 \leq f_2 \leq \ldots \leq f_n$
2: $S \leftarrow \emptyset$
3: **for** $i \leftarrow 1$ to n **do**
4:      **if** $[s_i, f_i)$ does not overlap any activity in S **then**
5:          $S \leftarrow S \cup \{[s_i, f_i)\}$
6:      **end if**
7: **end for**
8: **return** S

Theorem 4.1.4 *Algorithm 11 produces an optimal solution for the activity selection problem.*

Proof Let us prove it by induction on n. For $n = 1$, it is trivial.

Consider $n \geq 2$. Suppose $\{I_1^*, I_2^*, \ldots, I_k^*\}$ is an optimal solution. By Lemma 4.1.2, we may assume that $I_1^* = [s_1, f_1)$. By Lemma 4.1.3, $\{I_2^*, \ldots, I_k^*\}$ is an optimal solution for the activity selection problem on input $\{I_i \mid I_i \cap I_1^* = \emptyset\}$.

Note that after select $[s_1, f_1)$, if we ignore all iterations i with $[s_i, f_i) \cap [s_1, f_1) \neq \emptyset$, then the remaining part is the same as greedy algorithm running on input $\{I_i \mid I_i \cap I_1^* = \emptyset\}$. By induction hypothesis, it will produce an optimal solution for the activity selection problem on input $\{I_i \mid I_i \cap I_1^* = \emptyset\}$, which must contain $k - 1$ activities. Together with $[s_1, f_1)$, they form a subset of k non-overlapping activities, which should be optimal.   □

Next, we study another example.

Example 4.1.5 (Huffman Tree) Given n characters $a_1, a_2, \ldots, a_n$ with weights $f_1, f_2, \ldots, f_n$, respectively, find a binary tree with n leaves labeled by $a_1, a_2, \ldots, a_n$, respectively, to minimize

$$d(a_1) \cdot f_1 + d(a_2) \cdot f_2 + \cdots + d(a_n) \cdot f_n$$

where $d(a_i)$ is the depth of leaf a_i, i.e., the number of edges on the path from the root to a_i.

First, we show a property of optimal solutions.

Lemma 4.1.6 *In any optimal solution, every internal node has two children, i.e., every optimal binary tree is full.*

Proof If an internal node has only one child, then this internal node can be removed to reduce the objective function value.   □

We can also show an exchange property and a self-reducibility.

Lemma 4.1.7 (Exchange Property) *If $f_i > f_j$ and $d(a_i) > d(a_j)$, then exchanging a_i with a_j would make the objective function value decrease.*

Proof Let $d'(a_i)$ and $d(a_j)$ be the depths of a_i and a_j, respectively, after exchanging a_i with a_j. Then $d'(a_i) = d(a_j)$ and $d'(a_j) = d(a_i)$. Therefore, the difference of objective function values before and after exchange is

$$(d(a_i) \cdot f_i + d(a_j) \cdot f_j) - (d'(a_i) \cdot f_i + d'(a_j) \cdot f_j)$$
$$= (d(a_i) \cdot f_i + d(a_j) \cdot f_j) - (d(a_j) \cdot f_i + d(a_i) \cdot f_j)$$
$$= (d(a_i) - d(a_j))(f_i - f_j)$$
$$> 0$$

□

Lemma 4.1.8 (Self-Reducibility) *In any optimal tree T^*, if we assign the weight of an internal node u with the total weight w_u of its descendant leaves, then removal of the subtree T_u at the internal node results in an optimal tree T'_u for weights at remainder's leaves (Fig. 4.1).*

Proof Let $c(T)$ denote the objective function value of tree T, i.e.,

Fig. 4.1 A self-reducibility

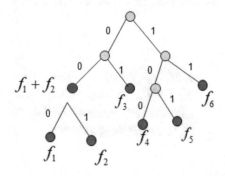

$$c(T) = \sum_{a \text{ over leaves of } T} d(a) \cdot f(a)$$

where $d(a)$ is the depth of leaf a and $f(a)$ is the weight of leaf a. Then we have

$$c(T^*) = c(T_u) + c(T_u').$$

If T_u' is not optimal for weights at leaves of T_u', then we have a binary tree T_u'' for those weights with $c(T_u'') < c(T_u')$. Therefore, $c(T_u \cup T_u'') < c(T^*)$, contradicting optimality of T^*.   □

By Lemmas 4.1.7 and 4.1.8, we can construct an optimal Huffman tree in the following:

- Sort $f_1 \le f_2 \le \cdots \le f_n$.
- By exchange property, there must exist an optimal tree in which a_1 and a_2 are sibling at bottom level.
- By self-reducibility, the problem can be reduced to construct optimal tree for leaves weights $\{f_1 + f_2, f_3, \ldots, f_n\}$.
- Go back to initial sorting step. This process continues until only two weights exist.

In Fig. 4.2, an example is presented to explain this construction. This construction can be implemented with min-priority queue (Algorithm 12)

The Huffman tree problem is raised from the study of Huffman codes as follows.

Problem 4.1.9 (Huffman Codes) Given n characters $a_1, a_2, \ldots, a_n$ with frequencies $f_1, f_2, \ldots, f_n$, respectively, find prefix binary codes $c_1, c_2, \ldots, c_n$ to minimize

$$|c_1| \cdot f_1 + |c_2| \cdot f_2 + \cdots + |c_n| \cdot f_n,$$

where $|c_i|$ is the length of code c_i, i.e., the number of symbols in c_i.

Actually, $c_1, c_2, \ldots, c_n$ are called *prefix* binary codes if no one is a prefix of another one. Therefore, they have a binary tree representation.

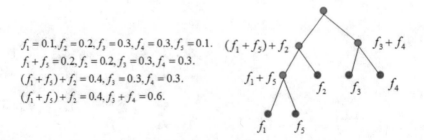

$f_1 = 0.1, f_2 = 0.2, f_3 = 0.3, f_4 = 0.3, f_5 = 0.1.$ $(f_1 + f_5) + f_2$
$f_1 + f_5 = 0.2, f_2 = 0.2, f_3 = 0.3, f_4 = 0.3.$
$(f_1 + f_5) + f_2 = 0.4, f_3 = 0.3, f_4 = 0.3.$
$(f_1 + f_5) + f_2 = 0.4, f_3 + f_4 = 0.6.$

Fig. 4.2 An example for construction of Huffman tree

Algorithm 12 Greedy algorithm for Huffman tree

Input: A sequence of leaf weights $\{f_1, f_2, \ldots, f_n\}$.
Output: A binary tree.
1: Put $f_1, f_2, \ldots, f_n$ into a min-priority queue Q
2: **for** $i \leftarrow 1$ to $n - 1$ **do**
3:    allocate a new node z
4:    $left[z] \leftarrow x \leftarrow$ Extract-Min(Q)
5:    $right[z] \leftarrow y \leftarrow$ Extract-Min(Q)
6:    $f[z] \leftarrow f[x] + f[y]$
7:    Insert(Q, z)
8: **end for**
9: **return** Extract-Min(Q)

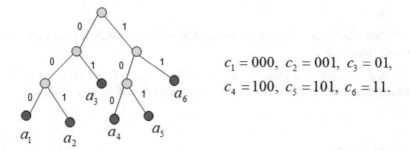

$$c_1 = 000, \ c_2 = 001, \ c_3 = 01,$$
$$c_4 = 100, \ c_5 = 101, \ c_6 = 11.$$

Fig. 4.3 Huffman codes

- Each edge is labeled with 0 or 1.
- Each code is represented by a path from the root to a leaf.
- Each leaf is labeled with a character.
- The length of a code is the length of corresponding path.

An example is as shown in Fig. 4.3. With this representation, the Huffman codes problem can be transformed exactly to the Huffman tree problem.

In Chap. 1, we see that the Kruskal greedy algorithm can compute the minimum spanning tree. Thus, we may have a question: Does the minimum spanning tree problem have an exchange property and self-reducibility? The answer is yes, and they are given in the following.

Lemma 4.1.10 (Exchange Property) *For an edge e with the smallest weight in a graph G and a minimum spanning tree T without e, there must exist an edge e' in T such that $(T \setminus e') \cup e$ is still a minimum spanning tree.*

Proof Suppose u and v are two endpoints of edge e. Then T contains a path p connecting u and v. On path p, every edge e' must have weight $c(e') = c(e)$. Otherwise, $(T \setminus e') \cup e$ will be a spanning tree with total weight smaller than $c(T)$, contradicting minimality of $c(T)$.

Now, select any edge e' in path p. Then $(T \setminus e') \cup e$ is a minimum spanning tree.
□

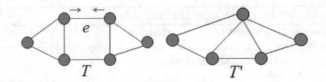

Fig. 4.4 Lemma 4.1.11

Lemma 4.1.11 (Self-Reducibility) *Suppose T is a minimum spanning tree of a graph G and edge e in T has the smallest weight. Let G' and T' be obtained from G and T, respectively, by shrinking e into a node (Fig. 4.4). Then T' is a minimum spanning tree of G'.*

Proof Note that T is a minimum spanning tree of G if and only if T' is a minimum spanning tree of G'.   □

With the above two lemmas, we are able to give an alternative proof for correctness of the Kruskal algorithm. We leave it as an exercise for readers.

4.2   Matroid

There is a combinatorial structure which has a close relationship with greedy algorithms. This is the matroid. To introduce matroid, let us first study independent systems.

Consider a finite set S and a collection $\mathcal{C}$ of subsets of S. $(S, \mathcal{C})$ is called an *independent system* if

$$A \subset B, B \in \mathcal{C} \Rightarrow A \in \mathcal{C},$$

i.e., it is *hereditary*. In the independent system $(S, \mathcal{C})$, each subset in $\mathcal{C}$ is called an independent set.

Consider a maximization problem as follows.

Problem 4.2.1 (Independent Set Maximization) Let c be a nonnegative cost function on S. Denote $c(A) = \sum_{x \in A} c(x)$ for any $A \subseteq S$. The problem is to maximize $c(A)$ subject to $A \in \mathcal{C}$.

Also, consider the greedy algorithm in Algorithm 13.

For any $F \subseteq E$, a subset I of F is called a *maximal* independent subset if no independent subset of E contains F as a proper subset. Define

$$u(F) = \max\{|I| \mid I \text{ is an independent subset of } F\},$$

$$v(F) = \min\{|I| \mid I \text{ is a maximal independent subset of } F\}.$$

Algorithm 13 Greedy algorithm for independent set maximization

---

Input: An independent system $(S, \mathcal{C})$ with a nonnegative cost function c on S.
Output: An independent set.
1: Sort all elements in S into ordering $c(x_1) \geq c(x_2) \geq \cdots \geq c(x_n)$
2: $A \leftarrow \emptyset$
3: **for** $i \leftarrow 1$ to n **do**
4:     **if** $A \cup \{x_i\} \in \mathcal{C}$ **then**
5:         $A \leftarrow A \cup \{x_i\}$
6:     **end if**
7: **end for**
8: **return** A

---

where $|I|$ is the number of elements in I. Then we have the following theorem to estimate the performance of Algorithm 13.

Theorem 4.2.2 *Let A_G be a solution obtained by Algorithm 13. Let A^* be an optimal solution for the independent set maximization. Then*

$$1 \leq \frac{c(A^*)}{c(A_G)} \leq \max_{F \subseteq S} \frac{u(F)}{v(F)}.$$

Proof Note that $S = \{x_1, x_2, \ldots, x_n\}$ and $c(x_1) \geq c(x_2) \geq \cdots \geq c(x_n)$. Denote $S_i = \{x_1, \ldots, x_i\}$. Then

$$c(A_G) = c(x_1)|S_1 \cap A_G| + \sum_{i=2}^{n} c(x_i)(|S_i \cap A_G| - |A_{i-1} \cap A_G|)$$

$$= \sum_{i=1}^{n-1} |S_i \cap A_G|(c(x_i) - c(x_{i+1})) + |A_n \cap A_G|c(x_n).$$

Similarly,

$$c(A^*) = \sum_{i=1}^{n-1} |S_i \cap A^*|(c(x_i) - c(x_{i+1})) + |S_n \cap A^*|c(x_n).$$

Thus,

$$\frac{c(A^*)}{c(A_G)} \leq \max_{1 \leq i \leq n} \frac{|A^* \cap S_i|}{|A_G \cap S_i|}.$$

We claim that $A_i \cap A_G$ is a maximal independent subset of S_i. In fact, for contradiction, suppose that $S_i \cap A_G$ is not a maximal independent subset of S_i. Then there exists an element $x_j \in S_i \setminus A_G$ such that $(S_i \cap A_G) \cup \{x_j\}$ is independent.

Thus, in the computation of Algorithm 2.1, $I \cup \{e_j\}$ as a subset of $(S_i \cap A_G)\{x_j\}$ should be independent. This implies that x_j should be in A_G, a contradiction.

Now, from our claim, we see that

$$|S_i \cap A_G| \geq v(S_i).$$

Moreover, since $S_i \cap A^*$ is independent, we have

$$|S_i \cap A^*| \leq u(S_i).$$

Therefore,

$$\frac{c(A^*)}{c(A_G)} \leq \max_{F \subseteq S} \frac{u(F)}{v(F)}.$$

□

The matroid is an independent system satisfying an additional property, called *augmentation property*:

$$A, B \in \mathcal{C} \text{ and } |A| > |B|$$

$$\Rightarrow \exists x \in A \setminus B : B \cup \{x\} \in \mathcal{C}.$$

This property is equivalent to some others.

Theorem 4.2.3 *An independent system $(S, \mathcal{C})$ is a matroid if and only if for any $F \subseteq S$, $u(F) = v(F)$.*

Proof For forward direction, consider two maximal independent sets A and B. If $|A| > |B|$, then there exists $x \in A \setminus B$ such that $B \cup \{x\} \in \mathcal{C}$, contradicting maximality of B.

For backward direction, consider two independent sets with $|A| > |B|$. Set $F = A \cup B$. Then every maximal independent set of F has size at least $|A|$ ($> |B|$). Hence, B cannot be a maximal independent set of F. Thus, there exists an element $x \in F \setminus B = A \setminus B$ such that $B \cup \{x\} \in \mathcal{C}$. □

Theorem 4.2.4 *An independent system $(S, \mathcal{C})$ is a matroid if and only if for any cost function $c(\cdot)$, Algorithm 13 gives a maximum solution.*

Proof For necessity, we note that when $(S, \mathcal{C})$ is matroid, we have $u(F) = v(F)$ for any $F \subseteq S$. Therefore, Algorithm 13 gives an optimal solution.

For sufficiency, we give a contradiction argument. To this end, suppose independent system $(S, \mathcal{C})$ is not a matroid. Then, there exists $F \subseteq S$ such that F has two maximal independent sets I and J with $|I| < |J|$. Define

$$c(e) = \begin{cases} 1 + \varepsilon & \text{if } e \in I \\ 1 & \text{if } e \in J \setminus I \\ 0 & \text{otherwise} \end{cases}$$

where ε is a sufficient small positive number to satisfy $c(I) < c(J)$. The greedy algorithm will produce I, which is not optimal. □

This theorem gives tight relationship between matroids and greedy algorithms, which is built up on all nonnegative objective function. It may be worth mentioning that the greedy algorithm reaches optimal for a certain class of objective functions may not provide any additional information to the independent system. The following is a counterexample.

Example 4.2.5 Consider a complete bipartite graph $G = (V_1, V_2, E)$ with $|V_1| = |V_2|$. Let $\mathcal{I}$ be the family of all matchings. Clearly, $(E, \mathcal{I})$ is an independent system. However, it is not a matroid. An interesting fact is that maximal matchings may have different cardinalities for some subgraph of G although all maximal matchings for G have the same cardinality.

Furthermore, consider the problem $\max\{c(\cdot) \mid I \in \mathcal{I}\}$, called the *maximum assignment* problem.

If $c(\cdot)$ is a nonnegative function such that for any $u, u' \in V_1$ and $v, v' \in V_2$,

$$c(u, v) \geq \max(c(u, v'), c(u', v)) \implies c(u, v) + c(u', v') \geq c(u, v') + c(u', v).$$

This means that replacing edges (u_1, v') and (u', v_1) in M^* by (u_1, v_1) and (u', v') will not decrease the total cost of the matching. Similarly, we can put all (u_i, v_i) into an optimal solution, that is, they form an optimal solution. This gives an exchange property. Actually, we can design a greedy algorithm to solve the maximum assignment problem. (We leave this as an exercise.)

Next, let us present some examples of the matroid.

Example 4.2.6 (Linear Vector Space) Let S be a finite set of vectors and $\mathcal{I}$ the family of linearly independent subsets of S. Then $(S, \mathcal{I})$ is a matroid.

Example 4.2.7 (Graph Matroid) Given a graph $G = (V, E)$ where V and E are its vertex set and edge set, respectively. Let $\mathcal{I}$ be the family of edge sets of acyclic subgraphs of G. Then $(E, \mathcal{I})$ is a matroid.

Proof Clearly, $(E, \mathcal{I})$ is an independent system. Consider a subset F of E. Suppose that the subgraph (V, F) has m connected components. Note that in each connected component, every maximal acyclic subgraph must be a spanning tree which has the number of edges one less than the number of vertices. Thus, every maximal acyclic subgraph of (V, E) has exactly $|V| - m$ edges. By Theorem 4.2.3, $(E, \mathcal{I})$ is a matroid. □

In a matroid, all maximal independent subsets have the same cardinality. They are also called *bases*. In a graph matroid obtained from a connected graph, every base is a spanning tree.

Let $\mathcal{B}$ be the family of all bases of a matroid $(S, \mathcal{C})$. Consider the following problem:

Problem 4.2.8 (Base Cost Minimization) Consider a matroid $(S, \mathcal{C})$ with base family $\mathcal{B}$ and a nonnegative cost function on S. The problem is to minimize $c(B)$ subject to $B \in \mathcal{B}$.

Algorithm 14 Greedy algorithm for base cost minimization

Input: A matroid $(S, \mathcal{C})$ with a nonnegative cost function c on S.
Output: A base.
1: Sort all elements in S into ordering $c(x_1) \leq c(x_2) \leq \cdots \leq c(x_n)$
2: $A \leftarrow \emptyset$
3: **for** $i \leftarrow 1$ to n **do**
4:    **if** $A \cup \{x_i\} \in \mathcal{C}$ **then**
5:       $A \leftarrow A \cup \{x_i\}$
6:    **end if**
7: **end for**
8: **return** A

Theorem 4.2.9 *An optimal solution of the base cost minimization can be computed by Algorithm 14, a variation of Algorithm 13.*

Proof Suppose that every base has the cardinality m. Let M be a positive number such that for any $e \in S$, $c(e) < M$. Define $c'(e) = M - c(e)$ for all $e \in E$. Then $c'(\cdot)$ is a positive function on S, and the non-decreasing ordering with respect to $c(\cdot)$ is the non-increasing ordering with respect to $c'(\cdot)$. Note that $c'(B) = mM - c(B)$ for any $B \in \mathcal{B}$. Since Algorithm 13 produces a base with maximum value of c', Algorithm 14 produces a base with minimum value of function c.   □

The correctness of greedy algorithm for the minimum spanning tree can also be obtained from this theorem.

Next, consider the following problem.

Problem 4.2.10 (Unit-Time Task Scheduling) Consider a set of n unit-time tasks, $S = \{1, 2, \ldots, n\}$. Each task i can be processed during a unit-time and has to be completed before an integer deadline d_i and, if not completed, will receive a penalty w_i. The problem is to find a schedule for S on a machine within time n to minimize total penalty.

A set of tasks is independent if there exists a schedule for these tasks without penalty. Then we have the following.

Lemma 4.2.11 *A set A of tasks is independent if and only if for any $t = 1, 2, \ldots, n$, $N_t(A) \leq t$ where $N_t(A) = |\{i \in A \mid d_i \leq t\}|$.*

Proof It is trivial for "only if" part. For the "if" part, note that if the condition holds, then tasks in A can be scheduled in order of nondecreasing deadlines without penalty. □

Example 4.2.12 Let S be a set of unit-time tasks with deadlines and penalties and $\mathcal{C}$ the collection of all independent subsets of S. Then, $(S, \mathcal{C})$ is a matroid. Therefore, an optimal solution for the unit-time task scheduling problem can be computed by a greedy algorithm (i.e., Algorithm 13).

Proof (Hereditary) Trivial.

(Augmentation) Consider two independent sets A and B with $|A| < |B|$. Let k be the largest k such that $N_t(A) \geq N_t(B)$. (A few examples are presented in Fig. 4.5 to explain the definition of k.) Then $k < n$ and $N_t(A) < N_t(B)$ for $k + 1 \leq t \leq n$. Choose $x \in \{i \in B \setminus A \mid d_i = k + 1\}$. Then

$$N_t(A \cup \{x\}) = N_t(A) \leq t \text{ for } 1 \leq t \leq k$$

and

$$N_t(A \cup \{x\}) \leq N_t(A) + 1 \leq N_t(B) \leq t \text{ for } k + 1 \leq t \leq n.$$

□

Example 4.2.13 Consider an independent system $(S, \mathcal{C})$. For any fixed $A \subseteq S$, define

$$\mathcal{C}_A = \{B \subseteq S \mid A \not\subseteq B\}.$$

deadline of task

Fig. 4.5 In proof of Example 4.2.12

Then, (S, C_A) is a matroid.

Proof Consider any $F \subseteq S$. If $A \nsubseteq F$, then F has unique maximal independent set, which is F. Hence, $u(F) = v(F)$.

If $A \subseteq F$, then every maximal independent subset of F is in the form $F \setminus \{x\}$ for some $x \in A$. Hence, $u(F) = v(F) = |F| - 1$.   □

4.3 Minimum Spanning Tree

Let us revisit the minimum spanning tree problem.

Consider a graph $G = (V, E)$ with nonnegative edge weight $c : E \to R_+$, and a spanning tree T. Let (u, v) be an edge in T. Removal (u, v) would break T into two connected components. Let U and W be vertex sets of these two components, respectively. The edges between U and V constitute a *cut*, denoted by (U, W). The cut (U, W) is said to be induced by deleting (u, v). For example, in Fig. 4.6, deleting $(3, 4)$ induces a cut $(\{1, 2, 3\}, \{4, 5, 6, 7, 8\})$.

Theorem 4.3.1 (Cut Optimality) *A spanning tree T^* is a minimum spanning tree if and only if it satisfies the cut optimality condition as follows:*

Cut Optimality Condition   *For every edge (u, v) in T^*, $c(u, v) \leq c(x, y)$ for every edge (x, y) contained in the cut induced by deleting (u, v).*

Proof Suppose, for contradiction, that $c(u, v) > c(x, y)$ for some edge (x, y) in the cut induced by deleting (u, v) from T^*. Then $T' = (T^* \setminus (u, v)) \cup (x, y)$ is a spanning tree with cost less than $c(T^*)$, contradicting the minimality of T^*.

Conversely, suppose that T^* satisfies the cut optimality condition. Let T' be a minimum spanning tree such that among all minimum spanning trees, T' is the one with the most edges in common with T^*. Suppose, for contradiction, that $T' \neq T^*$. Consider an edge (u, v) in $T^* \setminus T'$. Let p be the path from u to v in T'. Then p has at least one edge (x, y) in the cut induced by deleting (u, v) from T^*. Thus, $c(u, v) \leq c(x, y)$ by the cut optimality condition. Hence, $T'' = (T' \setminus (x, y)) \cup (u, v)$ is also a minimum spanning tree, contradicting the assumption on T'.   □

The following algorithm is designed based on cut optimality condition.

Fig. 4.6 A cut induced by deleting an edge from a spanning tree

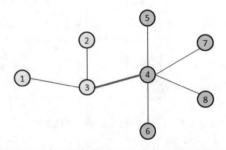

Prim Algorithm
input: A graph $G = (V, E)$ with nonnegative edge weight $c :\to R_+$.
output: A spanning tree T.
   $U \leftarrow \{s\}$ for some $s \in V$;
   $T \leftarrow \emptyset$;
   while $U \neq V$ **do**
          find the minimum weight edge (u, v) from cut $(U, V \setminus U)$
          and $T \leftarrow T \cup (u, v)$;
   return T.

An example for using Prim algorithm is shown in Fig. 4.7. The construction starts at node 1 and guarantees that the cut optimality conditions are satisfied at the end.

The min-priority queue can be used for implementing Prim algorithm to obtain the following result.

Theorem 4.3.2 *Prim algorithm can construct a minimum spanning tree in* $O(m \log m)$ *time where m is the number of edges in input graph.*

Proof Prim algorithm can be implemented by using min-priority queue in the following way:

- Keep to store all edges in a cut (U, W) in the min-priority queue S.
- At each iteration, choose the minimum weight edge (u, v) in the cut (U, W) by using operation Extract-Min(S) where $u \in U$ and $v \in W$.
- For every edge (x, v) with $x \in U$, delete (c, v) from S. This needs a new operation on min-priority queue, which runs $O(m)$ time.
- Add v to U.
- For every edge (v, y) with $y \in V \setminus U$, insert (v, y) into priority queue. This also requires $O(\log m)$ time.

In this implementation, Prim algorithm runs in $O(m \log m)$ time.   □

Prim algorithm can be considered as a local-information greedy algorithm. Actually, its correctness can also be established by an exchange property and a self-reducibility as follows.

Lemma 4.3.3 (Exchange Property) *Consider a cut (U, W) in a graph $G = (V, E)$. Suppose edge e has the smallest weight in cut (U, W). If a minimum spanning tree T does not contain e, then there must exist an edge e' in T such that $(T \setminus e') \cup e$ is still a minimum spanning tree.*

Lemma 4.3.4 (Self-Reducibility) *Suppose T is a minimum spanning tree of a graph G and edge e in T has the smallest weight in the cut induced by deleting e from T. Let G' and T' be obtained from G and T, respectively, by shrinking e into a node. Then T' is a minimum spanning tree of G'.*

We leave proofs of them as exercises.

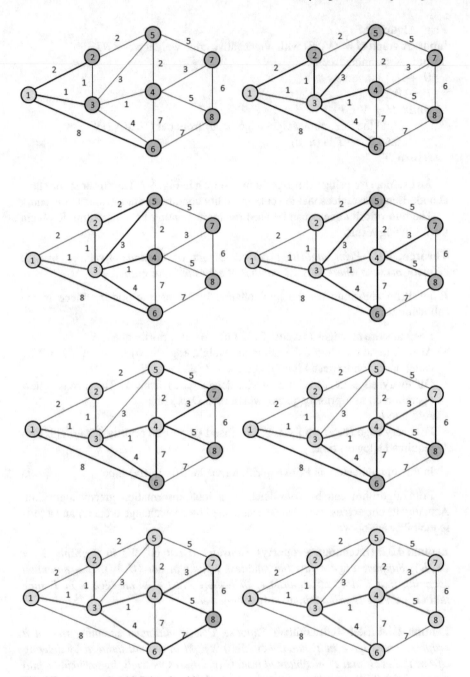

Fig. 4.7 An example with Prim algorithm

4.4 Local Ratio Method

The local ratio method is also a type of algorithm with self-reducibility. Its basic idea is as follows.

Lemma 4.4.1 *Let* $c(x) = c_1(x) + c_2(x)$. *Suppose* x^* *is an optimal solution of* $\min_{x \in \Omega} c_1(x)$ *and* $\min_{x \in Omega} c_2(x)$. *Then* x^* *is an optimal solution of* $\min_{x \in \Omega} c(x)$. *The similar statement holds for the maximization problem.*

Proof For any $x \in \Omega$, $c_1(x) \geq c_1(x^*)$, $c_2(x) \geq c_2(x^*)$, and hence $c(x) \geq c(x^*)$.
□

Usually, the objective function $c(x)$ is decomposed into $c_1(x)$ and $c_2(x)$ such that optimal solutions of $\min_{x \in \Omega} c_1(x)$ constitute a big pool so that the problem is reduced to find an optimal solution of $\min_{x \in \Omega} c_2(x)$ in the pool. In this section, we present two examples to explain this idea.

First, we study the following problem.

Problem 4.4.2 (Weighted Activity Selection) Given n activities each with a time period $[s_i, f_i)$ and a positive weight w_i, find a nonoverlapping subset of activities to maximize the total weight.

Suppose, without loss of generality, $f_1 \leq f_2 \leq \cdots \leq f_n$. First, we consider a special case that for every activity $[s_i, f_i)$, if $s_i < f_1$, i.e., activity $[s_i, f_i)$ overlaps with activity $[s_1, f_1)$, then $w_i = w_1 > 0$, and if $s_i \geq f_1$, then $w_i = 0$. In this case, every feasible solution containing an activity overlapping with $[s_1, f_1)$ is an optimal solution. Motivated from this special case, we may decompose the problem into two subproblems. The first one is in the special case, and the second one has weight as follows

$$w_i' = \begin{cases} w_i - w_1 & \text{if } s_i < f_1, \\ w_i & \text{otherwise.} \end{cases}$$

In the second subproblem obtained from the decomposition, some activity may have non-positive weight. Such an activity can be removed from our consideration because putting it in any feasible solution would not increase the total weight. This operation would simplify the problem by removing at least one activity. Repeat the decomposition and simplification until no activity is left.

To explain how to obtain an optimal solution, let A' be the set of remaining activities after the first decomposition and simplification and Opt' is an optimal solution for the weighted activity selection problem on A'. Since simplification does not effect the objective function value of optimal solution, Opt' is an optimal solution of the second subproblem in the decomposition. If Opt' contains an activity overlapping with activity $[s_1, f_1)$, then Opt' is also an optimal solution of the first subproblem, and hence by Lemma 4.4.1, Opt' is an optimal solution for the weighted activity selection problem on original input A. If Opt' does not contain an activity overlapping with $[s_1, f_1)$, then $Opt' \cup \{[s_1, f_1)\}$ is an optimal solution for the first subproblem and the second subproblem and hence also an optimal solution for the original problem.

Based on the above analysis, we may construct the following algorithm.

Local Ratio Algorithm for Weighted Activity Selection
input $A = \{[s_1, f_1), [s_2, f_2), \ldots, [s_n, f_n)\}$ with $f_1 \leq f_2 \leq \cdots \leq f_n$.
$B \leftarrow \emptyset$.
output Opt.
    while $A \neq \emptyset$ **do begin**
        $[s_j, f_j) \leftarrow \mathrm{argmin}_{[s_i, f_i) \in A} f_i$;
        $B \leftarrow B \cup \{[s_j, f_j)\}$;
        for every $[s_i, f_i) \in A$ **do**
            if $s_i < f_j$ **then** $w_i \leftarrow w_i - w_j$;
        end-for
        for every $[s_i, f_i) \in A$ **do**
            if $w_i \leq 0$ **then** $A \leftarrow A - \{[s_i, f_i)\}$;
        end-for
    end-while;
    $[s_k, f_k) \leftarrow \mathrm{argmax}_{[s_i, f_i) \in B} f_i$;
    $Opt \leftarrow \{[s_k, f_k)\}$;
    $B \leftarrow B - \{[s_k, f_k)\}$;
    while $B \neq \emptyset$ **do**
        $[s_h, f_h) \leftarrow \mathrm{argmax}_{[s_i, f_i) \in B} f_i$;
        if $s_k \geq f_h$,
            then $Opt \leftarrow Opt \cup \{[s_h, f_h)\}$
                and $[s_k, f_k) \leftarrow [s_h, f_h)$;
        end-if
        $B \leftarrow B - \{[s_h, f_h)\}$;
    end-while;
    return Opt.

Now, we run this algorithm on an example as shown in Fig. 4.8.

Next, we study the second example.

Consider a directed graph $G = (V, E)$. A subgraph T is called an *arborescence* rooted at a vertex r if T satisfies the following two conditions:

(a) If it ignores direction on every arc, then T is a tree.
(b) For any vertex $v \in V$, T contains a directed path from r to v.

Let T be an arborescence with root r. Then for any vertex $v \in V - \{r\}$, there is exactly one arc coming to v. This property is quite important.

Lemma 4.4.3 *Suppose T is obtained by choosing one incoming arc at each vertex $v \in V - \{r\}$. Then T is an arborescence if and only if T does not contain a directed cycle.*

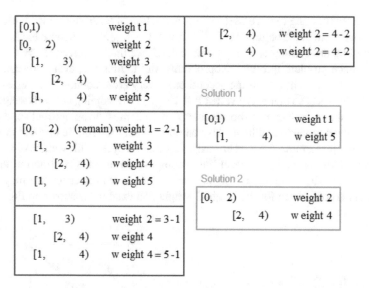

Fig. 4.8 An example for weighted activity selection

Proof Note that the number of arcs in T is equal to $|V| - 1$. Thus, condition (b) implies the connectivity of T when ignore direction, which implies condition (a). Therefore, if T is not an arborescence, then condition (b) does not hold, i.e., there exists $v \in V - \{r\}$ such that there does not exist a directed path from r to v. Now, T contains an arc (v_1, v) coming to v with $v_1 \neq r$, an arc (v_2, v_1) coming to v_1 with $v_2 \neq v$, and so on. Since the directed graph G is finite. The sequence $(v, v_1, v_2, \ldots)$ must contain a cycle.

Conversely, if T contains a cycle, then T is not an arborescence by the definition. This completes the proof of the lemma. □

Now, we consider the minimum arborescence problem.

Problem 4.4.4 (Minimum Arborescence) Given a directed graph $G = (V, E)$ with positive arc weight $w : E \to R^+$ and a vertex $r \in V$, compute an arborescence with root r to minimize total arc weight.

The following special case gives a basic idea for a local ratio method.

Lemma 4.4.5 *Suppose for each vertex $v \in V - \{r\}$ all arcs coming to v have the same weight. Then every arborescence with root r is optimal for the* MIN ARBORESCENCE *problem.*

Proof It follows immediately from the fact that each arborescence contains exactly one arc coming to v for each vertex $v \in V - \{r\}$. □

Since arcs coming to r are useless in construction of an arborescence with root r, we remove them at the beginning. For each $v \in V - \{r\}$, let w_v denote the minimum weight of an arc coming to v. By Lemma 4.4.5, we may decompose the minimum arborescence problem into two subproblems. In the first one, every arc coming to a vertex v has weight w_v. In the second one, every arc e coming to a vertex v has weight $w(e) - w_v$, so that every vertex $v \in V - \{r\}$ has a coming arc with weight 0. If all 0-weight arcs contain an arborescence T, then T must be an optimal solution for the second subproblem and hence also an optimal solution for the original problem. If not, then by Lemma 4.4.3, there exists a directed cycle with weight 0. Contract this cycle into one vertex. Repeat the decomposition and the contraction until an arborescence with weight 0 is found. Then in backward direction, we may find a minimum arborescence for the original weight. An example is shown in Fig. 4.9.

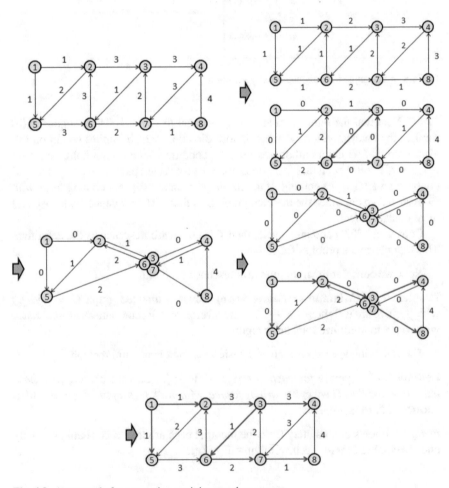

Fig. 4.9 An example for computing a minimum arborescence

According to above analysis, we may construct the following algorithm.

Local Ratio Algorithm for Minimum Arborescence
input a directed graph $G = (V, E)$ with arc weight $w : E \to R^+$,
         and a root $r \in V$.
output An arborescence T with root r.
   $C \leftarrow \emptyset$;
   repeat
      for every $v \in V \setminus \{r\}$ **do**
            let e_v be the one with minimum weight among arcs coming
            to v and $T \leftarrow T \cup \{e_v\}$;
            for every edge $e = (u, v)$ coming to v **do**
                  $w(e) \leftarrow w(e) - w_v$;
            end-for
      end-for
      if T contains a cycle C
         then $C \leftarrow C \cup \{C\}$ and
                  contract cycle C into one vertex in G and T;
      end-if
   until T does not contain a cycle;
   for every $C \in C$ **do**
         add C into T and properly delete an arc of C.
   end-for
   return T.

Exercises

1. Suppose that for every cut of the graph, there is a unique light edge crossing the cut. Show that the graph has a unique minimum spanning tree. Does the inverse hold? If not, please give a counterexample.
2. Consider a finite set S. Let $\mathcal{I}_k$ be the collection of all subsets of S with size at most k. Show that $(S, \mathcal{I}_k)$ is a matroid.
3. Solve the following instance of the unit-time task scheduling problem.

a_i	1	2	3	4	5	6	7
d_i	4	2	4	3	1	4	6
w_i	70	60	50	40	30	20	10

   Please solve the problem again when each penalty w_i is replaced by $80 - w_i$.
4. Suppose that the characters in an alphabet is ordered so that their frequencies are monotonically decreasing. Prove that there exists an optimal prefix code whose codeword length are monotonically increasing.
5. Show that if $(S, \mathcal{I})$ is a matroid, then $(S, \mathcal{I}')$ is a matroid, where

$$\mathcal{I}' = \{A' \mid S - A' \text{ contains some maximal } A \in \mathcal{I}\}.$$

That is, the maximal independent sets of $(S, \mathcal{I}')$ are just complements of the maximal independent sets of $(S, \mathcal{I})$.

6. Suppose that a set of activities are required to schedule in a large number of lecture halls. We wish to schedule all the activities using as few lecture halls as possible. Give an efficient greedy algorithm to determine which activity should use which lecture hall.

7. Consider a set of n files, $f_1, f_2, \ldots, f_n$, of distinct sizes $m_1, m_2, \ldots, m_n$, respectively. They are required to be recorded sequentially on a single tape, in some order, and retrieve each file exactly once, in the reverse order. The retrieval of a file involves rewinding the tape to the beginning and then scanning the files sequentially until the desired file is reached. The *cost* of retrieving a file is the sum of the sizes of the files scanned plus the size of the file retrieved. (Ignore the cost of rewinding the tape.) The *total cost* of retrieving all the files is the sum of the individual costs.

   (a) Suppose that the files are stored in some order $f_{i_1}, f_{i_2}, \ldots, f_{i_n}$. Derive a formula for the total cost of retrieving the files, as a function of n and the m_{i_k}'s.

   (a) Describe a greedy strategy to order the files on the tape so that the total cost is minimized, and prove that this strategy is indeed optimal.

8. In merge sort, the merge procedure is able to merge two sorted lists of lengths n_1 and n_2, respectively, into one by using $n_1 + n_2$ comparisons. Given m sorted lists, we can select two of them and merge these two lists into one. We can then select two lists from the $m - 1$ sorted lists and merge them into one. Repeating this step, we shall eventually end up with one merged list. Describe a general algorithm for determining an order in which m sorted lists $A_1, A_2, \ldots, A_m$ are to be merged so that the total number of comparisons is minimum. Prove that your algorithm is correct.

9. Let $G = (V, E)$ be a connected undirected graph. The distance between two vertices x and y, denoted by $d(x, y)$, is the number of edges on the shortest path between x and y. The *diameter* of G is the maximum of $d(x, y)$ over all pairs (x, y) in $V \times V$. In the remainder of this problem, assume that G has at least two vertices.

   Consider the following algorithm on G: Initially, choose arbitrarily $x_0 \in V$. Repeatedly, choose x_{i+1} such that $d(x_{i+1}, x_i) = \max_{v \in V} d(v, x_i)$ until $d(x_{i+1}, x_i) = d(x_i, x_{i-1})$.

   Can this algorithm always terminate? When it terminates, is $d(x_{i+1}, x_i)$ guaranteed to equal the diameter of G? (Prove or disprove your answer.)

10. Consider a graph $G = (V, E)$ with positive edge weight $c : E \to R^+$. Show that for any spanning tree T and the minimum spanning tree T^*, there exists a one-to-one onto mapping $\rho : E(T) \to E(T^*)$ such that $c(\rho(e)) \leq c(e)$ for every $e \in E(T)$ where $E(T)$ denotes the edge set of T.

11. Consider a point set P in the Euclidean plane. Let R be a fixed positive number. A steinerized spanning tree on P is a tree obtained from a spanning tree on P by putting some Steiner points on its edges to break them into pieces each of length at most R. Show that the steinerized spanning with minimum number of Steiner points is obtained from the minimum spanning tree.

12. Consider a graph $G = (V, E)$ with edge weight $w : E \to R^+$. Show that the spanning tree T which minimizes $\sum_{e \in E(T)} \|e\|^\alpha$ for any fixed $1 < \alpha$ is the minimum spanning tree, i.e., the one which minimizes $\sum_{e \in E(T)} \|e\|$.

13. Let $\mathcal{B}$ be the family of all maximal independent subsets of an independent system $(E, \mathcal{I})$. Then $(E, \mathcal{I})$ is a matroid if and only if for any nonnegative function $c(\cdot)$, Algorithm 14 produces an optimal solution for the problem $\min\{c(I) \mid I \in \mathcal{B}\}$.

14. Consider a complete bipartite graph $G = (U, V, E)$ with $|U| = |V|$. Let $c(\cdot)$ be a nonnegative function on E such that for any $u, u' \in V_1$ and $v, v' \in V_2$,

$$c(u, v) \geq \max(c(u, v'), c(u', v)) \implies c(u, v) + c(u', v') \geq c(u, v') + c(u', v).$$

   (a) Design a greedy algorithm for problem $\max\{c(\cdot) \mid I \in \mathcal{I}\}$.
   (b) Design a greedy algorithm for problem $\min\{c(\cdot) \mid I \in \mathcal{I}\}$.

15. Given n intervals $[s_i, f_i)$ each with weight $w_i \geq 0$, design an algorithm to compute the maximum weight subset of disjoint intervals.

16. Give a counterexample to show that an independent system with all maximal independent sets of the same size may not be a matroid.

17. Consider the following scheduling problem. There are n jobs, $i = 1, 2, \ldots, n$, and there is one super-computer and n identical PCs. Each job needs to be pre-processed first on the supercomputer and then finished by one of the PCs. The time required by job i on the supercomputer is p_i for $i = 1, 2, \ldots, n$; the time required on a PC for job i is f_i for $i = 1, 2, \ldots, n$. Finishing several jobs can be done in parallel since we have as many PCs as there are jobs. But the supercomputer processes only one job at a time. The input to the problem is the vectors $p = [p_1, p_2, \ldots, p_n]$ and $f = [f_1, f_2, \ldots, f_n]$. The objective of the problem is to minimize the completion time of last job (i.e., minimize the maximum completion time of any job). Describe a greedy algorithm that solves the problem in $O(n \log n)$ time. Prove that your algorithm is correct.

18. Consider an independent system $(S, \mathcal{C})$. For a fixed $A \in \mathcal{C}$, define $\mathcal{C}_A = \{B \subseteq S \mid A \setminus B \neq \emptyset\}$. Prove that $(S, \mathcal{C}_A)$ is a matroid.

19. Prove that every independent system is an intersection of several matroids, that is, for every independent system $(S, \mathcal{C})$, there exist matroids $(S, \mathcal{C}_1)$, $(S, \mathcal{C}_2)$, $\ldots (S, \mathcal{C}_k)$ such that $\mathcal{C} = \cap_{i=1}^k \mathcal{C}_i$.

20. Suppose that an independent system $(S, \mathcal{C})$ is the intersection of k matroids. Prove that for any subset $F \subseteq S$, $u(F)/v(F) \leq k$ where $u(F)$ is the cardinality of maximum independent subset of F and $v(F)$ is the minimum cardinality of maximal independent subset of F.

21. Design a local ratio algorithm to compute a minimum spanning tree.
22. Consider a graph $G = (V, E)$ with edge weight $w : E \to Z$ and a minimum spanning tree T of G. Suppose the weight of an edge $e \in T$ is increased by an amount $\delta > 0$. Design an efficient algorithm to find a minimum spanning tree of G after this change.
23. Consider a graph $G = (V, E)$ with distinct edge weights. Suppose that a minimum spanning tree T is already computed by Prim algorithm. A new edge (u, v) (not in E) is being added to the graph. Please write an efficient algorithm to update the minimum spanning tree. Note that no credit is given for just computing a minimum spanning tree for graph $G' = (V, E \cup \{(u, v)\})$.
24. Consider a matroid $\mathcal{M} = (X, \mathcal{I})$. Each minimal dependent set C is called a *circuit*. A *cut* D is a minimal set such that D intersects every base. Suppose that a circuit C intersects a cut D. Show that $|C \cap D| \geq 2$.

Historical Notes

The greedy algorithm is an important class of computer algorithms with self-reducibility, for solving combinatorial optimization problems. It uses the greedy strategy in construction of an optimal solution. There are several variations of greedy algorithms, e.g., Prim algorithm for minimum spanning tree in which greedy principal applies not globally but a subset of edges.

Could Prim algorithm be considered as a local search method? The answer is no. Actually, in a local search method, a solution is improved by finding a better one within a local area. Therefore, the greedy strategy applies to search for the best moving from a solution to another better solution. This can also be called as incremental method, which will be introduced in the next chapter.

The minimum spanning tree has been studied since 1926 [30]. Its history can be found a remarkable article [185]. The best known theoretical algorithm is due to Bernard Chazelle [49, 50]. The algorithm runs almost in $O(m)$ time. However, it is too complicated to implement and hence may not be practical.

Matroid was first introduced by Hassler Whitney in 1935 [406] and independently by Takeo Nakasawa [329]. It is an important combinatorial structure to describe the independence with axioms. Especially, those axioms provide an abstraction for common properties in linear algebra and graphs. Therefore, many concepts and terminologies are analogous in these two areas. The relationship between matroid and greedy algorithm is only a small portion in the theory of matroid [334, 384, 403]. Actually, the study of a matroid contains a much larger field, with connections to many topics [404], such as combinatorial geometry [37, 74, 405], unimodular matrices [171], projective geometry [308], electrical networks [316, 348], and software systems [254].

Chapter 5
Incremental Method and Maximum Network Flow

Change is incremental. Change is small.

—Theodore Melfi

In this chapter, we study the incremental method which is very different from those methods in the previous chapters. This method does not use the self-reducibility. It starts from a feasible solution, and in each iteration, computation moves from a feasible solution to another feasible solution by improving the objective function value. The incremental method has been used in the study of many problems, especially in the study of network flow.

5.1 Maximum Flow

Consider a *flow network* $G = (V, E)$, i.e., a directed graph with a nonnegative capacity $c(u, v)$ on each arc (u, v), and two given nodes, *source s* and *sink t*. An example of the flow network is shown in Fig. 5.1. For simplicity of description for flow, we may extend capacity $c(u, v)$ to every pair of nodes u and v by defining $c(u, v) = 0$ if $(u, v) \notin E$.

A *flow* in flow network G is a real function f on $V \times V$ satisfying the following three conditions:

1. (Capacity constraint) $f(u, v) \leq c(u, v)$ for every $u, v \in V$.
2. (Skew symmetry) $f(u, v) = -f(v, u)$ for all $u, v \in V$.
3. (Flow conservation) $\sum_{v \in V \setminus \{u\}} f(u, v) = 0$ for every $u \in V \setminus \{s, t\}$.

The flow has the following properties.

Lemma 5.1.1 *Let f be a flow of network $G = (V, E)$. Then the following holds:*

(a) If $(u, v) \notin E$ and $(v, u) \notin E$, then $f(u, v) = 0$.
(b) For any $x \in V \setminus \{s, t\}$, $\sum_{f(u,x)>0} f(u, x) = \sum_{f(x,v)>0} f(x, v)$.

© The Author(s), under exclusive license to Springer Nature Switzerland AG 2022
D.-Z. Du et al., *Introduction to Combinatorial Optimization*, Springer Optimization and Its Applications 196, https://doi.org/10.1007/978-3-031-10596-8_5

Fig. 5.1 A flow network

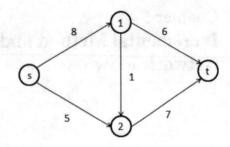

(c) $\sum_{f(s,v)>0} f(s, v) - \sum_{f(u,s)>0} f(u, s) = \sum_{f(u,t)>0} f(u, t) - \sum_{f(t,v)>0} f(t, v).$

Proof

(a) By capacity constraint, $f(u, v) \leq c(u, v) = 0$ and $f(v, u) \leq c(v, u) = 0$. By skew symmetric, $f(u, v) = -f(v, u) \geq 0$. Hence, $f(u, v) = 0$.

(b) By flow conservation, for any $x \in V \setminus \{s, t\}$,

$$\sum_{f(x,u)<0} f(x, u) + \sum_{f(x,v)>0} f(x, v) = \sum_{v \in V} f(x, v) = 0.$$

By skew symmetry,

$$\sum_{f(u,x)>0} f(u, x) = - \sum_{f(x,u)<0} f(x, u) = \sum_{f(x,v)>0} f(x, v).$$

(c) By (b), we have

$$\sum_{x \in V \setminus \{s,t\}} \sum_{f(u,x)>0} f(u, x) = \sum_{x \in V \setminus \{s,t\}} \sum_{f(x,v)>0} f(x, v).$$

For $(y, z) \in E$ with $y, z \in V \setminus \{s, t\}$, if $f(y, z) > 0$, then $f(y, z)$ appears in both the left-hand and the right-hand sides, and hence it will be cancelled. After cancellation, we obtain

$$\sum_{f(s,v)>0} f(s, v) + \sum_{f(t,v)>0} = \sum_{f(u,s)>0} f(u, s) + \sum_{f(u,t)>0} f(u, t).$$

$\square$

Now, the *flow value* of f is defined to be

$$|f| = \sum_{f(s,v)>0} f(s, v) - \sum_{f(u,s)>0} f(u, s) = \sum_{f(u,t)>0} f(u, t) - \sum_{f(t,v)>0} f(t, v).$$

In case that the source s does not have arc coming in, we have

$$|f| = \sum_{f(s,v)>0} f(s, v).$$

In general, we can also represent $|f|$ as

$$|f| = \sum_{v \in V \setminus \{s\}} f(s, v) = \sum_{u \in V \setminus \{t\}} f(u, t).$$

In Fig. 5.2, arc labels with underline give a flow. This flow has value 11.

The maximum flow problem is as follows.

Problem 5.1.2 (Maximum Flow) Given a flow network $G = (V, E)$ with arc capacity $c : V \times V \to R_+$, a source s, and a sink t, find a flow f with maximum flow value. Usually, assume that s does not have incoming arc and t does not have outgoing arc.

An important tool for study of the maximum flow problem is the residual network. The *residual network* for a flow f in a network $G = (V, E)$ with capacity c is the flow network with $G_f(V, E')$ with capacity $c'(u, v) = c(u, v) - f(u, v)$ for any $u, v \in V$ where $E' = \{(u, v) \in V \times V \mid c'(u, v) > 0\}$. For example, the flow in Fig. 5.2 has its residual network as shown in Fig. 5.3. Two important properties of the residual network are included in the following lemmas.

Lemma 5.1.3 *Suppose f' is a flow in the residual network G_f. Then $f + f'$ is a flow in network G and $|f + f'| = |f| + |f'|$.*

Fig. 5.2   A flow in network

Fig. 5.3   The residual network G_f of the flow f in Fig. 5.2

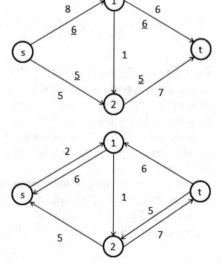

Proof For any $u, v \in V$, since $f'(u, v) \le c'(u, v) = c(u, v) - f(u, v)$, we have $f(u, v) + f'(u, v) \le c(u, v)$, that is, $f + f'$ satisfies the capacity constraint. Moreover, $f(u, v) + f'(u, v) = -f(v, u) - f'(v, u) = -(f(v, u) + f'(v, u))$ and for every $u \in V \setminus \{s, t\}$,

$$\sum_{v \in V \setminus \{u\}} (f + f')(u, v) = \sum_{v \in V \setminus \{u\}} f(u, v) + \sum_{v \in V \setminus \{u\}} f'(u, v) = 0.$$

This means that $f + f'$ satisfies the skew symmetry and the flow conservation conditions. Therefore, $f + f'$ is a flow. Finally,

$$|f + f'| = \sum_{v \in V \setminus \{s\}} (f + f')(s, v) = \sum_{v \in V \setminus \{s\}} f(s, v) + \sum_{v \in V \setminus \{s\}} f'(s, v) = |f| + |f'|.$$

$\square$

Lemma 5.1.4 *Suppose f' is a flow in the residual network G_f. Then $(G_f)_{f'} = G_{f+f'}$, i.e., the residual network of f' in network G_f is the residual network of $f + f'$ in network G.*

Proof The arc capacity of $(G_f)_{f'}$ is

$$c'(u, v) - f'(u, v) = c(u, v) - f(u, v) - f'(u, v) = c(u, v) - (f + f')(u, v)$$

which is the same as that in $G_{f+f'}$. $\square$

In order to get a flow with larger value, Lemmas 5.1.3 and. 5.1.4 suggest us to find a flow f' in G_f with $|f'| > 0$. A simple way is to find a path P from s to t and define f' by

$$f'(u, v) = \begin{cases} \min_{(x,y) \in P} c'(x, y) & \text{if } (u, v) \in P, \\ 0 & \text{otherwise.} \end{cases}$$

The following algorithm is motivated from this idea.

Using this algorithm, an example is shown in Fig. 5.4. The s-t path of the residual network is called an *augmenting path*, and hence Ford-Fulkerson algorithm is an augmenting path algorithm (Algorithm 15).

Now, we may have two questions: Can Ford-Fulkerson algorithm stop within finitely many steps? When Ford-Fulkerson algorithm stops, does output reach the maximum?

The answer for the first question is negative, that is, Ford-Fulkerson algorithm may run infinitely many steps. A counterexample can be obtained from the one as shown in Fig. 5.5 by setting $m = \infty$. However, with certain condition, Ford-Fulkerson algorithm will run within finitely many steps.

Algorithm 15 Ford-Fulkerson algorithm for maximum flow

Input: A flow network $G = (V, E)$ with capacity function c, a source s, and a sink t.
Output: A flow f.

1: $G \leftarrow G$;
2: $f \leftarrow 0$; (i.e., $\forall u, v \in V$, $f(u, v) = 0$)
3: **while** there exists a path P from s to t in G **do**
4:     $\delta \leftarrow \min\{c(u, v) \mid (u, v) \in P\}$ and
5:     send a flow f' with value δ from s to t along path P;
6:     $G \leftarrow G_{f'}$;
7:     $f \leftarrow f + f'$;
8: **end while**
9: **return** f.

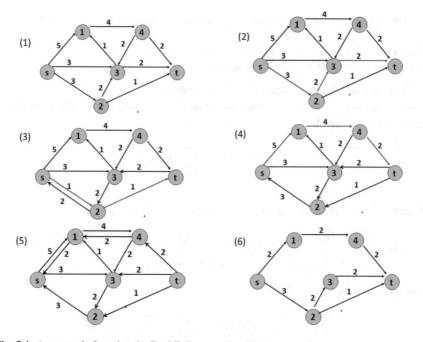

Fig. 5.4 An example for using the Ford-Fulkerson Algorithm

Theorem 5.1.5 *If every arc capacity is a finite integer, then Ford-Fulkerson algorithm runs within finitely many steps.*

Proof The flow value has upper bound $\sum_{(s,v) \in E} c(s, v)$. Since every arc capacity is an integer, in each step, the flow value will be increased by at least one. Therefore, the algorithm will run within at most $\sum_{(s,v) \in E} c(s, v)$ steps.   □

Remark Ford-Fulkerson algorithm may run with infinitely many augmentations even if all arc capacities are finite, but there is an irrational arc capacity. Such an example can be found in exercises.

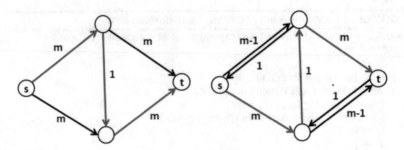

Fig. 5.5 Ford-Fulkerson algorithm runs not in polynomial time

The answer for the second question is positive. Actually, we have the following.

Theorem 5.1.6 *A flow f is maximum if and only if its residual network G_f does not contain a path from source s to sink t.*

To prove this theorem, let us first show a lemma.

A partition (S, T) of V is called an *s-t cut* if $s \in S$ and $t \in T$. The capacity of an *s-t* cut is defined by

$$\mathrm{CAP}(S, T) = \sum_{u \in S, v \in T} c(u, v).$$

Lemma 5.1.7 *Let (S, T) be an s-t cut. Then for any flow f,*

$$|f| = \sum_{f(u,v)>0, u \in S, v \in T} f(u, v) - \sum_{f(v,u)>0, u \in S, v \in T} f(v, u) \le \mathrm{CAP}(S, T).$$

Proof By Lemma 5.1.1(b),

$$\sum_{x \in S \setminus \{s\}} \sum_{f(u,x)>0} f(u, x) = \sum_{x \in S \setminus \{s\}} \sum_{f(x,v)>0} f(x, v).$$

Simplifying this equation, we will obtain

$$\sum_{f(s,x)>0, x \in S \setminus \{s\}} f(s, x) + \sum_{u \in T, x \in S \setminus \{s\}, f(u,x)>0} f(u, x)$$

$$= \sum_{f(x,s)>0, x \in S \setminus \{s\}} f(x, s) + \sum_{v \in T, x \in S \setminus \{s\}, f(x,v)>0} f(x, v).$$

Thus,

$$\sum_{f(s,x)>0} f(s,x) + \sum_{u\in T,x\in S, f(u,x)>0} f(u,x)$$

$$= \sum_{f(x,s)>0} f(x,s) + \sum_{v\in T,x\in S, f(x,v)>0} f(x,v),$$

that is,

$$|f| = \sum_{f(s,x)>0} f(s,x) - \sum_{f(x,s)>0} f(x,s)$$

$$= \sum_{v\in T,x\in S, f(x,v)>0} f(x,v) - \sum_{u\in T,x\in S, f(u,x)>0} f(u,x)$$

$$\le \sum_{v\in T,x\in S, f(x,v)>0} f(x,v)$$

$$\le \sum_{x\in S,v\in T} c(x,v).$$

□

Now, we prove Theorem 5.1.6.

Proof of Theorem 5.1.6 If residual network G_f contains a path from source s to sink t, then a positive flow can be added to f and hence f is not maximum. Next, we assume that G_f does not contain a path from s to t.

Let S be the set of all nodes each of which can be reached by a path from s. Set $T = V \setminus S$. Then (S, T) is a partition of V such that $s \in S$ and $t \in T$. Moreover, G_f has no arc from S to T. This fact implies two important facts:

(a) For any arc (u, v) with $u \in S$ and $v \in T$, $f(u, v) = c(u, v)$.
(b) For any arc (v, u) with $u \in S$ and $v \in T$, $f(v, u) = 0$.

Based on these two facts, by Lemma 5.1.7, we obtain that

$$|f| = \sum_{u\in S,v\in T} c(u, v).$$

Hence, f is a maximum flow. □

Corollary 5.1.8 *The maximum flow is equal to minimum s-t cut capacity.*

Finally, we remark that Ford-Fulkerson algorithm is not a polynomial-time. A counterexample is given in Fig. 5.5. On this counterexample, the algorithm runs in $2m$ steps. However, the input size is $O(\log m)$. Clearly, $2m$ is not a polynomial with respect to $O(\log m)$.

5.2   Edmonds-Karp Algorithm

To improve the running time of Ford-Fulkerson algorithm, a simple modification is found which works very well, that is, at each iteration, find a shortest augmenting path instead of an arbitrary augmenting. By the shortest, we mean the path contains the minimum number of arcs. This algorithm is called Admonds-Karp algorithm (Algorithm 16).

An example for using Edmonds-Karp algorithm is shown in Fig. 5.6. Compared with Fig. 5.4, we may find that input flow network is the same, but obtained maximum flows are different. Thus, for this input flow network, there are two different maximum flows. Actually, in this case, there are infinitely many maximum flows. The reader may prove it as an exercise.

To estimate the running time, let us study some properties of Edmonds-Karp algorithm.

Algorithm 16 Edmonds-Karp algorithm for maximum flow

Input: A flow network $G = (V, E)$ with capacity function c, a source s and a sink t.
Output: A flow f.
1: $G \leftarrow G$;
2: $f \leftarrow 0$; (i.e., $\forall u, v \in V, f(u, v) = 0$)
3: **while** there exists a path from s to t in G **do**
4:      find a shortest path P from s to t;
5:      set $\delta \leftarrow \min\{c(u, v) \mid (u, v) \in P\}$ and
6:      send a flow f' with value δ from s to t along path P;
7:      $G \leftarrow G_{f'}$;
8:      $f \leftarrow f + f'$;
9: **end while**
10: **return** f.

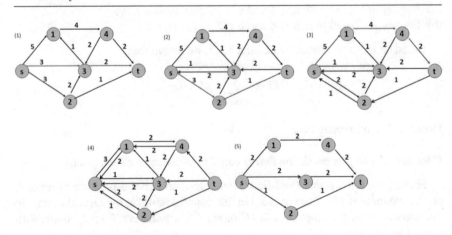

Fig. 5.6   An example for using the Edmonds-Karp algorithm

Let $\delta_f(x)$ denote the shortest path distance from source s to node x in the residual network G_f of flow f where each arc is considered to have unit distance.

Lemma 5.2.1 *When Edmonds-Karp algorithm runs, $\delta_f(x)$ increases monotonically with each flow augmentation.*

Proof For contradiction, suppose flow f' is obtained from flow f through an augmentation with path P and $\delta_{f'}(v) < \delta_f(v)$ for some node v. Without loss of generality, assume $\delta_{f'}(v)$ reaches the smallest value among such v, i.e.,

$$\delta_{f'}(u) < \delta_{f'}(v) \Rightarrow \delta_{f'}(u) \geq \delta_f(u).$$

Suppose arc (u, v) is on the shortest path from s to v in $G_{f'}$. Then $\delta_{f'}(u) = \delta_{f'}(v) - 1$ and hence $\delta_{f'}(u) \geq \delta_f(u)$. Next, let us consider two cases.

Case 1. $(u, v) \in G_f$. In this case, we have

$$\delta_f(v) \leq \delta_f(u) + 1 \leq \delta_{f'}(u) + 1 = \delta_{f'}(v),$$

a contradiction.

Case 2. $(u, v) \notin G_f$. Then arc (v, u) must lie on the augmenting path P in G_f (Fig. 5.7). Therefore,

$$\delta_f(v) = \delta_f(u) - 1 \leq \delta_{f'}(u) - 1 = \delta_{f'}(v) - 2 < \delta_{f'}(v),$$

a contradiction.   □

An arc (u, v) is *critical* in residual network G_f if (u, v) has the smallest capacity in the shortest augmenting path in G_f.

Lemma 5.2.2 *Each arc (u, v) can be critical at most $(|V| + 1)/2$ times.*

Proof Suppose arc (u, v) is critical in G_f. Then (u, v) will disappear in the next residual network. Before (u, v) appears again, (v, u) has to appear in augmenting path of a residual network $G_{f'}$. Thus, we have

$$\delta_{f'}(u) = \delta_{f'}(v) + 1.$$

Fig. 5.7 Proof of
Lemma 5.2.1

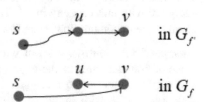

Since $\delta_f(v) \leq \delta_{f'}(v)$, we have

$$\delta_{f'}(u) = \delta_{f'}(v) + 1 \geq \delta_f(v) + 1 = \delta_f(u) + 2.$$

By Lemma 5.2.1, the shortest path distance from s to u will increase by $2(k-1)$ when arc (u, v) can be critical k times. Since this distance is at most $|V| - 1$, we have $2(k-1) \leq |V| - 1$, and hence $k \leq (|V| + 1)/2$.   □

Now, we establish a theorem on running time.

Theorem 5.2.3 *Edmonds-Karp algorithm runs in time* $O(|V| \cdot |E|^2)$.

Proof In each augmentation, there exists a critical arc. Since each arc can be critical $(|V| + 1)/2$ times, there are at most $O(|V| \cdot |E|)$ augmentations. In each augmentation, finding the shortest path takes $O(|E|)$ time, and operations on the augmenting path take also $O(|E|)$ time. Putting all together, Edmonds-Karp algorithm runs in time $O(|V| \cdot |E|^2)$.   □

Note that the above theorem does not require that all arc capacities are integers. Therefore, the modification of Edmonds and Karp is twofold: (1) Make the algorithm halt within finitely many iterations, and (2) the number of iterations is bounded by a polynomial.

5.3 Applications

The maximum flow has many applications. Let us show a few examples in this section.

Example 5.3.1 Given an undirected graph $G = (V, E)$ and two distinct vertices $s, t \in V$, please give an algorithm to determine the connectivity between s and t, i.e., the maximum number of s-to-t paths that are vertex-disjoint paths (other than at s and t).

For each vertex $v \in V$, create two vertices v^+ and v^- together with an arc (v^+, v^-). For each edge $(u, v) \in E$, create two arcs (u^-, v^+) and (v^-, u^+). Then, we obtain a directed graph G' from G (Fig. 5.8). Every path from s to t in G induces a path from s^- to t^+ in G', and a family of vertex-disjoint paths from s to t in G will induce a family of arc-disjoint paths from s^- to t^+, vice versa. Therefore, assign every arc with unit capacity in G'. Then the connectivity between s and t in G is equal to the maximum flow value from s^- to t^+ in G'.

Example 5.3.2 Consider a set of wireless sensors lying in a rectangle which is a piece of boundary area of the region of interest. The region is below the rectangle and outside is above the rectangle. The monitoring area of each sensor is a unit disk, i.e., a disk with radius of unit length. A point is said to be covered by a sensor if it lies in the monitoring disk of the sensor. The set of sensors is called

Fig. 5.8 Construct G' from G

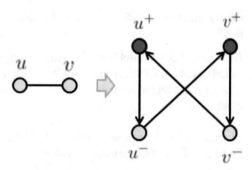

Fig. 5.9 Sensor barrier covers

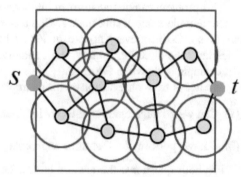

a *barrier cover* if they can cover a line (not necessarily straight) connecting two vertical edges (Fig. 5.9) of the rectangle. The barrier cover is used for protecting any intruder coming from outside. Sensors are powered with batteries and hence lifetime is limited. Assume that all sensors have unit lifetime. The problem is to find the maximum number of disjoint barrier covers so that they can be used in turn to maximize the lifetime of the system.

Use two points s and t to represent two vertical edges of the rectangle; we call them *vertical lines s and t*, respectively. Construct a graph G by setting the vertex set consisting of all sensors together with s and t (Fig. 5.9). The edge is constructed based on the following rules:

- If the monitoring disk of sensor u and the monitoring disk of sensor v have nonempty intersection, then add an edge (u, v).
- If vertical line s and the monitoring disk of sensor v have nonempty intersection, then add an edge (s, v).
- If the monitoring disk of sensor u and vertical line t have nonempty intersection, then add an edge (u, t).

In graph G, every path between s and t induces a barrier cover, and every set of vertex-disjoint paths between s and t will induce a set of disjoint barrier covers, vice versa. Therefore, we can further construct G' from G as above (Fig. 5.8), so the maximization of disjoint barrier is transformed to the maximum flow problem in G'.

Definition 5.3.3 (Matching) Consider a graph $G = (V, E)$. A subset of edges is called a *matching* if edges in the subset are not adjacent to each other. In other words, a matching is an independent edge subset. A *bipartite* matching is a matching in a bipartite graph.

Example 5.3.4 (Maximum Bipartite Matching) Given a bipartite graph (U, V, E), find a matching with maximum cardinality.

This problem can be transformed into a maximum flow problem as follows. Add a source node s and a sink node t. Connect s to every node u in U by adding an arc (s, u). Connect every node v in V to t by adding an arc (v, t). Add to every edge in E the direction from U to V. Finally, assign every arc with unit capacity. An example is shown in Fig. 5.10.

Motivated from observation on the example in Fig. 5.10, we may have questions:

(1) Can we do augmentation directly in bipartite graph without putting it in a flow network?
(2) Can we perform the first three augmentations in the same time?

For both questions, the answer is yes. Let us explain the answer in the next section.

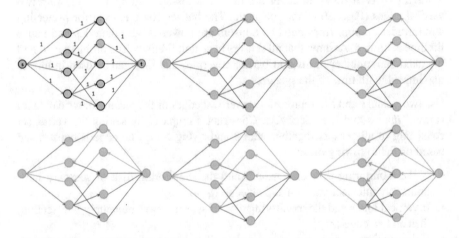

Fig. 5.10 Maximum bipartite matching is transformed to maximum flow

5.4 Matching

In this section, we study matching in a directed way. First, we define the augmenting path as follows.

Consider a matching M in a bipartite graph $G = (U, V, E)$. Let us call every edge in M as *matched* edge and every edge not in M as *unmatched* edge. A node v is called a *free node* if v is not an ending point of a matched edge.

Definition 5.4.1 (Augmenting Path) The *augmenting path* is now defined to be a path satisfying the following:

- It is an *alternating path*, that is, edges on the path are alternatively unmatched and matched.
- The path is between two free nodes.

There are totally odd number of edges in an augmenting path. The number of unmatched edges is one more than the number of matched edges. Therefore, on an augmenting path, turn all matched edges to unmatched and turn all unmatched edges to matched. Then considered matching will become a matching with one more edge. Therefore, if a matching M has an augmenting path, then M cannot be maximum. The following theorem indicates that the inverse holds.

Theorem 5.4.2 *A matching M is maximum if and only if M does not have an augmenting path.*

Proof Let M be a matching without augmenting path. For contradiction, suppose M is not maximum. Let M^* be a maximum matching. Then $|M| < |M^*|$. Consider $M \oplus M^* = (M \setminus M^*) \cup (M^* \setminus M)$, in which every node has degree at most two (Fig. 5.11).

Hence, it is disjoint union of paths and cycles. Since each node with degree two must be incident to two edges belonging to M and M', respectively. Those paths and cycles must be alternative. They can be classified into four types as shown in Fig. 5.12.

Note that in each of the first three types of connected components, the number of edges in M is not less than the number of edges in M^*. Since $|M| < |M^*|$, we have $|M \setminus M^*| < |M^* \setminus M|$. Therefore, the connected component of the fourth type must exist, that is, M has an augmenting path, a contradiction.   □

We now return to the question on augmentation of several paths at the same time. The following algorithm is the result of a positive answer.

We next analyze Hopcroft-Karp algorithm (Algorithm 17).

Lemma 5.4.3 *In each iteration, the length of the shortest augmenting path is increased by at least two.*

Proof Suppose matching M' is obtained from matching M through augmentation on a maximal set of shortest augmenting paths, $\{P_1, P_2, \ldots, P_k\}$, for M. Let P be a shortest augmenting path for M'. If P is disjoint from $\{P_1, P_2, \ldots, P_k\}$, then P is

$$M \qquad\qquad M^* \qquad\qquad M \oplus M^*$$

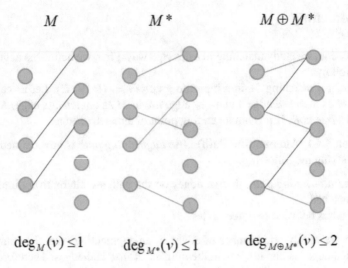

$$\deg_M(v) \le 1 \qquad \deg_{M^*}(v) \le 1 \qquad \deg_{M \oplus M^*}(v) \le 2$$

Fig. 5.11 $M \oplus M^*$

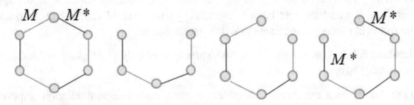

Fig. 5.12 Connected components of $M \oplus M^*$

Algorithm 17 Hopcroft-Karp algorithm for maximum bipartite matching

---

Input: A bipartite graph $G = (U, V, E)$.
Output: A maximum matching M.

1: $M \leftarrow$ any edge;
2: **while** there exists an augmenting path **do**
3:     find a maximal set of disjoint augmenting paths $\{P_1, P_2, \ldots, P_k\}$;
4:     $M \leftarrow M \oplus (P_1 \cup P_2 \cup \cdots P_k)$;
5: **end while**
6: **return** M.

---

also an augmenting path for M. Hence, the length of P is longer than the length of P_1. Note that the augmenting path must have odd length. Therefore, the length of P at-least-two longer than the length of P_1.

Next, assume that P has an edge lying in P_i for some i. Note that every augmenting path has two endpoints in U and V, respectively. Let u and v be two endpoints of P and u_i and v_i two endpoints of P_i where $u, u_i \in U$ and $v, v_i \in V$. Without loss of generality, assume that (x, y) is the edge lying on P and also on some P_i such that no such edge exists from y to v. Clearly,

$$\text{dist}_P(y, v) \geq \text{dist}_{P_i}(y, v_i), \tag{5.1}$$

where $\text{dist}_P(y, v)$ denotes the distance between y and v on path P. In fact, if $\text{dist}_P(y, v) < \text{dist}_{P_i}(y, v_i)$, then replacing the piece of P_i between y and v_i by the piece of P between y and v, we obtain an augmenting path for M, shorter than P_i, contradicting to shortest property of P_i. Now, we claim that the following holds.

$$\text{dist}_{P_i}(u_i, y) + 1 = \text{dist}_{P_i}(u_i, x) \leq \text{dist}_P(u, x) = \text{dist}_P(u, y) - 1. \tag{5.2}$$

To prove this claim, we may put the bipartite graph into a flow network as shown in Fig. 5.10. Then every augmenting path receives a direction from U to V, and the claim can be proved as follows.

Firstly, note that on path P, we assumed that the piece from y to v is disjoint from all $P_1, P_2, \ldots, P_k$. This assumption implies that edge (x, y) is in direction from x to y on P, so that $\text{dist}_P(u, x) = \text{dist}_P(u, y) - 1$.

Secondly, note that edge (x, y) also appears on P_i, and after augmentation, every edge in P_i must change its direction. Thus, edge (x, y) is in direction from y to x on P_i. Hence, $\text{dist}_{P_i}(u_i, y) + 1 = \text{dist}_{P_i}(u_i, x)$.

Thirdly, by Lemma 5.2.1, we have $\text{dist}_{P_i}(u_i, x) \leq \text{dist}_P(u, x)$.

Finally, putting (5.1) and (5.2) together, we obtain

$$\text{dist}_{P_i}(u_i, v_i) + 2 \leq \text{dist}_P(u, v).$$

$\square$

Theorem 5.4.4 *Hopcroft-Karp algorithm computes a maximum bipartite matching in time* $O(|E|\sqrt{|V|})$.

Proof In each iteration, it takes $O(|E|)$ time to find a maximal set of shortest augmenting paths and to perform augmentation on these paths. (We will give more explanation after the proof of this theorem.) Let M be the matching obtained through $\sqrt{|V|}$ iterations. Let M^* be the maximum matching. Then $M \oplus M^*$ contains $|M^*| \setminus |M|$ augmenting path, each of length at least $1 + 2\sqrt{|V|}$ by Lemma 5.4.3. Therefore, each takes at least $2 + 2\sqrt{|V|}$ nodes. This implies that the number of augmenting paths in $M \oplus M^*$ is upper bounded by

$$|V|/(2 + 2\sqrt{|V|}) < \sqrt{|V|}/2.$$

Thus, M^* can be obtained from M through at most $\sqrt{|V|}/2$ iterations. Therefore, M^* can be obtained within at most $\frac{3}{2} \cdot \sqrt{|V|}$ iterations. This completes the proof.

$\square$

There are two steps in finding a maximal set of disjoint augmenting paths for a matching M in bipartite graph $G = (U, V, E)$.

In the first step, employ the breadth-first search to put nodes into different levels as follows. Initially, select all free nodes in U and put them in the first level. Next,

Fig. 5.13 The breadth-first
search

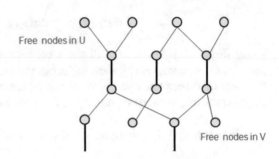

Free nodes in U

Free nodes in V

put in the second level all nodes each with an unmatched edge connecting to a node
in the first level. Then, put in the third level all nodes each with a matched edge
connecting to a node in the second level. Continue in this alternating ways, until a
free node in V is discovered, say in the kth level (Fig. 5.13). Let F be all free nodes
in the kth level and H the obtained subgraph. If the breadth-first search comes to an
end and still cannot find a free node in V, then this means that there is no augmenting
path, and a maximum matching has already obtained by Hopcroft-Karp algorithm.

In the second step, employ the depth-first search to find path from each node in
F to a node in the first level. Such paths will be searched one by one in H, and once
a path is obtained, all nodes on this depth-first-search path will be deleted from H,
until no more such path can be found.

Since both steps can work in $O(|E|)$ time, the total time for finishing this task is
$O(|E|)$.

The alternating path method can also be used for the maximum matching in
general graph.

Problem 5.4.5 (Maximum Graph Matching) Given a graph $G = (V, E)$, find a
matching with maximum cardinality.

The *augmenting path* is also defined to be a path satisfying the following:

- It is an *alternating path*, that is, edges on the path are alternatively unmatched
  and matched.
- The path is between two free nodes.

Now, the proof of Theorem. 5.4.2 can be applied to the graph matching without
any change, to show the following.

Theorem 5.4.6 *A matching M is maximum if and only if M does not have an
augmenting path.*

Therefore, we obtained Algorithm 18 for the maximum graph matching problem.

How to find an augmenting path for matching in a general graph $G = (V, E)$?
Let us introduce the Blossom algorithm of Edmonds. A *blossom* is an almost
alternating odd cycle as shown in Fig. 5.14.

Algorithm 18 Algorithm for maximum graph matching

Input: A graph $G = (V, E)$.
Output: A maximum matching M.
 1: $M \leftarrow$ {an arbitrary edge};
 2: **while** there exists an augmenting path P **do**
 3:     $M \leftarrow M \oplus P$;
 4: **end while**
 5: **return** M.

Fig. 5.14 A blossom shrinks into a node

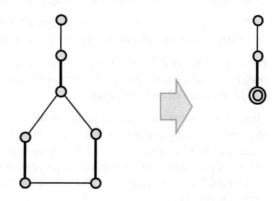

The blossom algorithm is similar to the first step of augmenting-path finding in Hopcroft-Karp algorithm, i.e., employ the breadth-first search by using unmatched edge and matched edge alternatively. However, start from one free node x at a time.

Definition 5.4.7 (Alternating Tree) The alternating tree has a root at a free node x. Its first level consists of unmatched edges, its second level consists of matched edges, and alternatively continue.

Definition 5.4.8 (Even and Odd Nodes) In an alternating tree, a node is called an *odd* node if its distance to the root has odd length. A node is called an *even* node if its distance to the root has even length, e.g., the root is an even node.

Now, let us describe the blossom algorithm.

- For each free node x, construct an alternating tree with root x in the breadth-first-search ordering.
- At an odd node y, if no matched edge is incident to y, then y is a free node, and an augmenting path from x to y is found. If there exists a matched edge incident to y, then such a matched edge is unique, and y can be extended uniquely to an even node.
- At an even node z, if no unmatched edge is incident to z, then z cannot be extended. If there exists an unmatched edge (u, z) incident to z, then consider another ending node u of this edge. If u is a known even node, then a blossom is found; shrink the blossom into an even node. If u is not a known even node, then u can be counted as an odd node to continue our construction.

- At a level consisting of even nodes, if none of them can be extended, then there is no augmenting path starting from free node x.
- Therefore, above construction of alternating tree, we can either find an augmenting path starting from free node x or determine not existing of such a path. As soon as an augmenting path is found, we can carry out an augmentation, matching is updated, and we restart to search for an augmenting path.
- If for all free nodes, no augmenting path can be found from construction of alternating trees, then current matching is maximum.

To show the correctness of the above algorithm, it is sufficient to explain why we can shrink a blossom into a node. An explanation is given in the following.

Lemma 5.4.9 *Let B be a blossom in graph G. Let G/B denote the graph obtained from G by shrinking B into a node. Then G contains an augmenting path if and only if G/B contains an augmenting path.*

Proof Note that the alternating path can be extended passing through a blossom out-reach to its any connection (Fig. 5.15). Therefore, if an augmenting path passes through a blossom, then after shrink the blossom into a node, the augmenting path is still an augmenting path. Conversely, if an augmenting path contains a node which is obtained from a blossom, then after de-shrink the blossom, we can still obtain an augmenting path.   □

Clearly, this algorithm runs in $O(|V| \cdot |E|)$ time. Thus, we have the following.

Theorem 5.4.10 *With blossom algorithm, the maximum cardinality matching in graph $G = (V, E)$ can be computed in $O(|V|^2 \cdot |E|)$ time.*

Proof To obtain a maximum matching, we can carry out at most $|V|$ augmentations. To find an augmenting path, we may spend $O(|V| \cdot |E|)$ time to construct alternating trees. Therefore, the total running time is $O(|V|^2 |E|)$.   □

For weighted bipartite matching and weighted graph matching, can we use the alternating path to deal with them? The answer is yes. However, it is more complicated. We can find a better way, which will be introduced in Sect. 6.8.

Fig. 5.15 An alternating path passes a blossom

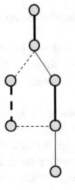 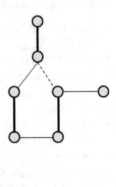

5.5 Dinitz Algorithm

In this and the next sections, we present more algorithms for the maximum flow problem. They have running time better than Edmonds-Karp algorithm.

First, we note that the idea in Hopcroft-Karp algorithm can be extended from matching to flow. This extension gives a variation of Edmonds-Karp algorithm, called Dinitz algorithm.

Consider a flow network $G = (V, E)$. The algorithm starts with a zero flow $f(u, v) = 0$ for every arc (u, v). In each substantial iteration, consider residual network G_f for flow f. Start from source node s to do the breadth-first search until node t is reached. If t cannot be researched, then algorithm stops, and the maximum flow is already obtained. If t is reached with distance ℓ from node s, then the breadth-first-search tree contains ℓ level, and its nodes are divided into ℓ classes $V_0, V_1, \ldots, V_\ell$ where V_i is the set of all nodes each with distance i from s and $\ell \leq |V|$. Collect all arcs from V_i to V_{i+1} for $i = 0, 1, \ldots, \ell - 1$. Let $L(s)$ be the obtained levelable subnetwork. Above computation can be done in $O(|E|)$ time.

Next, the algorithm finds augmenting paths to do augmentations in the following way.

Step 1. Iteratively, for $v \neq t$ and $u \neq s$, remove, from $L(s)$, every arc (u, v) with no coming arc at u or no outgoing arc at v. Denote by $\hat{L}(s)$ the obtained levelable network.

Step 2. If $\hat{L}(s)$ is empty, then this iteration is completed, and go to the next iteration. If $\hat{L}(s)$ is not empty, then it contains a path of length ℓ, from s to t. Find such a path P by using the depth-first search. Do augmentation along the path P. Update $L(s)$ by using $\hat{L}(s)$ and deleting all critical arcs on P. Go to Step 1.

This algorithm has the following property.

Lemma 5.5.1 *Let $\delta_f(s, t)$ denote the distance from s to t in residual graph G_f of flow f. Suppose flow f' is obtained from flow f through an iteration of Dinitz algorithm. Then $\delta_{f'}(s, t) \geq \delta_f(s, t) + 2$.*

Proof The proof is similar to the proof of Lemma 5.2.1.   □

The correctness of Dinitz algorithm is stated in the following theorem.

Theorem 5.5.2 *Dinitz algorithm produces a maximum flow in $O(|V|^2|E|)$ time.*

Proof By Lemma 5.5.1, Dinitz algorithm runs within $O(|V|)$ iterations. Let us estimate the running time in each iteration.

- The construction of $L(s)$ spends $O(|E|)$ time.
- It needs $O(|V|)$ time to find each augmenting path and to do augmentation. Since each augmentation will remove at least one critical arc, there are at most $O(|E|)$ augmentations. Thus, the total time for augmentations is $O(|V| \cdot |E|)$.
- Amortizing all time for removing arcs, it is at most $O(|E|)$.

Therefore, each iteration runs in $O(|V| \cdot |E|)$ time. Hence, Dinitz algorithm runs in $O(|V|^2|E|)$ time. At the end of the algorithm, G_f does not contain a path from s to t. Thus, f is a maximum flow.   □

5.6 Goldberg-Tarjan Algorithm

In this section, we study a different type of incremental method for maximum network flow. In this method, a valid label will play an important role. This valid label will be on each arc to guide the incremental direction.

Consider a flow network $G = (V, E)$ with capacity $c(u, v)$ for each arc $(u, v) \in E$; s and t are source and sink, respectively. As usual, for simplicity of description, we extend capacity $c(u, v)$ to every pair of nodes u and v by defining $c(u, v) = 0$ if $(u, v) \notin E$.

A function $f : V \times V \to R$ is called a *preflow* if

1. (Capacity constraint) $f(u, v) \leq c(u, v)$ for every $u, v \in V$.
2. (Skew symmetry) $f(u, v) = -f(v, u)$ for all $u, v \in V$.
3. For every $v \in V \setminus \{s, t\}$, $\sum_{v \in V \setminus \{u\}} f(u, v) \geq 0$, i.e., $\sum_{(u,v) \in E} f(u, v) \geq \sum_{(v,w) \in E} f(v, w)$.

Compared with those three conditions in the definition of flow, the first two are the same, and the third one is different. The flow conservation condition is relaxed to allow more flow coming than going out at any node other than s and t. This difference is called the *excess* at node v and denotes

$$e(v) = \sum_{(u,v) \in E} f(u, v) \geq \sum_{(v,w) \in E} f(v, w).$$

A node v is said to be *active* if $e(v) > 0$, $v \neq s$, and $v \neq t$. In preflow-relabel algorithm, the excess will be pushed from an active node toward the sink, relying on the valid distance label $d(v)$ for $v \in V$, satisfying the following conditions.

- $d(t) = 0$.
- $d(u) \leq d(v) + 1$ for $(u, v) \in E$.

An arc (u, v) is said to be *admissible* if $d(u) = d(v) + 1$ and $c(u, v) > 0$. Note that if we consider a residual graph, then $c(u, v)$ should be considered as updated capacity.

Lemma 5.6.1 *Let $dist(u, v)$ denote the minimum number of arcs on the path from u to v. Then $d(u) \leq dist(u, t)$.*

Proof It can be proved by induction on $dist(u, t)$. For $dist(u, t) = 0$, u must be t, and hence $d(t) \leq dist(t, t)$. For $dist(u, t) = k > 0$, suppose $(u, u_1, \ldots, u_k = t)$ is the shortest path from u to t. Then $dist(u_1, t) = k - 1$. By induction hypothesis,

$d(u_1) \le dist(u_1, t)$. Hence,

$$d(u, t) \le 1 + d(u_1) \le 1 + dist(u_1, t) = dist(u, t).$$

$\square$

Next, we explain two operations, push and relabel. Consider an active node v. Suppose that there exists an admissible arc (v, w). Then a flow $\min(e(v), c(v, w))$ will be pushed along arc (v, w). If $e(v) \le c(v, w)$, then it is called a *saturated push*. Otherwise, the push is called a *non-saturated* one.

Suppose that there does not exist an admissible arc (v, w). Then relabel $d(v)$ by setting

$$d(v) = 1 + \min\{d(w) \mid c(v, w) > 0\}.$$

An important observation is stated in the following lemma.

Lemma 5.6.2 *After push and relabel, the (residual) network and label $d(\cdot)$ are updated. However, $d(\cdot)$ is still a valid label for updated network. Hence, Lemma 5.6.1 holds. Moreover, each relabel for node v will increase its label at least one.*

Proof First, consider a push along arc (v, w). This push may add a new arc (w, v) to the residual network. Therefore, we need to make sure $d(w) \le d(v) + 1$. This is true because $d(v) = d(w) + 1$.

Next, consider relabel node v. By the rule, new label for v is upper bounded by $d(w) + 1$ for any arc (v, w) with $c(v, w) > 0$. Moreover, suppose $(v, w') = \operatorname{argmin}\{d(w) \mid c(v, w) > 0\}$. Since (v, w') is not admissible, we have $d(v) \le d(w')$. Hence, the new label for v is $d(w') + 1 \ge d(v) + 1$. $\square$

Now, we are ready to describe the push-relabel algorithm of Goldberg and Tarjan (see Algorithm 19). An example is shown as in Fig. 5.16.

We next analyze this algorithm.

Lemma 5.6.3 *Let f be a preflow appearing in computation process of Goldberg-Tarjan algorithm. Then, in residual network G_f, every active node v has a path connecting to s.*

Proof In flow decomposition of f, there is a path flow from s to active node v. This path flow will result in a path from v to s in residual network G_f. $\square$

Lemma 5.6.4 *For any node v, $d(v) \le 2n$ during computation, and there are at most $2n$ relabels at each node v, where n is the number of nodes. Moreover, all relabels need at most $O(mn)$ time of computation.*

Proof Note that the relabel occurs only at active nodes. If a node has never been active, then its label is at most $n - 1$. If a node v has been active, then at last time that v is active, there is a path from v to s. After push, this path still exists, and

---

Algorithm 19 Goldberg-Tarjan algorithm for maximum flow

Input: A flow network $G = (V, E)$ with source s, sink t, and capacity $c(u, v)$ for every $(u, v) \in E$.
Output: A flow $f : V \times V \rightarrow R$.
1: $f(u, v) \leftarrow 0$ for all $(u, v) \in V \times V$;
2: $d(v) \leftarrow dist(v, t)$;
3: $f(s, v) \leftarrow c(s, v)$ for $(s, v) \in E$;
4: $d(s) \leftarrow n; (n = |V|)$
5: $G \leftarrow G_f$;
6: **while** there is an active node v **do**
7:   **if** there is an admissible arc (v, w) **then**
8:     $f(v, w) \leftarrow \min(e(v), c(v, w))$
9:   **else**
10:     $d(v) \leftarrow \min\{d(w) \mid c(v, w) > 0\}$
11:   **end if**
12:   $G \leftarrow G_f$;
13: **end while**
14: **return** f

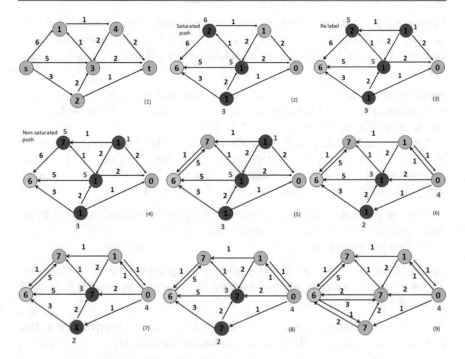

Fig. 5.16 An example for Goldberg-Tarjan algorithm (it contains three iterations from (5) to (6) and two iterations from (8) to (9))

hence, $d(v) \leq n - 1 + d(s) = 2n - 1$. Since each relabel makes a node's label increased at least once, a node can be relabeled at most $2n - 1$ times.

Let $deg(v)$ be the number of arcs at node v. Then each relabel spends time at most $deg(v)$. Therefore, all relabels need computational time at most

$$\sum_{v \in V} deg(v)(2n - 1) \le (2n - 1) \cdot 2m = O(mn)$$

where $m = |E|$. □

Lemma 5.6.5 *There are at most $O(mn)$ saturated pushes in computation of Goldberg-Tarjan algorithm, where $n = |V|$ and $m = |E|$.*

Proof Note that a push must be on an arc (u, v) in G_f, and hence $(u, v) \in E$ or $(v, u) \in E$ where E is the set of arcs in input flow network G. Between two consecutive saturated pushes on an arc (u, v), there must exist a relabel for v. The total number of relabels for v is at most $2n$. Therefore, the total number of saturated pushes is at most $O(mn)$. □

Lemma 5.6.6 *There are at most $O(mn^2)$ non-saturated pushes in computation of Goldberg-Tarjan algorithm, where $n = |V|$ and $m = |E|$.*

Proof Consider a potential function $\Phi = \sum_{v:\text{active}} d(v)$. For simplicity of speaking, let us call some operation making a "deposit" if it decreases the value of Φ and "withdraw" if it increases the values of Φ.

First, note that each node has its label at most $2n - 1$. Therefore, the relabel can make total deposit at a node at most $2n - 1$. Hence, totally, the relabel can make deposit at most $2n^2 - n$.

Now, consider the saturated push. Each saturated push may increase a new active node, which results in a deposit at most $2n - 1$. Since there are totally $O(mn)$ saturated pushes, the saturated push can make totally $O(mn^2)$ deposit.

Finally, consider the non-saturated push. Each non-saturated push will remove an active node while increasing an active node. Suppose the push is on arc (u, v). Then $d(u) = d(v) + 1$. When u is removed and v may be added, the non-saturated push will make a withdraw at least one. Therefore, the number of non-saturated pushes is at most

$$n^2 + O(n^2) + O(mn^2) = O(m^2)$$

where note that initially, Φ has value at most n^2. □

Theorem 5.6.7 *Goldberg-Tarjan algorithm must terminate at a maximum flow within time $O(mn^2)$.*

Proof The algorithm must terminate after all pushes and relabels are done. Therefore, by Lemmas 5.6.4, 5.6.5, and 5.6.6, the algorithm will terminate within time

$$O(mn) + O(mn) + O(mn^2) = O(mn^2).$$

Moreover, when the algorithm terminates, there is no active node, and hence the preflow becomes a normal flow. Moreover, the label of s is still $d(s) = n$, which means that there is no path from s to t in the residual graph. Therefore, the flow is maximum.   □

Note that in Goldberg-Tarjan algorithm, the selection of an active node is arbitrary. This gives an opportunity for improvement. There are two interesting rules for the selection of active node, which can improve the running time.

The First Rule (Excess Scaling) The algorithm is divided into phases, Δ-scaling phase for $\Delta = 2^{\lceil \log_2 C \rceil}, 2^{\lceil \log_2 C \rceil - 1}, \ldots, 1$ where $C = \max\{c(u, v) \mid (u, v) \in E\}$. At beginning of the Δ-scaling phase, $e(v) \leq \Delta$ for every active node v. At the end of the Δ-scaling phase, $e(v) \leq \Delta/2$ for every active node v. (When $\Delta = 1$, $\Delta/2$ is replaced by 0.) During the Δ-scaling phase, active node v is selected to be

$$v = \operatorname{argmin}\{d(u) \mid e(u) > \Delta/2\}.$$

In order to keep all active nodes with excess no more than Δ, a modification has to be made on flow amount in a push. Along an admissible arc (u, v), the flow of amount $\min(e(u), c(v, w), \Delta - e(w))$ is pushed.

Note that with the modification, there is no change on the relabel and the saturated push. Therefore, Lemmas 5.6.4 and 5.6.5 still hold. However, the non-saturated push occurs when pushed amount is either $e(v)$ or $\Delta - e(w)$. In either case, this amount is at least $\Delta/2$. In Fig. 5.17, an example is presented for computation in a Δ-scaling phase.

We next analyze the excess scaling algorithm, i.e., Goldberg-Tarjan algorithm with excess scaling rule.

Lemma 5.6.8 *In each Δ-scaling phase, the number of non-saturated pushes is at most $O(n^2)$.*

Proof Consider a potential function $\Phi = \sum_{v \in V} d(v) \cdot e(v)/\Delta$. For simplicity of speaking, let us call some operation making a "deposit" if it decreases the value of Φ and "withdraw" if it increases the values of Φ.

Note that $e(v)$ is nondecreasing during computation and $e(v) \leq 2n$. Therefore, the relabel can deposit at most $2n$ at each node v. Hence, the relabel deposits totally at most $2n^2$ for Φ.

Every push will withdraw from Φ since it moves a certain amount value from node v to w with $d(v) = d(w) + 1$. Especially, every non-saturated push will move at least $\Delta/2$ from $e(v)$ to $e(w)$, that is, it withdraws at least $1/2$ from Φ. Thus, the total number of non-saturated pushes is at most $4n^2$ during each Δ-scaling phase.
                                                                                                   □

Theorem 5.6.9 *The excess scaling algorithm must terminate at a maximum flow within time $O(mn + n^2 \log C)$.*

Proof By Lemmas 5.6.4 and 5.6.5, the relabel and the saturated push use totally $O(mn)$ time. By Lemma 5.6.8, the non-saturated push spends totally $n^2 \log C$ time. At the end of algorithm, there is no active node, i.e., the preflow becomes a flow.

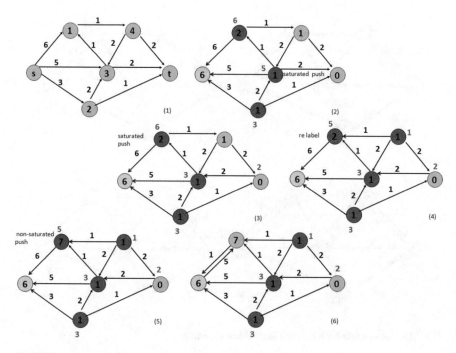

Fig. 5.17 An example for Δ-scaling phase ($\Delta = 8$)

Moreover, in its residual graph, there is no path from source s to sink t since $d(s) = n$. Thus, the flow is maximum.   □

The Second Rule (Highest-Level Pushing) A *level* is subset of nodes with the same label. In this rule, the active node v is selected from the highest level, i.e.,

$$v = \text{argmax}\{d(u) \mid u \text{ is active}\}.$$

In Fig. 5.18, an example is presented for computation in Goldberg-Tarjan algorithm with this rule for selection of active node.

We next analyze Goldberg-Tarjan algorithm with highest-level pushing.

Let a *phase* be a consecutive sequence of pushes all at the same level.

Lemma 5.6.10 *There are totally at most $O(n^2)$ phases.*

Proof Let v be selected active node. A phase ends only if one of the following two cases occurs:

(a) All active nodes at level $d(v)$ have become inactive.
(b) A relabel for v occurs before v becomes inactive.

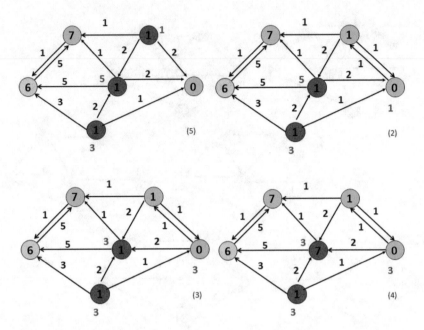

Fig. 5.18 An example for a phase of highest-level pushing

By Lemma 5.6.4, there are at most $O(n^2)$ relabels, and hence (b) occurs at most $O(n^2)$ times. Let $\Phi = \max\{d(v) \mid v$ is active$\}$. Then initially, $\Phi \leq n$. If (b) occurs, then Φ is increased. However, the total amount of increasing at each node is at most $2n$. Hence, Φ can be increased at most $2n^2$ by relabels. If (a) occurs, then Φ will be decreased by one. Therefore, (b) can occur at most $n + 2n^2 = O(n^2)$ times. Putting together, the number of phases is at most $O(n^2)$.   □

Lemma 5.6.11 *There are at most $O(n^2 m^{1/2})$ non-saturated pushes.*

Proof Let $k = \lceil m^{1/2} \rceil$. Call a phase as a *cheap one* if it contains at most k non-saturated pushes. Otherwise, call it as an *expensive phase*. Since there are at most $O(n^2)$ phases. The number of non-saturated pushes in all cheap phases is at most $O(n^2 k) = O(n^2 m^{1/2})$. In the following, we estimate the number of non-saturated pushes in all expensive phases.

To do so, define a potential function:

$$\Phi = \sum_{\text{active } v} z(v)$$

where $z(v)$ is the number of nodes each with label at most $d(v)$, i.e.,

$$z(v) = |\{u \in V \mid d(u) \leq d(v)\}|.$$

Again, for simplicity of speaking, let us call some operation making a "deposit" if it decreases the value of Φ and "withdraw" if it increases the values of Φ.

Each relabel makes an active node v have label increased, which will make $z(v)$ increased. However, the increase cannot exceed n, the total number of nodes. Therefore, each relabel makes a deposit with at most value n. $O(n^2)$ relabels will deposit with at most value $O(n^3)$.

Each saturated push may activate a node, which will deposit with value at most $2n$. Since there are totally $O(mn)$ saturated pushes, they can deposit with value at most $O(n^2)$.

Every non-saturated push on admissible arc (v, w) will make u inactive. Since $d(v) = d(w) + 1$, $z(v) - z(w)$ is equal to the number of nodes at the highest level in the phase containing the non-saturated push. Note that during a phase, an active node v at the level becomes inactive if and only if a non-saturated push occurs at active node v. Therefore, at beginning of an expensive phase, there must exist at least k active nodes at the highest level. This means that every non-saturated push will withdraw at least value k from Φ.

Summarized from above argument, we can conclude that the total number of non-saturated pushes in expensive phases is at most $O((n^3 + n^2 m)/k) = O(n^2 m^{1/2})$. Therefore, the total number of non-saturated pushes during whole computation is at most $O(n^2 m^{1/2})$.   □

Theorem 5.6.12 *The Goldberg-Tarjan algorithm with highest-level push must terminate at a maximum flow within time $O(n^2 m^{1/2})$.*

Proof It follows immediately from Lemmas 5.6.4, 5.6.5, and 5.6.11.   □

Exercises

1. A conference organizer wants to set up a review plan. There are m submitted papers and n reviewers. Each reviewer has made p papers as "prefer to review." Each paper should have at least q review reports. Find a method to determine whether such a review plan exists or not.
2. A conference organizer wants to set up a review plan. There are m submitted papers and n reviewers. Each reviewer is allowed to make at least p_1 papers as "prefer to review" and at least p_2 papers as "likely to review." Each paper should have at least q_1 review reports and at most q_2 review reports. Please give a procedure to make the review plan.
3. Let f be a flow of flow network G and f' a flow of residual network G_f. Show that $f + f'$ is a flow of G.
4. Let G be a flow network in which every arc capacity is a positive even integer. Show that its maximum flow value is an even integer.

5. Let G be a flow network in which every arc capacity is a positive odd integer. Can we conclude that its maximum flow value is an odd integer? If not, please give a counterexample.

6. Let G be a flow network. An arc (u, v) is said to be *critical* for a maximum flow f if $f(u, v) = c(u, v)$ where $c(u, v)$ is the capacity of (u, v). Show that an arc (u, v) is critical for every maximum flow if and only if decreasing it capacity by one will result in maximum flow value getting decreased by one.

7. Let A be an $m \times n$ matrix with non-negative real numbers such that for every row and every column, the sum of entries is an integer. Prove that there exists an $m \times n$ matrix B with non-negative integers and the same sums as in A, for every row and every column.

8. Suppose there exist two distinct maximum flows f_1 and f_2. Show that there exist infinitely many maximum flows.

9. Consider a directed graph G with a source s, a sink t and nonnegative arc capacities. Find a polynomial-time algorithm to determine whether G contains a unique s-t cut.

10. (This is an example on which Ford-Fulkerson algorithm runs with infinitely many augmentations.) Consider a flow network as shown in Fig. 5.19 where $x = \frac{\sqrt{5}-1}{2}$. Show by induction on k that the residual capacity $c(u, v) - f(u, v)$ on three vertical arcs can be $x^k, 0, x^{k+1}$ for every $k = 0, 1, 2, \ldots$. (Hint: The case of $k = 0$ is shown in Fig. 5.20. The induction step is as shown in Fig. 5.21.)

11. Consider a flow network $G = (V, E)$ with a source s, a sink t, and nonnegative capacities. Suppose a maximum flow f is given. If an arc is broken, find a fast algorithm to compute a new maximum flow based on f. A favorite algorithm will run in $O(|E| \log |V|)$ time.

12. Consider a flow network $G = (V, E)$ with a source s, a sink t, and nonnegative integer capacities. Suppose a maximum flow f is given. If the capacity of an

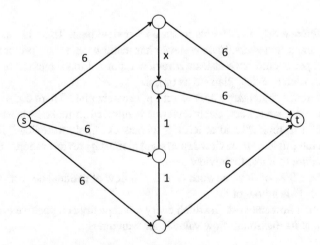

Fig. 5.19 An example for Ford-Fulkerson algorithm

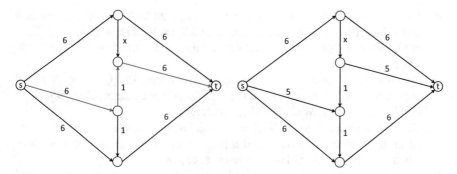

Fig. 5.20 Base step

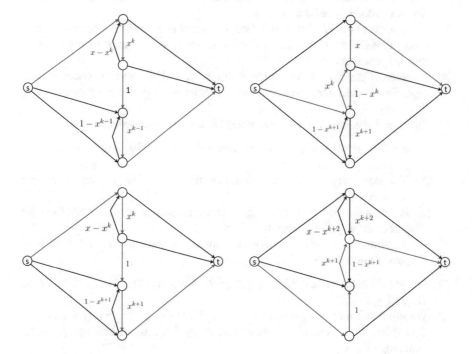

Fig. 5.21 Induction step

arc is increased by one, find a fast algorithm to update the maximum flow. A favorite algorithm runs in $O(|E| + |V|)$ time.

13. Consider a directed graph $G = (V, E)$ with a source s and a sink t. Instead of arc capacity, assume that there is the nonnegative integer node capacity $c(v)$ on each node $v \in V$, that is, the total flow passing node v cannot exceed $c(v)$. Show that the maximum flow can be computed in polynomial-time.

14. Show that the maximum flow of a flow network $G = (V, E)$ can be decomposed into at most $|E|$ path-flows.

15. Let $G = (V, E)$ be a undirected connected graph with two distinct nodes s and t. Find two disjoint node subsets S and T such that $s \in S$ and $t \in T$, to minimize $\delta(S) + \delta(T)$ where $\delta(X)$ denotes the number of edges between X and $V \setminus X$.

16. Suppose a flow network $G = (V, E)$ is symmetric, i.e., $(u, v) \in E$ if and only if $(v, u) \in E$ and $c(u, v) = c(v, u)$. Show that Edmonds-Karp algorithm terminates within at most $|V| \cdot |E|/4$ iterations.

17. Consider a directed graph G. A node-disjoint set of cycles is called a *cycle-cover* if it covers all nodes. Find a polynomial-time algorithm to determine whether a given graph G has a cycle-cover or not.

18. Consider a graph G. Given two nodes s and t, and a positive integer k, find a polynomial-time algorithm to determine whether there exist or not k edge-disjoint paths between s and t.

19. Consider a graph G. Given two nodes s and t and a positive integer k, find a polynomial-time algorithm to determine whether there exist or not k node-disjoint paths between s and t.

20. Consider a graph G. Given three nodes x, y, and z, find a polynomial-time algorithm to determine whether there exists a simple path from x to z passing through y.

21. Prove or disprove (by counterexample) the following statements.

    (a) If a flow network has unique maximum flow, then it has a unique minimum s-t cut.
    (b) If a flow network has unique minimum s-t cut, then it has a unique maximum flow.
    (c) A maximum flow must associate with a minimum s-t cut such that the flow passes through the minimum s-t cut.
    (d) A minimum s-t cut must associate with a maximum flow such that the flow passes through the minimum s-t cut.

22. Let M be a maximal matching of a graph G. Show that for any matching M' of G, $|M'| \leq 2 \cdot |M|$.

23. We say that a bipartite graph $G = (L, R, E)$ is d-regular if every vertex $v \in L \cup R$ has degree exactly d. Prove that every d-regular bipartite graph has a matching of size $|L|$.

24. There are n students who studied at a late-night study for final exam. The time has come to order pizzas. Each student has his own list of required toppings (e.g., mushroom, pepperoni, onions, garlic, sausage, etc.). Everyone wants to eat at least half a pizza, and the topping of that pizza must be in his required list. A pizza may have only one topping. How to compute the minimum number of pizzas to order to make everyone happy?

25. Consider bipartite graph $G = (U, V, E)$. Let $\mathcal{H}$ be the collection of all subgraphs H that for every $u \in U$, H has at most one edge incident to u. Let $E(H)$ denote the edge set of H and $\mathcal{I} = \{E(H) \mid H \in \mathcal{H}\}$. Show that (a) $(E, \mathcal{I})$ is a matroid and (b) all matchings in G form an intersection of two matroids.

26. Consider a graph $G = (V, E)$ with nonnegative integer function $c : V \to N$. Find an augmenting path method to compute a subgraph $H = (V, F)$ $(F \subseteq E)$ with maximum number of edges such that for every $v \in V$, $deg(v) \leq c(v)$.

27. A conference with a program committee of 30 members received 100 papers. Before making an assignment, the PC-chair first asked all PC-members each to choose 15 preferred papers. Based on what PC-members choose, the PC-chair wants to find an assignment such that each PC-member reviews 10 papers among 15 chosen ones and each paper gets 3 PC-members to review. How do we figure out whether such an assignment exists? Please design a maximum flow formulation to answer this question.

28. Let $U = \{u_1, u_2, \ldots, u_n\}$ and $V = \{v_1, v_2, \ldots, v_n\}$. A bipartite graph $G = (U, V, E)$ is *convex* if (u_i, v_k), $(u_j, v_k) \in E$ with $i < j$ imply $(u_h, v_k) \in E$ for all $h = i, i+1, \ldots, j$. Find a greedy algorithm to compute the maximum matching in a convex bipartite graph.

29. Consider a bipartite graph $G = (U, V, E)$ and two node subsets $A \subseteq U$ and $B \subseteq V$. Show that if there exist a matching M_A covering A and a matching M_B covering B, then there exists a matching $M_{A \cup B}$ covering $A \cup B$.

30. For a graph G, let $odd(G)$ denote the number of connected components of odd size in G. Prove the following.

(a) In any graph $G = (V, E)$, the minimum number of free nodes in any matching is

$$\max_{U \subseteq V} (odd(G \setminus U) - |U|).$$

(b) In any graph G, the maximum size of a matching is

$$\min_{U \subseteq V} \frac{1}{2} \cdot (|V| + |U| - odd(G \setminus U)).$$

(c) A graph $G = (V, E)$ has a perfect matching if and only if for any $U \subseteq V$, $odd(G \setminus U) \leq |U|$.

Historical Notes

Maximum flow problem was proposed by T. E. Harris and F. S. Ross in 1955 [204, 360] and was first solved by L.R. Ford and D.R. Fulkerson in 1956 [145]. However, Ford-Fulkerson algorithm is a pseudo polynomial-time algorithm when all arc capacities are integers. If arc capacities may not be integers, the termination of the algorithm may meet a trouble. The first strong polynomial-time algorithm was designed by Edmonds and Karp [123]. Later, various designs appeared in the literature, including Dinitz algorithm [88, 89], Goldberg-Tarjan push-relabel algorithm [178], Goldberg-Rao algorithm [175], Sherman algorithm [365], and the algorithm of Kelner, Lee, Orecchia, and Sidford [240]. Currently, the best running time is $O(|V||E|)$. This record is kept by Orlin algorithm [331] for approximation solution, running time can be further improved [239].

Matching is a classical subject in graph theory. Both maximum (cardinality) matching and minimum cost perfect matching problems in bipartite graphs can be easily transformed to maximum flow problems. However, they can also be solved with alternating path methods. So far, Hopcroft-Karp algorithm [215] is the fastest algorithm for the maximum bipartite matching. In general graph, they have to be solved with alternating path method since currently, no reduction has been found to transform matching problem to flow problem. Those algorithms were designed by Edmonds [118]. An extension of Hopcroft-Karp algorithm was made by Micali and Vazirani [313], which runs in $O(\sqrt{|E|}|V|)$ time.

For maximum weight matching, nobody has found any method to transform it to a flow problem. Therefore, we have to employ the alternating path and cycle method [118], too.

Chinese postman problem was proposed by Kwan [269], and the first polynomial-time algorithm was given by Edmonds and Johnson [122] with minimum cost perfect matching in complete graph with even number of nodes.

Chapter 6
Linear Programming

Our intuition about the future is linear.

—Ray Kurzweil

Linear programming (LP) is an important combinatorial optimization problem, and in addition, it is an important tool to design and to understand algorithms for other problems. In this chapter, we introduce LP theory starting from Simplex Algorithm, which is an incremental method.

6.1 Simplex Algorithm

An LP is an optimization problem with linear objective function and a constraint system of equalities and inequalities. The following is an example:

$$\text{maximize } z = 4x + 5y$$
$$\text{subject to } 2x + 3y \leq 60$$
$$x \geq 0, y \geq 0.$$

This example can be explained in the Euclidean plane as shown in Fig. 6.1. Each of three inequalities gives a half plane. Their intersection is a triangle, which is called a *feasible domain*. In general, the feasible domain of an LP is the set of all points satisfying all constraints. For different value of z, $z = 4x + 5y$ gives different lines which form a family of parallel lines. When z increases, line $z = 4x + 5y$ moves from left to right, and at point $(30, 0)$, it is the last moment to intersect the feasible domain. Hence, $(30, 0)$ is the point at which $z = 4x + 5y$ reaches its maximum value, i.e., 120.

In general, an LP may contain a large number of variables and a large number of constraints and hence cannot be solved geometrically as above. However, above example gives us a hint to find a general method. An important observation is that

© The Author(s), under exclusive license to Springer Nature Switzerland AG 2022
D.-Z. Du et al., *Introduction to Combinatorial Optimization*, Springer Optimization
and Its Applications 196, https://doi.org/10.1007/978-3-031-10596-8_6

Fig. 6.1 An example of LP

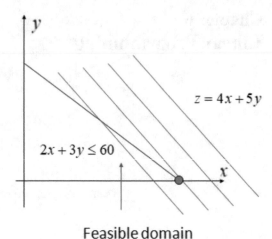

the maximum value of objective function is achieved at a vertex of the feasible
domain. This observation suggests an incremental method as follows: Start from
a vertex of the feasible domain and move from a vertex to another vertex with
improvement on objective function value.

Before we implement this idea, let us look at a standard form of LP. Every LP
can be transformed into the following form:

$$\max \ z = cx$$

$$\text{s.t. } Ax = b$$

$$x \geq 0,$$

where c is an n-dimensional row vector; b is an m-dimensional column vector; A
is an $m \times n$ coefficient matrix with rank m, i.e., $rank(A) = m$; and x is a column
vector with n variables as components. Thus, the above example can be transformed
into the following:

$$\text{maximize } z = 4x + 5y$$

$$\text{subject to } 2x + 3y + w = 60$$

$$x \geq 0, y \geq 0, w \geq 0.$$

The feasible domain of this LP is in the three-dimensional space as shown in
Fig. 6.2. It is still a triangle with three vertices: $(30, 0, 0)$, $(0, 20, 0)$, and $(0, 0, 60)$.

In general, what is the vertex of the feasible domain? A point in a convex domain
Ω is a *vertex* if

$$x = \frac{y+z}{2}, y, z \in \Omega \Rightarrow x = y = z.$$

Fig. 6.2 The feasible domain
of an LP in standard form

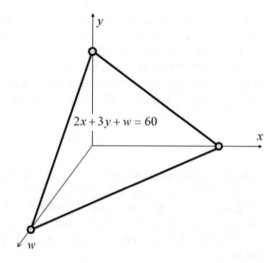

Theorem 6.1.1 (Fundamental Theorem) *Let* $\Omega = \{x \mid Ax = b, x \geq 0\}$. *If* $\max cx$ *over* $x \in \Omega$ *has an optimal solution, then it can be found in vertices of* Ω.

Proof Consider an optimal solution x^* with minimum number of zero components among all optimal solutions. We will show that x^* is a vertex of Ω. By contradiction, suppose x^* is not, that is, there exist $y, z \in \Omega$ such that $x^* = (y+z)/2$ and x^*, y, z are distinct. Since $cx^* \geq cy$, $cx^* \geq cz$, and $cx^* = (cy + cz)/2$, we must have $cx^* = cy = cz$. This means that y and z are also optimal solutions. It follows that all feasible points on line $x^* + \alpha(y - x^*)$ are optimal solutions. However, Ω does not contain any line. Thus, the line must contain a point x' not in Ω, that is, x' violates at least one constraint.

Note that for any α, $A(x^* + \alpha(y - x^*)) = b$. Thus, x' cannot violate constraint $Ax = b$. Suppose $x_i^* = 0$. Since $y_i \geq, z_i \geq 0$, and $x_i^* = (y_i + z_i)/2$, we have $z_i = y_i = x_i^* = 0$. Therefore, the ith component of $x^* + \alpha(y - x^*)$ is equal to 0 for any α. Hence, x' must violate some constraint $x_j \geq 0$ with $x_j^* > 0$, that is, $x_j' < 0$. Now, we can easily find an optimal solution on the line segment between x^* and x', which has one more zero component than x^* has, a contradiction. □

Let us consider our example again and start with $(0, 0, 60)$ to look at how to move from a vertex to another vertex.

$(0, 0, 60)$ is special because the objective function $z = 4x + 5y$ does not contain w so that the value of w does not affect the objective function value directly. Clearly, if we want to increase the objective function value, then we should increase value of x or y. Let us choose to increase y. To keep $2x + 3y + w = 60$, we may set $y = 20$ and $x = w = 0$, i.e., we move from vertex $(0, 0, 60)$ to vertex $(0, 20, 0)$. The objective function value is increased from 0 to 100.

Working with $(0, 20, 0)$, we get a little problem with objective function $z = 4x + 5y$. From this representation, we know that increasing x would increase the objective function value z; however, this would bring down y, in order to keep $2x + 3y + w = 60$, and hence bring down objective function value z. Can we give a representation of the objective function which does not contain y? The answer is yes. Substituting $y = 20 - \frac{2}{3}x - \frac{1}{3}w$ into $z = 4x + 5y$, we would obtain

$$z = 100 + \frac{2}{3}x - \frac{2}{3}w.$$

From this representation, we can easily see that increasing x would increase objective function value z although y would be brought down. As long as y is kept being nonnegative, it is ok. When y is brought down to 0, x can be increased to 30, that is, we move from vertex $(0, 20, 0)$ to vertex $(30, 0, 0)$.

At vertex $(30, 0, 0)$, we substitute $x = 30 - \frac{3}{2}y - \frac{1}{2}w$ into $z = 4x + 5y$ and obtain

$$z = 120 - y - w.$$

Since $y \geq 0$ and $w \geq 0$, $z = 120 - y - w \geq 120$ and equality sign holds when $y = w = 0$. This means that $(30, 0, 0)$ is the maximum point.

In above example, we were lucky to have all vertices each of which has only one nonzero variable and objective function can always be represented in a form not containing this nonzero variable. Can we always be so lucky? The answer is yes. Why? Let us consider the general standard form of LP.

Since $rank(A) = m$, there are m linearly independent columns. Suppose their indices form a set I, called a *basis*, and the rest of indices form a set $\bar{I}$, i.e., $\bar{I} = \{1, 2, \ldots, n\} - I$. Then we can write

$$A = (A_I, A_{\bar{I}}),$$

$$x = \begin{pmatrix} x_I \\ x_{\bar{I}} \end{pmatrix},$$

$$c = (c_I, c_{\bar{I}}).$$

Thus,

$$Ax = A_I x_I + A_{\bar{I}} x_{\bar{I}} = b, \tag{6.1}$$

and

$$z = c_I x_I + c_{\bar{I}} x_{\bar{I}}. \tag{6.2}$$

From (6.1), we obtain

$$x_I = A_I^{-1}b - A_I^{-1}A_{\bar{I}}x_{\bar{I}}.$$

Substituting it into (6.2), we obtain

$$z = c_I A_I^{-1}b + (c_{\bar{I}} - c_I A_I^{-1}A_{\bar{I}})x_{\bar{I}}.$$

This means that if we can have a feasible solution $x_{\bar{I}} = 0$, $x_I = A_I^{-1}b$, then we can have a representation of z, which does not contain any nonzero variable.

Actually, if $A_I^{-1}b \geq 0$, then I is called a *feasible basis*, which induces a *basic feasible solution* $x_{\bar{I}} = 0$, $x_I = A_I^{-1}b$. This basic feasible solution reaches the maximum if $c'_{\bar{I}} = c_{\bar{I}} - c_I A_I^{-1}A_{\bar{I}} \leq 0$.

At this point, we may have some feeling that the basic feasible solution has something to do with the vertex of the feasible domain. Let us first explain their relationship and then continue our discussion.

Lemma 6.1.2 *A feasible solution is a vertex of the feasible domain if and only if it is a basic feasible solution.*

Proof Denote $\Omega = \{x \mid Ax = b, x \geq 0\}$. First, we show that every basic feasible solution is a vertex. Consider a basic feasible solution x with feasible basis I. Then $x_{\bar{I}} = 0$ and $x_I = A_I^{-1}b$. Suppose $x = (y + z)/2$ for $y, z \in \Omega$. Then $x_{\bar{I}} = (y_{\bar{I}} + z_{\bar{I}})/2 = 0$ and $y \geq 0$, $z \geq 0$. It follows that $y_{\bar{I}} = z_{\bar{I}} = 0$. Therefore, $A_I y_I = b$ and $A_I z_I = b$. This means that $y_I = z_I = A_I^{-1}b = x_I$, implying that $y = z = x$.

Next, suppose x is a vertex. We claim that $\{a_j \mid x_j > 0\}$ is a set of linearly independent vectors where a_j is the jth column vector of A. For contradiction, suppose that those column vectors are not linearly independent. Then there exist α_j for j with $x_j > 0$ such that $\sum_{j:x_j>0} \alpha_j a_j = 0$ and not all α_j are zero. Define an n-dimensional column vector $d = (d_j)$ by setting

$$d_j = \begin{cases} \alpha_j & \text{if } x_j > 0, \\ 0 & \text{if } x_j = 0. \end{cases}$$

Then $d \neq 0$ and for sufficiently small $\varepsilon > 0$, $y = x + \varepsilon d \in \Omega$, $z = x - \varepsilon d \in \Omega$, and $x = (y + z)/2$. This means that x is not a vertex, a contradiction. Thus, our claim is true.

Now, note that every linearly independent subset of column vectors can be enlarged into a maximum linearly independent subset with m column vectors. Let I be the set of indices of those column vectors. Then $x_I = A_I^{-1}b \geq 0$, $x_{\bar{I}} = 0$, that is, x is a basic feasible solution with basis I. $\square$

Now, we return to our discussion and consider a feasible basis I with $c'_{\bar{I}} = c_{\bar{I}} - c_I A_I^{-1}A_{\bar{I}}$. If $c'_{\bar{I}} \leq 0$, then the basic feasible solution $(x_I, x_{\bar{I}}) = (A_I^{-1}b, 0)$ is optimal.

If $c'_{j*} > 0$ for some $j^* \in \bar{I}$, then it means that increasing x_{j*} will increase the objective function value z. How much can x_{j*} be increased? Denote $(a_{ij}) = A_I^{-1}A = (I_m, A_I^{-1}A_{\bar{I}})$ and $b' = A_I^{-1}b$ where I_m is the identity matrix of order m. We want to increase x_{j*} and keep $x_j = 0$ for all $j \in \bar{I} - \{j^*\}$. Thus, for every $j \in I$, $x_j = b'_{i_j} - a_{i_j j*}x_{j*}$ where i_j is the row index such that $a_{i_j j} = 1$. If $a_{i_j j*} \leq 0$, then $x_j \geq 0$ for any $x_{j*} \geq 0$. However, if $a_{i_j j*} > 0$, then $x_j \geq 0$ only for $x_{j*} \leq b'_{i_j}/a_{i_j j*}$. This means that x_{j*} can be increased at most to

$$\frac{b'_{i*}}{a_{i*j*}} = \min\{\frac{b'_i}{a_{ij*}} \mid a_{ij*} > 0\}, \tag{6.3}$$

and when x_{j*} is increased to this value, $x_{j'}$ for $j' \in I$ with $a_{i*j'} = 1$ would become 0. This means that such $j' \in I$ can be replaced by j^*.

Above process of replacing j^* by j' is called a *pivot*. Let us summarize the process of a pivot.

- Choose j^* from outside of feasible basis I because $c'_j > 0$.
- Choose row index i^* by (6.3).
- Choose j' if $a_{i*j'} = 1$.

Those choices imply that the new basis $I' = (I \setminus \{j'\}) \cup \{j^*\}$ must be feasible.

Wait for a moment! (6.3) requires the existence i such that $a_{ij*} > 0$. What happens if $a_{ij*} \leq 0$ for all $i = 1, 2, \ldots, m$? In such a case, x_{j*} can be increased to any large number so that the objective function value gets an increment $c'_{j*}x_{j*}$, which is going to ∞ as x_{j*} goes to ∞. Therefore, the LP has no optimal solution.

Now, let us come back to the case that there exists i such that $a_{ij*} > 0$. When x_{j*} is increased to b'_{i*}/a_{i*j*}, the objective function value gets an increment $c'_{j*}b'_{i*}/a_{i*j*}$, which is positive if $b'_{i*} > 0$. At this moment, we would like to make an assumption that the following condition holds:

Nondegeneracy Condition *For every feasible basis I, $A_I^{-1}b > 0$.*

Under this condition, we can always have $b'_{i*} > 0$ so that **the objective function value is increased from a feasible basis I to another feasible basis $I' = (I \setminus \{j'\}) \cup \{j^*\}$.**

We summarize results of above discussion into Algorithm 20, where a_i denotes the ith row of A, and Theorem 6.1.3.

Theorem 6.1.3 *Under nondegeneracy condition, simplex method starting from a basic feasible solution can find an optimal solution or no optimal solution in finitely many iterations.*

Now, we work on an example with simplex algorithm.

We will use a table designed as follows:

Algorithm 20 Simplex algorithm

Input: A LP $\max\{cx \mid Ax = b, x \geq 0\}$ and a feasible basis I.
Output: An optimal solution x or "optimal value $= +\infty$".
1: $A \leftarrow A_I^{-1} A$;
2: $b \leftarrow A_I^{-1} b$;
3: $c \leftarrow c - c A_I^{-1} A$;
4: $c_0 \leftarrow -c A_I^{-1} b$;
5: **while** $c_0 < +\infty$ and there exists $c_{j*} > 0$ for $j^* \in \bar{I}$ **do**
6:      **if** $a_{ij*} \leq 0$ for all i **then**
7:          $c_0 \leftarrow +\infty$
8:      **else**
9:          select i^* such that $\frac{b_{i*}}{a_{i*j*}} = \min\{\frac{b_i}{a_{ij*}} \mid a_{ij*} > 0\}$;
10:         select j' such that $a_{i*j'} = 1$;
11:         $I \leftarrow (I \setminus \{j'\}) \cup \{j^*\}$;
12:         $(b_i, a_i) \leftarrow (b_i, a_i) - (b_{i*}, a_{i*}) \cdot \frac{a_{ij*}}{a_{i*j*}}$;
13:         $(c_0, c) \leftarrow (c_0, c) - (b_{i*}, a_{i*}) \cdot \frac{c_{j*}}{a_{i*j*}}$;
14:     **end if**
15: **end while**
16: **return** c_0 and x if $c_0 < +\infty$

$$\begin{array}{c|c} -z & x^T \\ \hline 0 & c \\ b & A \end{array}$$

With initial feasible basis I, this initial table can be changed to

$$\begin{array}{c|c} -z & x^T \\ \hline -cA_I^{-1}b & c - cA_I^{-1}A \\ A_I^{-1}b & A_I^{-1}A \end{array}$$

Consider this example:

$$\max \quad z = 3x_1 + x_2 + 2x_3$$
$$\text{s.t.} \quad x_1 + x_2 + 3x_3 + x_4 = 30$$
$$2x_1 + 2x_2 + 5x_3 + x_5 = 24$$
$$4x_1 + x_2 + 2x_3 + x_6 = 36$$
$$x_1, x_2, x_3, x_4, x_5, x_6 \geq 0.$$

Its initial simplex table with feasible basis $I_0 = \{4, 5, 6\}$ is as follows:

$-z$	x_1	x_2	x_3	x_4	x_5	x_6
0	3	1	2			
30	1	1	3	1		
24	2	2	5		1	
36	**4**	1	2			1

Since $c_1' = 3 > 0$, we may move x_1 into basis. Note that $36/4 = \min(30/1, 24/2, 36/4)$. We choose the black **4** as pivoting element. This means that $I_1 = (I_0 - \{6\}) \cup \{1\} = \{4, 5, 1\}$. The simplex table with I_1 is as follows:

$-z$	x_1	x_2	x_3	x_4	x_5	x_6
-27		1/4	1/2			$-3/4$
21		3/4	5/2	1		$-1/4$
6		**3/2**	4		1	$-1/2$
9	1	1/4	1/2			1/4

Since $c_2' > 0$ and $6/(3/2) = \min(21/(3/4), 6/(3/2), 9/(1/4))$, we choose the black **3/2** as pivoting element and set $I_2 = (I_1 - \{5\}) \cup \{2\} = \{4, 2, 1\}$. The simplex table with I_2 is as follows:

$-z$	x_1	x_2	x_3	x_4	x_5	x_6
-28			$-1/6$		$-3/8$	$-2/3$
18			1/2	1	$-9/8$	
4		1	8/3		3/2	$-1/3$
8	1		$-1/6$		$-3/8$	1/3

Now, $c_j' \leq 0$ for all $j \in \bar{I}_2$. Therefore, the basic feasible solution given by I_2, $(x_1 = 8, x_2 = 4, x_4 = 18, x_3 = x_5 = x_6 = 0)$, is optimal, which achieves the objective function value 28.

6.2 Lexicographical Ordering

When the nondegeneracy condition does not hold, the simplex method may fail into a cycle. The following is an example provided in [22]:

$-z$	x_1	x_2	x_3	x_4	x_5	x_6	x_7
1	3/4	-20	1/2	-6	0	0	0
0	1/4	-8	-1	9	1	0	0
0	1/2	-12	$-1/2$	3	0	1	0
0	0	0	1	0	0	0	1

After the first pivot, the following is obtained:

$-z$	x_1	x_2	x_3	x_4	x_5	x_6	x_7
1	0	4	7/2	−33	−3	0	0
0	1	−32	−4	36	4	0	0
0	0	4	3/2	−15	−2	1	0
0	0	0	1	0	0	0	1

After the second pivot, the following is obtained:

$-z$	x_1	x_2	x_3	x_4	x_5	x_6	x_7
1	0	0	2	−18	−1	−1	0
0	1	0	8	−84	−12	8	0
0	0	1	3/8	−15/4	−1/2	1/4	0
0	0	0	1	0	0	0	1

After the third pivot, the following is obtained:

$-z$	x_1	x_2	x_3	x_4	x_5	x_6	x_7
1	−1/4	0	0	3	2	−3	0
0	1/8	0	1	−21/2	−3/2	1	0
0	−3/64	1	0	**3/16**	1/16	−1/8	0
0	−1/8	0	0	21/2	3/2	−1	1

After the fourth pivot, the following is obtained:

$-z$	x_1	x_2	x_3	x_4	x_5	x_6	x_7
1	1/2	−16	0	0	1	−1	0
0	−3/2	56	1	0	2	−6	0
0	−1/4	16/3	0	1	1/3	−2/3	0
0	5/2	−56	0	0	−2	6	1

After the fifth pivot, the following is obtained:

$-z$	x_1	x_2	x_3	x_4	x_5	x_6	x_7
1	7/4	−44	−1/2	0	0	2	0
0	−5/4	28	1/2	0	1	−3	0
0	1/6	−4	−1/6	1	0	1/3	0
0	0	0	1	0	0	0	1

After the sixth pivot, the following is obtained:

$-z$	x_1	x_2	x_3	x_4	x_5	x_6	x_7
z	$3/4$	-20	$1/2$	-6	0	0	0
0	$1/4$	-8	-1	9	1	0	0
0	$1/2$	-12	$-1/2$	3	0	1	0
0	0	0	1	0	0	0	1

Therefore, we return to original table.

There are two ways to deal with the degeneracy, i.e., avoiding the cycle occurrence. The first one is called the *lexicographical ordering method*.

Consider two vectors $x = (x_1, x_2, \ldots, x_n)$ and $y = (y_1, y_2, \ldots, y_n)$. x is said to be lexicographically larger than y, written as $x >_L y$ if $x_1 = y_1, \ldots, x_{i-1} = y_{i-1}, x_i > y_i$ for some $1 \leq i \leq n$. A vector x is lexicographically positive if $x >_L 0$.

Now, we modify the simplex method as follows:

Initially, rearrange the ordering of n columns such that the initial feasible basis is placed at the first m columns as shown in the following simplex table:

$$
\begin{array}{c|cc}
-z & x_I^T & x_{\bar{I}}^T \\
\hline
c_0 & 0 & c_{\bar{I}} \\
b & I_m & A_{\bar{I}}
\end{array}
$$

This arrangement makes m rows, other than the top row, lexicographically positive. Moreover, they are distinct since they are distinct within identity matrix I_m.

If $c_{\bar{I}} \leq 0$, then I is an optimal basis. Otherwise, choose j^* such as $c_{j^*} > 0$. Denote $A = (I_m, A_{\bar{I}})$ and $c = (0, c_{\bar{I}})$. If $a_{ij^*} \leq 0$, then algorithm stops and we can conclude that the LP has no optimal solution. If there exists some i such that $a_{ij^*} > 0$, then choose i^* such that

$$
\left(\frac{b_{i^*}}{a_{i^*j^*}}, \frac{a_{i^*1}}{a_{i^*j^*}}, \ldots, \frac{a_{i^*n}}{a_{i^*j^*}} \right)
$$

is the lexicographically smallest one among

$$
\left(\frac{b_i}{a_{ij^*}}, \frac{a_{i1}}{a_{ij^*}}, \ldots, \frac{a_{in}}{a_{ij^*}} \right)
$$

for $a_{ij^*} > 0$.

Next, choose j' with $a_{i^*j'} = 1$ and then carry out pivot $I \leftarrow (I \setminus \{j'\}) \cup \{j^*\}$ by operations

$$
(b_i, a_i) \leftarrow (b_i, a_i) - (b_{i^*}, a_{i^*}) \cdot \frac{a_{ij^*}}{a_{i^*j^*}}
$$

and

$$(c_0, c) \leftarrow (c_0, c) - (b_{i*}, a_{i*}) \cdot \frac{c_{j*}}{a_{i*j*}}.$$

This means that after a pivot, the new simplex table becomes

$-z$	x^T
$c_0 - b_{i*} \cdot \dfrac{a_{ij*}}{a_{i*j*}}$	$c - a_{i*} \cdot \dfrac{c_{j*}}{a_{i*j*}}$
$b_i - b_{i*} \cdot \dfrac{a_{ij*}}{a_{i*j*}}$	$a_i - a_{i*} \cdot \dfrac{a_{ij*}}{a_{i*j*}}$

This table has the following two properties:

(a) By choice of i^*, all rows, other than the top row, are lexicographically positive and distinct.
(b) Since $c_{j*} > 0$, $a_{i*j*} > 0$, and $(b_{i*}, a_{i*}) >_L 0$, the top row gets strictly lexicographically decreasing.

Property (a) guarantees the continuation of this method. Property (b) implies that each feasible basis can appear at most once, and hence, if the optimal value is not infinity, then an optimal solution will be reached within finitely many iterations.

Theorem 6.2.1 *Simplex algorithm runs in finitely many iterations. It can either determine the optimal value is infinity or return an optimal solution.*

Next, we work on the example at the beginning of this section with lexicographical ordering method. Initially, we have

$-z$	x_5	x_6	x_7	x_1	x_2	x_3	x_4
1	0	0	0	3/4	-20	1/2	-6
0	1	0	0	1/4	-8	-1	9
0	0	1	0	1/2	-12	$-1/2$	3
0	0	0	1	0	0	1	0

After the first pivot, we obtain

$-z$	x_5	x_6	x_7	x_1	x_2	x_3	x_4
1	0	$-3/2$	0	0	-2	5/4	$-21/2$
0	1	$-1/2$	0	0	-2	$-3/4$	15/2
0	0	2	0	1	-24	-1	6
0	0	0	1	0	0	1	0

After the second pivot, we obtain

$-z$	x_5	x_6	x_7	x_1	x_2	x_3	x_4
1	0	$-3/2$	$-5/4$	0	-2	0	$-21/2$
0	1	$-1/2$	$3/4$	0	-2	0	$15/2$
0	0	2	1	1	-24	0	6
0	0	0	1	0	0	1	0

Thus, the algorithm ends after two iterations.

6.3   Bland's Rule

The second method for dealing with degeneracy is to modify the simplex algorithm by Bland's rule as follows:

- Choose the entering column index j^* satisfying

$$j^* = \min\{j \in \bar{I} \mid c_j > 0\}.$$

- Choose the row index i^* which is the smallest one if there are more than one i^* satisfying

$$\frac{b_{i^*}}{a_{i^*j^*}} = \min\left\{\frac{b_i}{a_{ij^*}} \mid a_{ij^*} > 0\right\}.$$

Theorem 6.3.1 *With Bland's rule, simplex algorithm will not run into a cycle, so that within finitely many iterations, the algorithm is able to determine whether the optimal value goes to infinity or not, and if the optimal value is finite, then the algorithm will obtain an optimal solution.*

Proof It is sufficient to show that with Bland's rule, simplex algorithm will not run into a cycle. For contradiction, suppose a cycle exists. For simplicity of discussion, we delete all constraints with row indices not selected in the cycle. Thus, for remaining row index i, $b_i = 0$ since objective function value cannot be changed during computation of the cycle. In this cycle, there also exist some column indices entering the feasible basis and then leaving or vice versa. Let t be the largest column index among them. For simplicity of discussion, we also delete all columns with index $j > t$ since we will always assign 0 to variable x_j for $j > t$. Next, let us consider two moments in this cycle.

At the first moment, t leaves the feasible basis. Assume column index s enters the feasible basis. Denote by a_{ij} and c_j coefficients of constraints and cost, respectively, at this moment.

At the second moment, t enters the feasible basis. Denote by a'_{ij} and c'_j coefficients of constraints and cost, respectively.

After deletion, suppose there are m rows left. Assume that at the first moment, $j_1, j_2, \ldots, j_m$ are base indices such that $a_{1j_1} = a_{2j_2} = \cdots = a_{mj_m} = 1$. Consider a variable assignment $\hat{x}$ with the following values:

$$
\hat{x}_j = \begin{cases} -1 & \text{if } j = s, \\ a_{is} & \text{if } j = j_i \text{ for } i = 1, 2, \ldots, m, \\ 0 & \text{otherwise} \end{cases}
$$

Clearly, $\hat{x}$ satisfies all constraints. Note that (c'_j) can be obtained from (c_j) through elementary row operations. Therefore, at the first moment and at the second moment, the cost function value should be the same at $\hat{x}$, that is,

$$
-c_s = -c'_s + \sum_{i=1}^{m} c'_{j_i} a_{is}.
$$

Since at the first moment, s enters in the feasible basis, we have $c_s > 0$. Note that $s < t$ and at the second moment, t enters in the feasible basis. By Bland's rule, we have $c'_s \leq 0$. Therefore,

$$
\sum_{i=1}^{m} c'_{j_i} a_{is} < 0.
$$

It follows that for some i, $c'_{j_i} a_{is} < 0$. By Bland's rule, $c'_{j_i} \leq 0$. Therefore, $a_{is} > 0$, contradicting that t is the leaving index at the first moment. □

Now, we apply Bland's rule to the following LP:

$-z$	x_1	x_2	x_3	x_4	x_5	x_6	x_7
1	3/4	−20	1/2	−6	0	0	0
0	1/4	−8	−1	9	1	0	0
0	1/2	−12	−1/2	3	0	1	0
0	0	0	1	0	0	0	1

After the seventh pivot, the following is obtained:

$-z$	x_1	x_2	x_3	x_4	x_5	x_6	x_7
1	0	4	7/2	−33	−3	0	0
0	1	−32	−4	36	4	0	0
0	0	4	3/2	−15	−2	1	0
0	0	0	1	0	0	0	1

After the seventh pivot, the following is obtained:

$-z$	x_1	x_2	x_3	x_4	x_5	x_6	x_7
1	0	0	2	-18	-1	-1	0
0	1	0	8	-84	-12	8	0
0	0	1	3/8	$-15/4$	$-1/2$	1/4	0
0	0	0	1	0	0	0	1

After the third pivot, the following is obtained:

$-z$	x_1	x_2	x_3	x_4	x_5	x_6	x_7
1	$-1/4$	0	0	3	2	-3	0
0	1/8	0	1	$-21/2$	$-3/2$	1	0
0	$-3/64$	1	0	**3/16**	1/16	$-1/8$	0
0	$-1/8$	0	0	21/2	3/2	-1	1

After the fourth pivot, the following is obtained:

$-z$	x_1	x_2	x_3	x_4	x_5	x_6	x_7
1	1/2	-16	0	0	1	-1	0
0	$-3/2$	56	1	0	2	-6	0
0	$-1/4$	16/3	0	1	1/3	$-2/3$	0
0	**5/2**	-56	0	0	-2	6	1

After the fifth pivot, the following is obtained:

$-z$	x_1	x_2	x_3	x_4	x_5	x_6	x_7
1	0	$-24/5$	0	0	7/5	$-11/5$	$-1/5$
0	0	112/5	1	0	**4/5**	$-12/5$	3/5
0	0	$-4/15$	0	1	2/15	$-1/15$	1/10
0	1	$-112/5$	0	0	$-4/5$	12/5	2/5

After the sixth pivot, the following is obtained:

$-z$	x_1	x_2	x_3	x_4	x_5	x_6	x_7
1	0	-44	0	0	0	2	$-4/5$
0	0	28	5/4	0	1	-3	3/4
0	0	-4	$-1/6$	1	0	**1/3**	0
0	1	0	1	0	0	0	1

After the seventh pivot, the following is obtained:

$-z$	x_1	x_2	x_3	x_4	x_5	x_6	x_7
1	0	-20	1	-6	0	0	$-4/5$
0	0	-8	$-1/4$	0	1	0	$3/4$
0	0	-12	$-1/2$	3	0	1	0
0	1	0	1	0	0	0	1

After the eighth pivot, the following is obtained:

$-z$	x_1	x_2	x_3	x_4	x_5	x_6	x_7
1	-1	-20	0	-6	0	0	$-9/5$
0	$1/4$	-8	0	0	1	0	1
0	$1/2$	-12	0	3	0	1	$1/2$
0	1	0	1	0	0	0	1

Now, computation stops.

6.4   Initial Feasible Basis

How do we find the initial feasible basis? A popular way is to introduce artificial variables $y = (y_1, y_2, \ldots, y_m)^T$ and solve the following LP:

$$\max \ w = -ey$$
$$\text{subject to} \quad Ax + I_m y = b$$
$$x \geq 0, y \geq 0,$$

where $e = (1, 1, \ldots, 1)$ and I_m is the identity matrix of order m. In this LP, those artificial variables form a feasible basis. There are three possible outcomes resulting from solving this LP.

(1) The cost function value w is reduced to 0 and all artificial variables are removed from the feasible basis. In this case, the final feasible basis can be used as initial feasible basis in original LP.
(2) The cost function reaches a negative maximum value. In this case, the original LP has no feasible solution.
(3) The cost function value w is reduced to 0; however, there is an artificial variable y_i in the feasible basis. Let b_i and a_{ij} denote coefficients of constraints at the last moment. In this case, we must have $y_i = b_i = 0$; otherwise, $w = ey > 0$. Note that there exists a variable x_j such that $a_{ij} \neq 0$ since $rank(A) = m$. This means that we may take a_{ij} as pivot element to move y_i out from feasible basis and to move in x_j, preserving cost function value 0. When all artificial variables are moved out from the feasible basis, this case is reduced to case (1).

We next show an example as follows:

$$\max z = -2x_1$$
$$\text{subject to} \quad x_1 \quad -x_3 \quad = 3,$$
$$x_1 -x_2 \quad -2x_4 = 1,$$
$$2x_1 \quad +x_4 \leq 7,$$
$$x_1, \quad x_2, \quad x_3, \quad x_4 \geq 0.$$

First, we transform it into a standard form:

$$\max z = -2x_1$$
$$\text{subject to} \quad x_1 \quad -x_3 \quad = 3,$$
$$x_1 -x_2 \quad -2x_4 \quad = 1,$$
$$2x_1 \quad +x_4 +x_5 = 7,$$
$$x_1, \quad x_2, \quad x_3, \quad x_4, \quad x_5, \geq 0.$$

To find an initial feasible basis, we introduce two artificial variables y_1 and y_2 and solve the following LP:

$$\max \quad w = \quad\quad\quad\quad\quad\quad -y_1 \quad -y_2$$
$$\text{subject to} \quad x_1 \quad\quad -x_3 \quad\quad\quad y_1 \quad\quad = 3,$$
$$x_1 \quad -x_2 \quad\quad -2x_4 \quad\quad\quad y_2 = 1,$$
$$2x_1 \quad\quad\quad\quad +x_4 \quad +x_5 \quad\quad\quad = 7,$$
$$x_1, \quad x_2, \quad x_3, \quad x_4, \quad x_5, \quad y_1, \quad y_2 \geq 0.$$

The following tables are obtained with simplex algorithm with lexicographical rule:

$-w$	y_1	y_2	x_5	x_1	x_2	x_3	x_4
4				2	-1	-1	
3	1			1		-1	
1		1		1	-1		-2
7			1	2			1

$-w$	y_1	y_2	x_5	x_1	x_2	x_3	x_4
2		-2			1	-1	4
2	1	-1			1	-1	2
1		1		1	-1		-2
5		-2	1		2		5

$-w$	y_1	y_2	x_5	x_1	x_2	x_3	x_4
0	-1	-1					2
2	1	-1			1	-1	2
3	1			1		-1	
1	-2		1			2	1

At this point, we may stop the algorithm since w has been reduced to 0 and all artificial variables are already moved out of feasible basis. An feasible basis $\{x_5, x_1, x_2\}$ is obtained for the original LP. Deleting columns of artificial variables and putting back original cost function, we obtain the following:

$-z$	x_5	x_1	x_2	x_3	x_4
0		-2			
2			1	-1	2
3		1		-1	
1	1			2	1

$-z$	x_5	x_1	x_2	x_3	x_4
6				-2	
2			1	-1	2
3		1		-1	
1	1			2	1

It is pretty lucky that this basis already reaches the optimal. Therefore, we obtain optimal solution ($x_1 = 3, x_2 = 2, x_3 = x_4 = 0$) with maximum objective function value -6.

Now, we can summarize what we obtained on the LP as follows:

Theorem 6.4.1 *There are three possible outcomes for solving LP* $\max\{cx \mid Ax = b, x \geq 0\}$.

(a) *There is no feasible solution.*
(b) *The maximum value of objective function is* $+\infty$.
(c) *There is a maximum solution with finite objective function value. Then, there exists a maximum solution which is a basic feasible solution associated with a feasible basis I such that $c - c_I A_I^{-1} A \leq 0$. Moreover, if a basic feasible solution is associated with the feasible basis I satisfying $c - c_I A_I^{-1} A \leq 0$, then it is a maximum solution.*

6.5   Duality

Consider the following two LPs:

$$(P): \quad \max \quad z = cx$$
$$\text{subject to } Ax = b$$
$$x \geq 0,$$

and

$$(D): \quad \min \quad w = yb$$
$$\text{subject to } yA \geq c,$$

where c is an n-dimensional row vector, b is an m-dimensional column vector, and M is a $m \times n$ matrix with $rank(A) = m$. LP (P) is called *primal* LP while LP (D) is called *dual* LP. x is said to be *primal-feasible* if x lies in the feasible domain of (P). y is said to be *dual-feasible* if y lies in the feasible domain of (D).

Lemma 6.5.1 *If x is primal-feasible and y is dual-feasible, then $yb \geq cx$.*

Proof $yb = yAx \geq cx$. $\qquad\qquad\qquad\qquad\qquad\qquad\qquad\qquad\qquad\qquad$ □

Theorem 6.5.2 (Duality)

(a) *The primal LP (P) has no feasible solution if and only if the dual LP (D) has minimum value $-\infty$.*
(b) *The primal LP (P) has maximum value $+\infty$ if and only if the dual LP has no feasible solution.*
(c) *If both the primal LP (P) and the LP (D) have feasible solutions, then they both have optimal solutions, say x^* and y^*, respectively. Moreover, $cx^* = y^*b$.*

Proof First two statements (a) and (b) follow from Lemma 6.5.1. To show (c), by Theorem 6.4.1, an optimal feasible basis I satisfies

$$c - c_I A_I^{-1} A \leq 0.$$

Set $y = c_I A_I^{-1}$. Then $yA \geq c$, i.e., y is dual-feasible. Suppose x is a maximum basic feasible solution associated with feasible basis I, i.e., $x_I = A_I^{-1} b$ and $x_{\bar{I}} = 0$. then

$$yb = yax = c_I A_I^{-1} Ax = (c_I, c_I A_I^{-1} A_{\bar{I}})x = c_I x_I = cx.$$

$\qquad\qquad\qquad\qquad\qquad\qquad\qquad\qquad\qquad\qquad\qquad\qquad\qquad\qquad\qquad$ □

Corollary 6.5.3 (Complementary Slackness) *Consider a primal-feasible solution x and a dual-feasible solution y. Then both x and y are optimal if and only if* $(yA - c)x = 0$.

Proof By the duality theorem, both x and y are optimal if and only if $cx = yb$. Since $b = Ax$, $cx = yb$ if and only if $cx = yAx$, i.e., $(yA - c)x = 0$. □

The condition $(yA - c)x = 0$ is called the *complementary slackness condition*. Consider a pair of primal and dual LPs in symmetric form as follows:

$$(P): \quad \max \quad z = cx$$
$$\text{subject to } Ax \le b,$$
$$x \ge 0,$$

and

$$(D): \quad \min \quad w = yb$$
$$\text{subject to } yA \ge c,$$
$$y \ge 0.$$

The duality theorem still holds for them and the complementary slackness condition has a different expression.

Corollary 6.5.4 (Complementary-Slackness) *Consider a primal-feasible solution x and a dual-feasible solution y in a pair of primal LP and dual LP in symmetric form. Then both x and y are optimal if and only if* $(yA-c)x = 0$ *and* $y(b-Ax) = 0$.

Proof Note that

$$cx \le yAx \le yb.$$

By the duality theorem, both x and y are optimal if and only if $cx = by$, that is, $cx = yAx$ and $yAx = yb$. These two equalities are equivalent $(yA - c)x = 0$ and $y(b - Ax) = 0$, respectively. □

Another important corollary of the duality theorem is about separating hyperplane.

Corollary 6.5.5 (Separating Hyperplane Theorem)

(a) *There does not exist $x \ge 0$ such that $Ax = b$ if and only if there exists y such that $yA \ge 0$ and $yb < 0$.*

(b) *There does not exist $x \ge 0$ such that $Ax \le b$ if and only if there exists $y \ge 0$ such that $yA \ge 0$ and $yb < 0$.*

Proof First, we prove (a). Consider the following pair of primal and dual LPs:

$$(P): \quad \max \quad z = 0$$
$$\text{subject to } Ax = b$$
$$x \geq 0,$$

and

$$(D): \quad \min \quad w = yb$$
$$\text{subject to } yA \geq 0.$$

By the duality theorem, (P) has no feasible solution if and only if (D) approaches to $-\infty$. Note that if y is dual-feasible, so is αy for any $\alpha > 0$. Therefore, (D) approaches to $-\infty$ if and only if there is a dual-feasible solution y such that $yb < 0$.

Similarly, we can show (b).   □

Let us give a little explanation for separating hyperplane. If the feasible domain $\{x \mid Ax = b, x \geq 0\}$ is nonempty, then b is located in the cone generated by $a_1, a_2, \ldots, a_n$, i.e., $\{\sum_{i=1}^{n} \alpha_i a_i \mid \alpha_i \geq 0 \text{ for } i = 1, 2, \ldots, n\}$. The separating hyperplane theorem (a) says that if b does not lie in this cone, then there exists a hyperplane separating the core and b. (b) has a similar background. The separating hyperplane theorem is quite useful in design of approximation algorithms in later chapters.

The duality gives the possibility to design other algorithms for LP, which may have advantage in some cases. The dual simplex algorithm is useful when initial dual-feasible solution is easily obtained.

Consider a basis I. I is called a dual-feasible basis if $c - c_I A_I^{-1} A \leq 0$, i.e., $y = c_I A_I^{-1}$ is a dual-feasible solution. Clearly, the dual-feasible basis is optimal if and only $A_I^{-1} b \geq 0$. In the following example, a primal-feasible basis is not explicitly appeared. However, a dual-feasible basis $\{x_4, x_5, x_6\}$ is easy to be found:

$$\max \quad z = -3x_1 - x_2 - 2x_3$$
$$\text{subject to } x_1 + x_2 + 3x_3 + x_4 = 30$$
$$2x_1 - 2x_2 + 5x_3 + x_5 = -24$$
$$4x_1 + x_2 + 2x_3 + x_6 = 36$$
$$x_1, x_2, x_3, x_4, x_5, x_6 \geq 0.$$

Its initial table is as follows. Note that all coefficients of cost, $c_j \leq 0$:

$-z$	x_1	x_2	x_3	x_4	x_5	x_6
0	-3	-1	-2			
30	1	1	3	1		
-24	2	-2	5		1	
4	1	2				1

In dual simplex algorithm, the computation moves from a dual-feasible basis to another dual-feasible basis through a pivot operation. To describe such a pivot, let a_{ij}, b_i, and c_j be coefficients of constraints and cost in current table. Note that all coefficients of cost, $c_j \leq 0$. However, some $b_j < 0$. First, we choose row index i^* such that $b_{i^*} < 0$. Then, choose column index j^* such that

$$\frac{c_{j^*}}{a_{i^*j^*}} = \min\{\frac{c_j}{a_{i^*j}} \mid a_{i^*j} < 0\}.$$

This condition yields that all c_j is keep nonpositive after pivot. According to these rules, the pivot element is $a_{22} = -2$ in this example. After, the first pivot, we obtain the following:

$-z$	x_1	x_2	x_3	x_4	x_5	x_6
12	-4	0	$-9/2$		$-1/2$	
18	2	0	$11/2$	1	$1/2$	
12	-1	1	$-5/2$		$-1/2$	
24	5	0	$9/2$		$1/2$	1

This table gives an optimal solution $x_1 = x_3 = 5 = 0$, $x_2 = 12$, $x_4 = 18$, $x_6 = 24$ with optimal value -12.

At the end of this section, let us list the correspondence between constraints in primal and dual LPs. This would be very helpful to write down the dual LP based on primal LP.

$$\text{Primal LP} \longleftrightarrow \text{Dual LP}$$

$$\text{max} \longleftrightarrow \text{min}$$

$$\sum_j a_{ij}x_j = b_i \longleftrightarrow y_i \text{ has no restriction}$$

$$\sum_j a_{ij}x_j \leq b_i \longleftrightarrow y_i \geq 0$$

$$\sum_j a_{ij}x_j \geq b_i \longleftrightarrow y_i \leq 0$$

$$x_j \text{ has no restriction} \longleftrightarrow \sum_i a_{ij}y_i = c_j$$

$$x_j \geq 0 \longleftrightarrow \sum_i a_{ij} y_i \geq c_j$$

$$x_j \leq 0 \longleftrightarrow \sum_i a_{ij} y_i \leq c_j.$$

For example, the maximum flow problem can be formulated as the following LP:

$$\max \sum_{(s,u)\in E} x_{su}$$

$$\text{subject to } 0 \leq x_{uv} \leq c(u, v) \text{ for } (u, v) \in E$$

$$\sum_{(u,v)\in E} x_{uv} = \sum_{(v,w)\in E} \text{ for } v \in V \setminus \{s, t\}.$$

Its dual LP is as follows:

$$\min \sum_{(u,v)\in E} c(u, v) z_{uv}$$

$$\text{subject to } y_u - y_v + z_{uv} \geq 0 \text{ for } (u, v) \in E, u \neq s \text{ and } v \neq t$$

$$-y_v \geq 1 \text{ for } (s, v) \in E$$

$$y_u \geq 0 \text{ for } (u, t) \in E$$

$$z_{uv} \geq 0 \text{ for } (u, v) \in E.$$

6.6 Primal-Dual Algorithm

In this section, we introduce an algorithm motivated from the complementary slackness condition. Consider the following two LPs:

$$(P): \quad \max \quad z = cx$$

$$\text{subject to } Ax = b$$

$$x \geq 0,$$

and

$$(D): \quad \min \quad w = yb$$

$$\text{subject to } yA \geq c,$$

where c is an n-dimensional row vector, b is an m-dimensional column vector, and M is an $m \times n$ matrix with $rank(A) = m$. Then the complementary slackness condition can be described equivalently as the following:

$$ya_j > c_j \Rightarrow x_j = 0,$$

or

$$x_j > 0 \Rightarrow ya_j = c_j$$

where a_j is the jth column of A. Let y be a dual-feasible solution. Denote $J(y) = \{i \mid ya_j = c_j\}$. Then, y is optimal if and only if there exists a primal-feasible solution x satisfying the complementary slackness condition with y, i.e., the following LP has optimal value:

$$(RP): \quad \max \quad -\sum_{i=1}^{m} u_i$$

$$\text{subject to } \sum_{j \in J(y)} a_{ij}x_j + u_i = b_i \text{ for } i = 1, 2, \ldots, m,$$

$$x_j \geq 0 \text{ for } j \in J(y),$$

$$u_i \geq 0 \text{ for } i = 1, 2, \ldots, m.$$

If this LP does not have optimal value 0, then solve its dual LP:

$$(RD): \quad \min \quad vb$$

$$\text{subject to } va_j \geq 0 \text{ for } j \in J(y),$$

$$v_i \geq -1 \text{ for } i = 1, 2, \ldots, m.$$

Let us give (RD) another explanation. Consider applying the feasible direction method to solve LP. At dual-feasible point y, we want to find next dual-feasible point $y + \lambda v$ ($\lambda > 0$) such that $yb > yb + \lambda yv$ where v is a descend feasible direction. We may like v to satisfy (RD). In fact, when (RP) does not have maximum value 0, it must have a negative maximum value. By the duality theorem, (RD) must have negative minimum value and hence v is a descend direction. Next, we can determine λ by

$$\lambda = \min\{\frac{c_j - ya_j}{va_j} \mid va_j < 0\}.$$

Here, note that if there does not exist j such that $va_j < 0$, then $\lambda = +\infty$ and hence (P) does not have a feasible solution. Now, we summarize the primal-dual algorithm.

Algorithm 21 Primal-dual algorithm

Input: A LP $\max\{cx \mid Ax = b, x \geq 0\}$ and an initial dual-feasible solution y.
Output: An optimal solution x or "no primal-feasible solution".
1: $\lambda \leftarrow 0$;
2: **while** $\lambda < +\infty$ and (RP) have maximum value < 0 **do**
3:     solve (RD) to obtain v;
4:     **if** there exists j such that $va_j < 0$ **then**
5:         compute $\lambda \leftarrow \min\{\frac{c_j - ya_j}{va_j} \mid va_j < 0\}$;
6:         set $y \leftarrow y + \lambda v$;
7:     **else**
8:         set $\lambda \leftarrow +\infty$;
9:     **end if**
10: **end while**
11: **if** $\lambda < +\infty$ **then**
12:     Solve (RP) to obtain x;
13: **end if**
14: **return** x or "(P) does not have a feasible solution" if $\lambda = +\infty$.

Can this algorithm terminate within finitely many iterations? We cannot find a conclusion in the literature. However, it can be known in nonlinear programming that this feasible direction method has the global convergence, that is, if it generates an infinite sequence, then every cluster point of the sequence is an optimal solution.

6.7   Interior Point Algorithm

Although three algorithms have been presented in previous sections for LP, they all are running not in polynomial-time. The reason is that in general, the number of extreme points (i.e., vertices) of feasible domain is exponential. In this section, we present a polynomial-time algorithm which moves from a feasible point to another feasible point in the interior of the feasible domain. Hence, it is called the interior point algorithm.

First, we assume that the LP is in the following form:

$$\min \ cx$$

$$\text{subject to} \ \ Ax = b$$

$$x \geq 0$$

where without loss of generality, assume

- A is $m \times n$ matrix with full rank m, i.e., $(AA^T)^{-1}$ exists,
- the feasible domain is bounded,
- the minimum value of objective function is zero, and
- an initial feasible solution x^0 is available.

Actually, from LP (P) and its dual (D') in Sect. 6.6, we can obtain LP as follows:

$$\min \quad w - z = yb - cx$$

$$\text{subject to} \quad yA \geq c$$

$$Ax = b$$

$$x \geq 0.$$

This LP has zero as the objective function value of optimal solution. Modify it into standard form. Then we will obtain an LP satisfying our assumptions.

In order to keep moving in the interior of feasible domain, we need to replace our linear objective function by a nonlinear one, called the potential function,

$$f(x) = q \log(cx) - \sum_{i=1}^{n} \log x_i$$

which contains a barrier terms $\sum_{i=1}^{n} \log x_i$ to keep moving away from boundaries. Moreover, for simplicity of notation, we assume the base of log is 2 in this section.

Next, we present the interior point algorithm and then explain and analyze it.

Interior Point Algorithm for LP
input: a LP described as above.
output: a minimum solution x^k.
  $k \leftarrow 0$;
  while $cx^k \geq \varepsilon$ **do**
    Scaling: $D \leftarrow \text{diag}\left(\frac{1}{x_1^k}, \ldots, \frac{1}{x_n^k}\right)$;
          $\bar{A} \leftarrow AD^{-1}$;
          $\bar{c} \leftarrow cD^{-1}$;
          $y^k \leftarrow Dx^k$;
    Update: $h \leftarrow P(-\nabla \bar{f}(y^k))$,
              where $P = I - \bar{A}^T(\bar{A}\bar{A}^T)^{-1}\bar{A}$
              and $\bar{f}(y) = q \log(\bar{c}y) - \sum_{i=1}^{n} \log y_i$ and $q > 0$;
            $y^{k+1} \leftarrow y^k + \lambda h$, where $\lambda = 0.3/\|h\|$;
            $k \leftarrow k + 1$;
    Scale Back $x^k \leftarrow D^{-1}y^k$;
  end-while
  return x^k.

In this algorithm, each iteration is divided into three stages, scaling, update, and scale back. In the scaling stage, the point x^k is moved to $y^k = \vec{1}$ where $\vec{1}$ is the vector with 1 for every component, which is away from boundaries $y_i = 0$ with same distance. Let $y = Dx$. Then

$$Ax = \bar{A}y,$$

$$cx = \bar{c}y,$$

and

$$\bar{f}(y) = q \log(\bar{c}y) - \sum_{i=1}^{n} \log y_i$$

$$= q \log(cx) - \sum_{i=1}^{n} \log(D_i x_i)$$

$$= f(x) - \sum_{i=1}^{n} \log D_i.$$

Therefore, to decrease $f(x)$, it suffices to decrease $\bar{f}(y)$.

In the update stage, the search direction h is obtained by considering the decreasing direction of $\bar{f}(y^k)$, i.e., the opposite of gradient, $-\nabla \bar{f}(y^k)$. To keep the feasibility of the search direction, the algorithm project $-\nabla \bar{f}(y^k)$ into plane $\bar{A}y = b$. Actually,

$$P = I - \bar{A}^T (\bar{A}\bar{A}^T)^{-1} \bar{A}$$

is the projection operator since we have

$$\bar{A}P = 0$$

and for any vector y,

$$(y - Py)(Py)^T = 0.$$

There are three parameters, q, λ, and ε. How do we choose q and ε? Why do we choose $\lambda = 0.3/\|h\|$? We will explain them in the following analysis:

Lemma 6.7.1 *For any $k \geq 0$, $Ax^k = b$ and $x^k \geq 0$.*

Proof We prove it by induction on k. For $k = 0$, x^0 is a feasible point and hence $Ax^0 = b$ and $x^0 \geq 0$. Assume that $Ax^k = b$ and $x^k \geq 0$. Then, we have

$$Ax^{k+1} = \bar{A}y^{k+1} = \bar{A}(y^k + \lambda h) = \bar{A}y^k = Ax^k = b.$$

and

$$y^{k+1} = y^k + \lambda h = \vec{1} + 0.3 \cdot \frac{h}{\|h\|} \geq 0.$$

□

Lemma 6.7.2 $\bar{f}(\vec{1} + \lambda h) \leq \bar{f}(\vec{1}) + \lambda \nabla \bar{f}(\vec{1})^T h + 2\lambda^2 \|h\|^2.$

Proof Let $\bar{f}(y) = f_1(y) - f_2(y)$ where

$$f_1(y) = q \log(\bar{c}y),$$

$$f_2(y) = \sum_{i=1}^{n} \log y_i.$$

Then

$$\frac{d}{d\lambda} f_1(\vec{1} + \lambda h) = \nabla f_1(\vec{1} + \lambda h)^T h,$$

$$\frac{d^2}{d\lambda^2} f_1(\vec{1} + \lambda h) = q \frac{-(\bar{c}h)^2}{(\bar{c}(\vec{1} + \lambda h))^2} \leq 0.$$

Thus, $f_1(\lambda)$ is concave. Hence,

$$f_1(\vec{1} + \lambda h) \leq f_1(\vec{1}) + \lambda \nabla f_1(\vec{1})^T h.$$

Note that $\log(1 + \delta) \geq \delta - 2\delta^2$. Hence,

$$f_2(\vec{1} + \lambda h) \geq \sum_{i=1}^{n} \lambda h_i - 2 \sum_{i=1}^{n} \lambda^2 h_i^2$$

$$= f_2(\vec{1}) + \lambda \nabla f_2(\vec{1})^T h - 2\lambda^2 \|h\|,$$

where h_i is the ith component of h. Putting together two inequalities, respectively about f_1 and f_2, we obtain

$$\bar{f}(\vec{1} + \lambda h) \leq \bar{f}(\vec{1}) + \lambda \nabla \bar{f}(\vec{1})^T h + 2\lambda^2 \|h\|.$$

□

Lemma 6.7.3 *Select* $q = n + \sqrt{n}$. *Then* $\|h\| \geq 1$.

Proof Let y^* be the optimal solution, i.e., $\bar{c}y^* = 0$. Then, $\bar{A}(y^* - \vec{1}) = 0$. Therefore,

$$h^T(y^* - \vec{1}) = -(P(\nabla \bar{f}(\vec{1})))^T(y^* - \vec{1})$$
$$= -(\nabla \bar{f}(\vec{1}))^T(y^* - \vec{1})$$
$$= \left(-\frac{q\bar{c}}{\bar{c}\vec{1}} + (\vec{1})^T\right)(y^* - \vec{1})$$
$$= -\frac{q}{\bar{c}\vec{1}} \cdot \bar{c}y^* + (\vec{1})^T y^* + q - n$$
$$= \sum_{i=1}^{n} y_i^* + \sqrt{n}$$
$$\geq \|y^*\| + \sqrt{n}.$$

Moreover, $h^T(y^* - \vec{1}) \leq \|h\| \cdot \|y^* - \vec{1}\|$. Therefore,

$$\|h\| \geq \frac{\|y^*\| + \sqrt{n}}{\|y^* - \vec{1}\|} \geq \frac{\|y^*\| + \sqrt{n}}{\|y^*\| + \|\vec{1}\|} = 1.$$

$\square$

Lemma 6.7.4 *For any $k \geq 0$, $f(x^k) - f(x^{k+1}) \geq 0.1$.*

Proof Note that

$$f(x^k) - f(x^{k+1}) = \bar{f}(y^k) - \bar{f}(y^{k+1}) = \bar{f}(\vec{1}) - \bar{f}(\vec{1} + \lambda h).$$

By Lemma 6.7.2,

$$\bar{f}(\vec{1} + \lambda h) \leq \bar{f}(\vec{1}) + \lambda \nabla \bar{f}(\vec{1})^T h + 2\lambda^2 \|h\|^2.$$

Since $h \in \text{Null}(\bar{A})$, we have $\nabla \bar{f}(\vec{1})^T h = -\|h\|^2$. Therefore,

$$\bar{f}(\vec{1} + \lambda h) - \bar{f}(\vec{1}) \leq -\lambda \|h\|^2 + 2\lambda^2 \|h\|^2$$
$$= -0.3 \cdot \|h\| + 0.18$$
$$< -0.1$$

since $\|h\| \geq 1$ by Lemma 6.7.3. $\square$

Let L be the number of bits which are required to represent all numbers in the input, i.e., $L = \log |(\text{product of all input numbers in } A \text{ and } c)|$.

Lemma 6.7.5 *Initial feasible solution x^0 can be assumed to satisfy $f(x^0) \leq 2nL$.*

Proof For any vertex x of the feasible domain, $\log(cx) \leq L$. By the fundamental theorem of LP, the maximum value of cx is achieved by a vertex, for every feasible solution x, $\log cx \leq L$. Thus, in order to have $f(x^0) \leq 2nL$, it suffices to have

$\log x_i^0 \leq L$. From content of Sect. 6.4, we may assume that initially, x^0 is a vertex. Next, compute a x' to satisfy $Ax' = b$ with $\log |x_i| \leq L$ for $1 \leq i \leq n$. Since the feasible domain is compact, we can find two boundary points y and y' on line passing through x^0 and x'. Consider $z = (y + y')/2$. Then z satisfies the condition. $\square$

Lemma 6.7.6 *If $cx^k < 2^{-L}$, then x^k can be rounded into an exact optimal solution within $O(n^3)$ time.*

Proof If x^k is not a vertex, then we can find a line ℓ passing through x_k such that ℓ contains two boundary points y and z of the feasible domain and $x^k \in (y, z)$. c^x is nonincreasing in either direction $(x^k, y]$ or $(x^k, z]$. Thus, we will find either $cy \leq cx^k$ or $cz \leq cx^k$. However, y and z have one more active constraint than x^k. In this way, we can find a vertex x' of the feasible domain such that $cx' \leq cx^k < 2^{-L}$. Note that for any vertex x', each component is a rational number with denominator at most 2^L since all numbers in the input are integers. Therefore, if x' is not optimal, then we must have $cx' \geq 2^{-L}$. Hence, x' is optimal. Above operation may be performed $O(n)$ time and in each operation, computing boundary points may take $O(n^2)$ time. Therefore, the total running time is $O(n^3)$. $\square$

Theorem 6.7.7 *Select $\varepsilon = 2^{-L}$. The interior point algorithm will be terminated within $O(nL)$ steps.*

Proof Since the feasible domain is compact, $\sum_{i=1}^n x_i$ has an upper bound M on the feasible domain. Therefore, right before the algorithm terminates, $f(x^k) \geq -L - M$. Therefore, by Lemmas 6.7.4 and 6.7.5, the number of iterations is upper-bounded by

$$\frac{2nL + L + M}{0.1} = O(nL).$$

$\square$

Corollary 6.7.8 *An optimal solution of LP can be obtained by the interior point algorithm in $O(n^4 L)$ time.*

Proof It follows immediately from Theorem 6.7.7, Lemma 6.7.6, and the fact that each iteration runs in $O(n^3)$ time. $\square$

6.8 Polyhedral Techniques

A *polyhedron* is a set of all points bounded by a system of linear inequalities and linear equalities. For example, the feasible domain of each LP is a polyhedron. Since LP is polynomial-time solvable, we may use LP as a tool for solving other combinatorial optimization problems in the following way: Find a polyhedron P such that every vertex of P is a feasible solution of considered combinatorial optimization problem, so that the problem is transformed into an LP. This method is called the *polyhedral technique*. In this section, we introduce this technique through a few examples.

The first example is the maximum weight bipartite matching.

Problem 6.8.1 (Maximum Weight Bipartite Matching) Given a bipartite graph (V_1, V_2, E) with nonnegative edge weight $w : E \to R_+$, find a matching with maximum total edge weight.

The polyhedron of bipartite matching is defined as follows:

Definition 6.8.2 (Polyhedron of Bipartite Matching) For each matching M, define $\chi_M \in \{0, 1\}^{|E|}$ by

$$\chi_M(e) = \begin{cases} 1 & \text{if } e \in M, \\ 0 & \text{otherwise.} \end{cases}$$

Define the polyhedron of bipartite match P_{bmatch} to be the convex hull of χ_M for M over all matchings that is

$$P_{bmatch} = \{ \sum_{M \in \mathcal{M}} \alpha_M \chi_M \mid \alpha_M \geq 0, \sum_{M \in \mathcal{M}} \alpha_M = 1 \}$$

where $\mathcal{M}$ the set of all matchings.

Note that a bounded polyhedron is also called a *polytope*. Thus, the convex hull of a finite number of vectors must be a bounded region. Therefore, P_{bmatch} is also called the polytope of bipartite matching.

Let $\delta(v)$ denote the set of edges incident to vertex v.

Theorem 6.8.3 $x \in P_{bmatch}$ *if and only if*

$$\sum_{e \in \delta(v)} x_e \leq 1 \quad \text{for every } v \in V_1 \cup V_2,$$

$$x_e \geq 0 \quad \text{for every } e \in E.$$

Proof First, we show the necessity. Note that for any $M \in \mathcal{M}$ and $v \in V_1 \cup V_2$, $\sum_{e \in \delta(v)} \chi_M(e) \leq 1$. Therefore, for $x \in P_{bmatch}$,

$$\sum_{e \in \delta(v)} x_e = \sum_{e \in \delta(v)} \left(\sum_{M \in \mathcal{M}} \alpha_M \chi_M(e) \right)$$

$$= \sum_{M \in \mathcal{M}} \alpha_M \sum_{e \in \delta(v)} \chi_M(e)$$

$$\leq \sum_{M \in \mathcal{M}} \alpha_M$$

$$\leq 1.$$

Next, we show the sufficiency. For $x \in P_{bmatch}$, define

$$\text{supp}(x) = \{e \in E \mid x_e > 0\}.$$

The proof is by induction on $|\text{supp}(x)|$. For $|\text{supp}(x)| = 0$, we have $x = \chi_\emptyset \in P_{bmatch}$ since $\emptyset \in \mathcal{M}$. For induction step, consider $|\text{supp}(x)| \geq 1$. We divide the proof into three cases.

Case 1. $\text{supp}(x)$ is a matching. Assume $\text{supp}(x) = \{e_1, e_2, \ldots, e_k\}$ in ordering $x_{e_1} \leq x_{e_2} \leq \cdots \leq x_{e_k}$. Denote $M_i = \{e_i, e_{i+1}, \ldots, e_k\}$. Then

$$x = x_{e_1}\chi_{M_1} + (x_{e_2} - x_{e_1})\chi_{M_2} + \cdots + (x_{e_k} - x_{e_{k-1}})\chi_{M_k} + (1 - x_{e_k})\chi_\emptyset.$$

Case 2. $\text{supp}(x)$ contains a cycle $C = (e_1, e_2, \ldots, e_k)$. Since G is bipartite, k must be even. Define a vector d by setting

$$d_{e_i} = 1 \quad \text{if } i \text{ is odd,}$$

$$d_{e_i} = -1 \quad \text{if } i \text{ is even,}$$

$$d(e) = 0 \quad \text{if } e \text{ is not on the cycle } C.$$

Let ε_1 be the maximum $\varepsilon > 0$ such that $x + \varepsilon d \geq 0$ and ε_2 the maximum $\varepsilon > 0$ such that $x - \varepsilon d \geq 0$. Denote $y = x + \varepsilon_1 d$ and $z = x - \varepsilon_2 d$. Then, we have

$$|\text{supp}(y)| \leq |\text{supp}(c)| - 1, \quad |\text{supp}(z)| \leq |\text{supp}(c)| - 1,$$

and

$$x = \frac{\varepsilon_2}{\varepsilon_1 + \varepsilon_2}y + \frac{\varepsilon_1}{\varepsilon_1 + \varepsilon_2}z.$$

By induction hypothesis, $y \in P_{bmatch}$ and $z \in P_{bmatch}$. Hence, $x \in P_{bmatch}$.

Case 3. $\text{supp}(x)$ is a forest, but not a matching. In this case, there exists a leaf-to-leaf path $(e_1, \ldots, e_k)$ with $k \geq 2$. Define d by setting

$$d_{e_i} = 1 \quad \text{if } i \text{ is odd,}$$

$$d_{e_i} = -1 \quad \text{if } i \text{ is even,}$$

$$d(e) = 0 \quad \text{if } e \text{ is not on the path.}$$

Note that $x_{e_1} + x_{e_2} \leq 1$ and $x_2 > 0$. Hence, $1 - x_{e_1} \geq x_{e_2} > 0$. Similarly, $1 - x_{e_k} \geq x_{e_{k-1}} > 0$. Let ε_1 be the maximum $\varepsilon > 0$ such that $x + \varepsilon d \geq 0$. Then, we must have $x_{e_1} + \varepsilon_1 d_{e_1} \le 1$ and $x_{e_k} + \varepsilon_1 d_{e_k} \le 1$. Therefore, $y = x + \varepsilon_1 d$ is a vector satisfying constraints in the theorem. Similarly, let ε_2 be the maximum $\varepsilon > 0$ such that $x - \varepsilon d \ge 0$. Denote $z = x - \varepsilon_2 d$. then z satisfies constraints in the theorem. Moreover,

$$|\text{supp}(y)| \le |\text{supp}(c)| - 1, \quad |\text{supp}(z)| \le |\text{supp}(c)| - 1,$$

By induction hypothesis, $y \in P_{bmatch}$ and $z \in P_{bmatch}$. Hence, $x \in P_{bmatch}$ since

$$x = \frac{\varepsilon_2}{\varepsilon_1 + \varepsilon_2} y + \frac{\varepsilon_1}{\varepsilon_1 + \varepsilon_2} z.$$

□

Now let us look at applications of above theorem.

Corollary 6.8.4 *The maximum weight bipartite matching problem is polynomial-time solvable.*

Proof By Theorem 6.8.3, the maximum weight bipartite matching problem can be formulated as an LP problem $\max\{w^T x \mid x \in P_{bmatch}\}$. Since LP can be solved in polynomial-time, so does the weight bipartite matching problem.   □

Corollary 6.8.5 (König Theorem) *In any bipartite graph, the cardinality of the maximum matching is equal to the cardinality of the minimum vertex cover.*

Proof By duality of LP,

$$\max\{\vec{1}^T x \mid x \in P_{bmatch}\}$$
$$= \max\{\vec{1}^T x \mid x \ge 0 \text{ and } \forall v \in V, \sum_{e \in \delta(v)} x_e \le 1\}$$
$$= \min\{y\vec{1} \mid y \ge 0 \text{ and } \forall (u, v) \in E, y_u + y_v \ge 1\},$$

where $\vec{1}$ is a column vector in which every component is equal to 1.   □

Next, we study the matching in general graphs.

Problem 6.8.6 (Maximum Weight Matching) Given a graph (V, E) with non-negative edge weight $w : E \to R_+$, find a matching with maximum total edge weight.

This problem can be reduced to the maximum weight perfect matching problem.

Problem 6.8.7 (Maximum Weight Perfect Matching) Consider a graph (V, E) with nonnegative edge weight $w : E \to R_+$. Assume that G has a perfect matching. The problem is to find a perfect matching with maximum total edge weight.

Fig. 6.3 Construction of graph $\hat{G}$

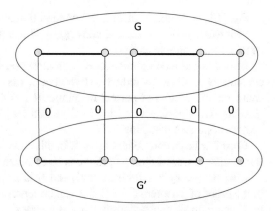

Lemma 6.8.8 *If the maximum weight perfect matching can be computed in polynomial-time, so does the maximum weight matching.*

Proof Let G be input graph of the maximum weight matching problem. Make a copy G' of G. Connect each pair of corresponding vertices by an edge with weight 0 (Fig. 6.3), and obtained graph is denoted by $\hat{G}$. Then, the maximum weight perfect matching in $\hat{G}$ will induce a maximum weight matching in G and vice versa. □

Theorem 6.8.9 *Consider a graph $G = (V, E)$. Let $\Omega(G)$ be the feasible region defined by the following constraints:*

$$\sum_{e \in \delta(v)} x_e = 1 \quad \text{for every } v \in V,$$

$$\sum_{e \in \delta(U)} x_e \geq 1 \quad \text{for every } U \subseteq V \text{ with odd } |U|,$$

$$x_e \geq 0 \quad \text{for every } e \in E,$$

where $\delta(U) = \{(u, v) \in V \mid u \in U \text{ and } v \notin U\}$. Then

$$\Omega(G) = conv\{\chi_M \mid M \text{ is a perfect matching in } G\}.$$

Proof First, it can be verified that for every perfect matching, χ_M satisfies all constraints. Hence,

$$\Omega(G) \supseteq conv\{\chi_M \mid M \text{ is a perfect matching in } G\}.$$

Next, we show by induction on $|E|$ that every $x \in \Omega$ belongs to $conv\{\chi_M \mid M$ is a perfect matching in $G\}$.

For $|E| = 1$, Ω contains only one point χ_M where M is the perfect matching consisting of the unique edge.

For $|E| \geq 2$, note that Ω is a bounded domain. Therefore, it suffices to show that for every vertex of Ω, $x \in \mathrm{conv}\{\chi_M \mid M$ is a perfect matching in $G\}$. First, we consider three simple cases.

Case 1. x has a component $x_e = 0$. In this case, we can delete edge e from G and consider $G' = G \backslash e$. By induction hypothesis, x is a convex combination of χ_M, i.e., characteristic vectors of perfect matching M's in G'. Since every perfect matching in G' is also a perfect matching in G, x is a convex combination of characteristic vectors of perfect matching in G.

Case 2. x has a component $x_e = 1$. In this case, we can delete e and its endpoints. It is similar to Case 1 that the proof can be completed by using induction hypothesis.

Case 3. $0 < x_e < 1$ for all $e \in E$, and G is a cycle. It is similar to the argument in the proof of Theorem 6.8.3 that x can be represented as a convex combination of two vectors in Ω, contradicting that x is a vertex.

Now, we can assume, without of generality, that above three cases do not occur. Therefore, for every $e \in E$, $0 < x_e < 1$, and every vertex has degree at least two; moreover, there exists a vertex with degree at least three. It follows that $|E| > |V|$. Since x is a vertex, there are at least $|E|$ active constraints. Hence, there exists $U \subseteq V$ with odd $|U|$ and $|U| \geq 3$ such that $\sum_{e \in \delta(U)} x_e = 1$. Denote $\bar{U} = V \subseteq U$. We can also assume $|\bar{U}| \geq 2$. In fact, if $\bar{U}$ is a singleton, say $\bar{U} = \{v\}$. Then, the constraint $\sum_{e \in \delta(U)} x_e = 1$ will be identical to the constraint $\sum_{e \in \delta(v)} x_e = 1$.

Let G/U denote the graph obtained by contracting U into a vertex u. Let x' be the restriction of x to those edges not disappearing in the contraction. Let $G/\bar{U}$ denote the graph obtained by contracting $\bar{U}$ into a vertex w and x'' the restriction of x to those edges not disappearing in the contraction. Note that $\sum_{e \in \delta(\bar{U})} x_e = \sum_{e \in \delta(U)} x_e = 1$. This fact implies that $\sum_{e \in \delta(u)} x_e = 1$ and $\sum_{e \in \delta(w)} x_e = 1$. Since $|U|$ is odd, we have that $x' \in \Omega(G/U)$ and $x'' \in \Omega(G/\bar{U})$. By induction hypothesis,

$$x' = \sum_{M'} \alpha_{M'} \chi_{M'}$$

$$x'' = \sum_{M''} \alpha_{M''} \chi_{M''},$$

where M' is over perfect matching in G/U and M'' is over perfect matching in $G/\bar{U}$. Since all constraints for $\Omega(G)$ have integer coefficients, every vertex of $\Omega(G)$ has rational components. Thus, x is rational and so are x' and x''. By choosing a common denominator m, we can write

$$x' = \frac{1}{m} \sum_{i=1}^{m} \chi_{M'_i}$$

$$x'' = \frac{1}{m} \sum_{i=1}^{m} \chi_{M''_i},$$

where M_i' and M_i'' are perfect matchings in G/U and $G/\bar{U}$, respectively (they may not be distinct). Since x' and x'' agree on $\delta(U) = \delta(\bar{U})$, we are able to pair up M_i' and M_i'' such that M_i' and M_i'' use the same edge in $\delta(U)$, so that $M_i = M_i' \cup M_i''$ is a perfect matching in G. Thus,

$$x = \frac{1}{m} \sum_{i=1}^{m} \chi_{M_i}.$$

This completes the proof of the theorem. □

The following corollary can immediately appear in front of us:

Corollary 6.8.10 *The minimum weight perfect matching can be computed in $O(n^2)$ time.*

However, this corollary cannot be driven immediately from the theorem. In fact, to show this corollary, we may study LP, $\min\{w^T x \mid x \in \Omega(G)\}$, i.e.,

$$\min\ w^T x$$

$$\text{subject to } \sum_{e \in \delta(v)} x_e = 1 \text{ for every } v \in V,$$

$$\sum_{e \in \delta(U)} x_e \geq 1 \text{ for every } U \subseteq V \text{ with odd } |U|,$$

$$x_e \geq 0 \text{ for every } e \in E.$$

This LP contains exponentially many constraints. Therefore, we cannot solve it in polynomial-time by employing the polynomial-time algorithm described in the last section. Therefore, we have to employ some techniques to overcome this difficulty.

In the following, we are going to design a primal-dual algorithm. Note that the dual LP is as follows:

$$\max \sum_{|U|=odd} y_U$$

$$\text{subject to } \sum_{U: e \in \delta(U), |U|=odd} y_U \leq w_e \text{ for all } e \in E$$

$$y_U \geq 0 \text{ for all } U \subseteq V \text{ with odd } |U| \geq 3.$$

In this algorithm, we do not need to write down all constraints; instead, only consider those U in a laminar family.

Definition 6.8.11 (Laminar Family) A family of sets, $\mathcal{F}$ is a laminar family if for any two sets $A, B \in \mathcal{F}$, $A \cap B = \emptyset$ or A or B, i.e., if $A \cap B \neq \emptyset$, then $A \subseteq B$ or $B \subseteq A$.

Lemma 6.8.12 *A laminar family $\mathcal{F}$ of distinct sets can have at most $2n-1$ members where n is the total number of elements.*

Proof Suppose $\mathcal{F}$ contains all singletons, i.e., for every element x, $\{x\} \in \mathcal{F}$. Construct a graph T with node set $\mathcal{F}$ and edge (A, B) exists if and only if $A \subset B$ and there is no third set $C \in \mathcal{F}$ such that $A \subset C \subset B$. Then, G is a forest, and a vertex is a leaf if and only if it is a singleton. Moreover, each connected component of G is a rooted tree such that the root is a largest set in this component. Since every singleton is in $\mathcal{F}$, every internal node of G has at least two children. Let us give a proof by induction on the number of internal node.

For basis step, consider G without internal node. Then the number of nodes in G is $n \le 2n - 1$.

For induction step, consider G with at least one internal node. Suppose r is the root of connected component which has at least one internal node. Then r has at least two children. Removal r will result in at least two subtrees. Suppose that T_1 is one of them and T_1 contains n_1 leaves. Let T_2 be the remaining part after removing T_1. Then T_2 has $n - n_1$ leaves. By induction hypothesis, T_1 has at most $2n_1 - 1$ nodes and T_2 has at most $2(n - n_1)-!$ nodes. Therefore, the number of nodes in G is at most

$$1 + 2n_1 - 1 + 2(n - n_1) - 1 = 2n - 1.$$

If $\mathcal{F}$ does not contain all singletons, then we can add those missing singletons to $\mathcal{F}$, which will enlarge $\mathcal{F}$. However, the enlarged $\mathcal{F}$ can have at most $2n - 1$ members. □

In this primal-dual algorithm, a matching M will be growing through augmentations. Each augmentation is on an alternating path with two free nodes. A node is *free* if it is not covered by matching M. To find an augmenting path, we may grow an alternating tree starting from each free node. On this tree, a node is called an *even node* if it has even distance from the free node; otherwise, it is called an *odd node*.

In this algorithm, we assume that input graph $G = (V, E)$ is simple and contains a perfect matching. Then, the algorithm consists of five steps as follows:

Step 0. Initially, set $M = \emptyset$, $\Gamma = \{\{v\} \mid v \in V\}$, and $y_U = 0$ for every $U \in \Gamma$.
Step 1. Let F be the set of free nodes with respect to matching M and

$$E_y = \{e \mid \sum_{U:e\in\delta(U),|U|=odd} y_U = w_e\}.$$

If $F = \emptyset$, then M is a perfect matching and the algorithm stops. If $F \ne \emptyset$, then construct an alternating tree in E_y, starting from each free node $v \in F$. If construction meets an edge between two even nodes, then a cycle of odd length is generated. This cycle is called a *blossom*. Shrink the blossom into an even node and put the union of nodes in the blossom, U into Γ. Continue the construction.

Step 2. If alternating tree rooted at free node v meets another free node u, then an augmenting path between v and u is found, perform an augmentation, and go back to Step 1.

Step 3. If no blossom can be found, no augmenting path can be found, and no alternating tree can be extended for all alternating tree, then for each alternating tree, modify dual solution by increasing y_U for all even nodes U and decreasing y_U for all odd nodes U at the same rate until stucking at boundary of the feasible domain of the dual LP (Fig. 6.4). If the process does not get stuck, then the primal is not feasible. (Since we assume that G contains a perfect matching, this cannot occur.)

Step 4. When the process in Step 3 gets stuck, there are two possibilities. (a) The first possibility is that a constraint $\sum_{e:e\in\delta(U),|U|=odd} y_U \le w_e$ becomes active, i.e., the equality is reached. In this case, add e to E_y and go back to Step 1.

(b) The second possibility is that the constraint $y_U \ge 0$ becomes active for $|U| = odd$ and $|U| \ge 3$. In this case, de-shrink the blossom U, add $\lfloor |U|/2 \rfloor$ matching edges back to M, remove U from Γ, and go back to Step 3 (Fig. 6.5). Please note that case (b) cannot occur if every node U is singleton. Therefore, finally (a) occurs and E_y will be increased.

Next, we analyze this algorithm.

Fig. 6.4 Increase y_U for all even nodes and decrease y_U for all odd nodes at the same rate

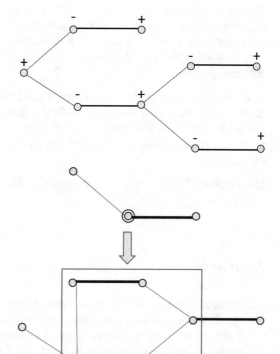

Fig. 6.5 De-shrink a blossom

Lemma 6.8.13 *Between a set U which is added to Γ and is removed, an augmentation must occur.*

Proof When a set U is added to Γ, U must be shrunk and y_U becomes an even node. However, when U is removed from Γ, y_U must be an odd node. Therefore, there must exist an augmentation which changes the parity of y_U.   □

Lemma 6.8.14 *The algorithm must stop within $O(n^3)$ time.*

Proof By Lemma 6.8.13, between two augmentations, if a set U is added to Γ, then it cannot be removed. Therefore, only those sets existing in Γ in the beginning can be removed. Since Γ is a laminar family, we have $|\Gamma| = O(n)$. Therefore, de-shrinking can occur at most $O(n)$ times. Also by Lemma 6.8.13, a set U added to Γ cannot leave Γ before an augmentation takes place. Therefore, shrinking can occur at most $O(n)$ time. Moreover, E_y can be increased at most $O(n)$ edges. Since each operation of de-shrinking, shrinking, and E_y-increasing takes $O(n)$ time, the running time between two augmentation is $O(n^2)$. Finally, note that there are $O(n)$ augmentations. Therefore, the algorithm runs in $O(n^3)$ time.   □

Lemma 6.8.15 *The algorithm terminates only when M is a minimum weight perfect matching.*

Proof First, we note that the algorithm is an extension of the blossom algorithm for maximum matching. Therefore, if M is not a perfect matching, then the algorithm cannot terminate. Moreover, note that every edge in M belongs to E_y, i.e., corresponds to an active dual constraint. Therefore, the complementary slackness condition holds. Hence, M gives an optimal solution for minimum weight perfect matching problem.   □

It follows immediately from above three lemmas that the minimum weight perfect matching can be computed in $O(n^3)$ time. Since all perfect matching have the same cardinality, the minimum weight perfect matching problem is equivalent to the maximum perfect matching problem. Furthermore, by Lemma 6.8.8, we have the following.

Corollary 6.8.16 *The maximum weight matching in graph can be computed in $O(n^3)$ time.*

Exercises

1. Linda plans to put \$12,000 in investment for two stocks. The history shows that the first stock earns 6% interests and the second stock earns 8% interests. If Linda wants to spend the money in the first stock at least twice as much as that in the second stock, but must not be greater than \$9000, then how can she buy these two stocks in order to maximize her profit?

2. A teacher plans to rent buses from a company for a trip of 200 students. The company has nine drivers and two types of buses. The first type bus has 50 seats; the rental cost is $800. The second type bus has 40 seats; rental cost is $600. The company has ten buses of the first type and eight buses of the second type. What plan can get the lowest total rental cost?

3. Transform the following LP into an equivalent LP in standard form:

$$\min \ x_1 - x_2$$

$$\text{subject to} \ 2x_1 + x_2 \geq 3$$

$$3x_1 - x_3 \leq 7$$

$$x_1 \geq 0.$$

4. Please formulate the following problem into an LP: Given n lines $a_i x + b_i y = c_i$ for $i = 1, 2, \ldots, n$ in a plane, find a point to minimize the total distance from the point to these lines.

5. Transform the following problem into an LP:

$$\min \max(c_1 x, c_2 x, \ldots, c_k x),$$

i.e., minimize the maximum of k linear functions.

6. Suppose in the following

$$\max \ z = cx$$

$$\text{subject to} \ Ax = b$$

$$x \geq 0,$$

A is an $m \times n$ coefficient matrix with rank $rank(A) < m$. Could you transform it into the standard form of LP? What cases would occur during your transformation?

7. Solve the following LPs:

(a)

$$\max \ 5x_1 + 7x_2$$

$$\text{subject to} \ x_1 + x_2 \leq 30$$

$$2x_1 + x_2 \leq 50$$

$$4x_1 + 3x_2 \geq 60$$

$$2x_1 \geq x_2$$

$$x_1 \geq 0, x_2 \geq 0.$$

(b)

$$\max\ x + y$$
$$\text{subject to}\ x + 2y \le 10$$
$$2x + y \le 16$$
$$-x + y \le 3$$
$$x \ge 0, y \ge 0.$$

(c)

$$\max\ x_1 + x_2 - 2x_3 + 2x_4$$
$$\text{subject to}\ x_1 - x_2 - x_3 - 2x_4 \ge 2$$
$$x_1 + x_2 + x_4 \le 8$$
$$x_1 + 2x_2 - x_3 = 4$$
$$x_1, x_2, x_3, x_4 \ge 0.$$

(d)

$$\max\ 3x + cy + 2z$$
$$\text{subject to}\ 2x + 4y + 2z \le 200$$
$$x + 3y + 2z \le 100$$
$$x, y, z \ge 0.$$

(e)

$$\max\ u + v$$
$$\text{subject to}\ -u - 2v - w = 2$$
$$3u + v \le -1$$
$$v \ge 0, w \ge 0.$$

(f)

$$\min\ 3x_1 + 4x_2 + 6x_3 + 7x_4 + x_5$$
$$\text{subject to}\ 2x_1 - x_2 + x_3 + 6x_4 - 5x_5 - x_6 = 6$$
$$x_1 + x_2 + 2x_3 + x_4 + 2x_5 - x_6 = 3$$
$$x_1, \ldots, x_6 \ge 0.$$

8. Can you find a simple method to solve the LP with only one equality constraint as follows?

$$\max \quad cx$$

$$\text{subject to} \quad Ax = b$$

$$x \geq 0$$

where A is a coefficient matrix of $1 \times n$.

9. Show that under nondegeneracy condition, a feasible basis is optimal if and only if cost coefficients $c_j \leq 0$ in the corresponding simplex table.

10. Give a counterexample to show that generally, in the simplex table corresponding a feasible basis I, all $c_j \leq 0$ may not be necessary for I to be optimal.

11. Show that in the simplex algorithm, if a variable leaves the feasible basis at a pivot, it cannot return to the feasible basis at next pivot.

12. Show that in the simplex algorithm, if a cycling occurs, there must exist an iteration in which the choice for a variable leaving feasible basis is not unique.

13. When using the following

$$\max \quad w = -ey$$

$$\text{subject to} \quad Ax + I_m y = b$$

$$x \geq 0, y \geq 0,$$

to find initial feasible basis for LP $\min\{cx \mid Ax = b, x \geq 0\}$, it is found that the maximum value of W is negative. Can you figure out what happens? Does $Ax = b$ not have a solution or does $Ax = b$ not have a nonnegative solution?

14. Consider the following relaxation of an LP, which allows a violation of constraints with an upper bound ε for total violations:

$$\min \quad yb$$

$$\text{subject to} \quad ya_j \geq c_j - \varepsilon_j \text{ for } j = 1, 2, \ldots, n$$

$$\sum_{j=1}^{n} \varepsilon_j \leq \varepsilon$$

$$y \geq 0, \varepsilon_j \geq 0 \text{ for } j = 1, 2, \ldots, n.$$

Find its dual LP and give an interpretation.

15. Show that if the LP

$$\max \quad cx$$

$$\text{subject to} \quad Ax \leq b$$

$$x \geq 0$$

is unbounded, then the LP

$$\max \ ex$$
$$\text{subject to} \ \ Ax \leq b$$
$$x \geq 0$$

is unbounded where $e = (1, 1, \ldots, 1)$. Is the inverse true?

16. Show that if the LP

$$\max \ cx$$
$$\text{subject to} \ \ Ax \leq b$$
$$x \geq 0$$

is unbounded, then for any $\hat{b}$, the LP

$$\max \ cx$$
$$\text{subject to} \ \ Ax \leq \hat{b}$$
$$x \geq 0$$

is either infeasible or unbounded.

17. Design Bland's rule for the dual simplex algorithm and prove the correctness of your design.

18. Show that the following two problems are equivalent:

$$\min \ cx$$
$$\text{subject to} \ \ b \leq Ax \leq \hat{b}$$
$$x \geq 0$$

and

$$\min \ cx$$
$$\text{subject to} \ \ Ax + y = b$$
$$x \geq 0$$
$$0 \leq y \leq \hat{b} - b.$$

19. Show that the following LP has either optimal value 0 or no feasible solution:

$$\min \ cx - yb$$
$$\text{subject to} \ Ax \geq b$$
$$yb \leq c$$
$$x \geq 0, y \geq 0,$$

where c is an n-dimensional row vector, b is an m-dimensional column vector, and A is an $m \times n$ matrix.

20. Let A be an $n \times n$ symmetric matrix. Suppose that x satisfies $Ax = c^T$ and $x \geq 0$. Prove that x is an optimal solution of the following LP:

$$\min \ cx$$
$$\text{subject to} \ Ax \geq c^T$$
$$x \geq 0.$$

21. Using the primal-dual algorithm, please solve the following LP:

$$\max \ x_2 - 10x_3 - 14x_4 - x_5$$
$$\text{subject to} \ x_2 + x_3 + x_4 + x_5 = 5$$
$$x_2 + 2x_3 + x_4 - x_5 = 4$$
$$x_1 + x_2 + 3x_3 + 2x_4 = 4$$
$$x_1, \ldots, x_5 \geq 0.$$

22. Consider a pair of primal LP (P) and dual LP (D). Suppose that both (P) and (D) have feasible solutions. Show that there exist optimal solutions x^* and y^* for (P) and (D), respectively, such that

$$x_j^* = 0 \text{ if and only if } y^* a_j > c_j$$

where a_j is the jth column of constraint coefficient matrix A and c_j is cost coefficient of x_j.

23. Consider a polyhedron P and $v \in P$. Show that the following statements are equivalent:

(a) v is a vertex of P.
(b) There exists a hyperplane $H = \{x \mid a^T x = \alpha\}$ such that $a^T x \leq \alpha$ for all $x \in P$ and $P \cap H = \{v\}$.
(c) There are n constraints $a_j^T x \leq b_j$ valid for P, which are active at v, i.e., $a_j^T v = b_j$ for $1 \leq j \leq n$, and $a_1, \ldots, a_n$ are linear independent.

24. Show that in a bipartite graph $G = (V_1, V_2, E)$,

$$x \in \text{conv}\{\chi_M \mid M \text{ is a perfect matching in } G\}.$$

if and only if

$$\sum_{e \in \delta(v)} x_e = 1 \quad \text{for every } v \in V_1 \cup V_2,$$

$$x_e \geq 0 \quad \text{for every } e \in E.$$

25. Show that in a bipartite graph, every perfect matching cannot be represented as a convex combination of other matchings.

26. (Doubly stochastic matrix) An $n \times n$ matrix $A = (a_{ij})$ is *doubly stochastic matrix* if

$$\text{for } 1 \leq i \leq n, 1 \leq j \leq n, \ a_{ij} \geq 0,$$

$$\text{for } 1 \leq i \leq n, \ \sum_{j=1}^{n} a_{ij} = 1,$$

$$\text{for } 1 \leq j \leq n, \ \sum_{i=1}^{n} a_{ij} = 1.$$

Show that every doubly stochastic matrix is a convex combination of permutation matrices.

27. (Unimodular Matrix) A matrix A is said to be *totally unimodular* if every square submatrix has determinant 0, 1, or -1. Show that for every bipartite graph G, the incidence matrix A is totally unimodular.

28. Consider a graph $G = (V, E)$. For $b \in Z_+^V$, a b-matching is a subset M of edges such that for every $v \in V$, $|M \cap \delta(v)| \leq b_v$. For $c \in Z_+^E$, a c-cover is a mapping $y : V \to Z_+$ such that every $(u, v) \in E$, $y_u + y_v \geq c_{(u,v)}$. Show that in any bipartite graph, for $b \in Z_+^V$ and $c \in Z_+^E$, the maximum c-weighted b-matching is equal to the minimum b-weighted c-vertex cover.

29. Consider a graph $G = (V, E)$. Let Ω be the feasible region defined by the following constraints:

$$\sum_{e \in \delta(v)} x_e \leq 1 \quad \text{for every } v \in V,$$

$$\sum_{e \in E(U)} x_e \leq \lfloor |U|/2 \rfloor \quad \text{for every } U \subseteq V \text{ with odd } |U|,$$

$$x_e \geq 0 \quad \text{for every } e \in E,$$

where $E(U) = \{(u, v) \in E \mid u, v \in U\}$. Show that

$$\Omega = \text{conv}\{\chi_M \mid M \text{ is a matching in } G\}.$$

30. Every d-regular bipartite graph has an edge-coloring with d colors, such that every vertex is incident to all d colors.

31. Let $G = (U, V, E)$ be a bipartite graph. Suppose that there exists a matching covering U and there exists a matching covering V. Show that a perfect matching exists.

32. Consider a graph G. Let $P_{match}(G)$ denote the matching polytope of G, i.e., $P_{match}(G) = \text{conv}\{\chi_M \mid M \text{ is a matching in } G\}$. Prove that $P_{match} \cap \{x \mid \vec{1}^T x = k\}$ is the convex hull of all matchings of size exactly k.

33. Consider a graph G and its perfect matching polytope $P = \text{conv}\{\chi_M \mid M \text{ is a perfect matching of } G\}$. An *edge* is a line segment between two vertices $s = [\chi_M, \chi_{M'}]$ such that $s = H \cap P$ for some hyperplane $H = \{x \mid w^T x = \alpha\}$ and $w^T x \le \alpha$ for all $x \in P$. Prove that $[\chi_M, \chi_{M'}]$ is an edge if and only if $M \oplus M'$ is a single cycle.

Historical Notes

The LP was initially proposed independently by Leonid Kantorovich [233] and T. C. Koopmans with a large number of applications. Hence, they shared a Nobel Prize in 1945. The simplex algorithm was designed by George B. Dantzig during 1946–1947 [79]. This algorithm may fall into cycling [22]. The first method for dealing with degeneracy was proposed by Charnes [47] in 1952. Lexicographical ordering method is motivated from it [80]. In 1977, Bland found a completely new one [32].

Klee and Minty [256] gave an example on which simplex algorithm runs in exponential number of iterations. The first polynomial-time algorithm for LP was discovered by Khachiyan in 1979 [241] using ellipsoid method. In 1984, Karmarkar [235] found the interior-point method, which has been developed extensively [181, 311, 375], to form a big class of efficient algorithms for LP currently. The interior point algorithm in Section 7.6 is a variation in [179], which is simpler to describe and analyze, and however, to have $O(nL)$ iterations. In the literature, there are several algorithms using $O(\sqrt{n}L)$ iterations [21, 180, 261, 321, 349, 376, 385].

Polyhedral technique is an important tool in the study of combinatorial optimization. The technique is initiated from a naive attempt to solve other combinatorial optimization problems by transforming them into LP [78, 79]. This method was quite successful for the transportation, maximum flow, as assignment problems [145–147, 214, 267, 268]. The pioneering work of Edmonds [116–121] opened a new page for polyhedral methods since he characterized several basic polytopes, such as the matching polytope and the perfect matching polytope, and found elegant solutions for weighted version of corresponding combinatorial optimization problems. Nowadays, the polyhedral technique also becomes an important tool for finding efficient approximation solutions [357, 358].

Since LP has been a subject of teaching in many areas, such as operations research, industrial engineering, management science, and computer science, for many years, many textbooks are already published in the literature. Here, we would like to mention one by Zhang and Xu [440] because the book provides many good exercises which do not appear in other publications. We adopt several of them in this chapter.

Chapter 7
Primal-Dual Methods and Minimum Cost Flow

Education today, more than ever before, must see clearly the dual objectives: education for living and educating for making a living.

—James Wood-Mason

There are three types of incremental methods, primal, dual, and primal-dual. In Chap. 6, we touched all of them for linear programming (LP). This chapter is contributed specially to primal-dual methods for further exploring techniques about primal-dual with a special interest in the minimum cost flow. Actually, the minimum cost flow is a fundamental optimization problem on networks. The shortest path problem and the assignment problem can be formulated as its special cases. We begin with the study of the assignment problem.

7.1 Hungarian Algorithm

Let us study the assignment problem as follows:

Problem 7.1.1 (Assignment) *Consider a set of n workers, $S = \{s_1, \ldots, s_n\}$; a set of n jobs, $T = \{t_1, t_2, \ldots, t_n\}$; and with an $n \times n$ table in which each entry $c(i, j)$ is the cost required by worker i finishing job j. The problem is to find an assignment for every worker to receive exactly one job with the minimum total cost.*

This problem is equivalent to the minimum cost perfect matching in a complete bipartite graph between node sets S and T. There is a very well-known primal-dual algorithm for the problem. Actually, in the history, it is the first algorithm of primal-dual type, called Hungarian algorithm, because the algorithm was initially designed by two Hungarian mathematicians.

To describe this algorithm, each node v is given a label $d(v)$. This label is *valid* if $d(u) + d(v) \leq c(u, v)$ for every $u \in S$ and $v \in T$. Clearly, if M is a perfect matching, then

$$c(M) = \sum_{(u,v) \in M} c(u,v) \geq \sum_{u \in S} d(u) + \sum_{v \in T} d(v).$$

It implies the following immediately:

Lemma 7.1.2 *If for a perfect matching M, there exists a valid label d such that for every $(u,v) \in M$, $d(u) + d(v) = c(u,v)$, then M has the minimum cost.*

In Hungarian algorithm, initially, set $d(v) = 0$ for every $v \in V$ and set matching $M = \emptyset$. During the computation, label d and matching M will be updated and finally, the condition in Lemma 7.1.2 will be reached. The success of this algorithm will indicate that this condition is necessary and sufficient. Actually, the valid label $d(v)$ plays the role of dual solution. To see this, let us consider an LP formulation of the assignment problem as follows:

$$\min \sum_{u=1}^{n} \sum_{v=1}^{n} c(u,v) x_{uv}$$

$$\text{subject to } \sum_{v \in T} x_{uv} = 1 \text{ for all } u \in S,$$

$$\sum_{u \in S} x_{uv} = 1 \text{ for all } v \in T,$$

$$x_{uv} \geq 0 \text{ for all } u \in S, v \in T.$$

The following is its dual LP:

$$\max \sum_{u \in S} d(u) + \sum_{v \in T} d(v)$$

$$\text{subject to } d(u) + d(v) \leq c(u,v) \text{ for all } u \in S, v \in T.$$

For simplicity of description, an edge (u,v) is said to be *tight* if $d(u) + d(v) = c(u,v)$. Let us also assign an orientation to each edge between S and T. We intend to use the orientation from T to S for representing edge in M, that is, an edge is in M if and only if it has orientation from T to S. Therefore, initially, every edge is assigned with orientation from S to T since $M = \emptyset$.

To update M, consider subgraph G_d consisting of all tight edges with their orientation. Let R_S be the set of free nodes in S, i.e., all nodes without incoming arc. Let R_T be the set of free nodes in T, i.e., all nodes without outgoing arc. If there is a path from R_S to R_T in G_d, then this path is an alternating path for matching M. Therefore, we can reverse the orientation on this path in order to increase members of M. Otherwise, if M is not a perfect matching, then we can consider to update valid label d as follows:

Let Z be the set of nodes each of which can be reached from R_S through a directed path in G_d. In this case, $Z \cap R_T = \emptyset$. Define

$$\Delta = \min\{c(u, v) - d(u) - d(v) \mid u \in Z \cap S, v \in T \setminus Z\}.$$

Lemma 7.1.3 *If M is not a perfect matching and $Z \cap R_T = \emptyset$, then $\Delta > 0$.*

Proof Since M is not maximum, there must exist an alternating path P from R_S to R_T. Since $Z \cap R_T = \emptyset$. P must contain loose (i.e., untight) arcs. Since every arc in M is tight, every loose arc (u, v) in P has orientation from S to T, i.e., $u \in S$ and $v \in T$. We first claim that there exists a loose arc (u, v) such that $u \in Z \cap S$. In fact, suppose that (u, v) is the first loose arc. Then we must have $u \in Z \cap S$.

Next, consider the last loose arc (u, v) in P such that $u \in Z \cap S$. We claim that $v \in T \setminus Z$. In fact, if not, i.e., $v \in Z$, then there exists a path P' from R_S to v in G_d. Replace the part of P, from R_S to v, by P'. The resulting path P'' is still an alternating path. The first loose arc (x, y) on P'' will have $x \in Z \cap S$. However, (x, y) is also a loose arc in P and appears later than (u, v), contradicting the choice of (u, v). Therefore, our second claim is true and hence $\Delta > 0$. □

Now, we update the valid label by setting

$$d(u) \leftarrow d(u) + \Delta \text{ for } u \in Z \cap S,$$

$$d(v) \leftarrow d(v) - \Delta \text{ for } v \in Z \cap T.$$

Lemma 7.1.4 *After update, d is still a valid label, G_d still contains M, and Z gets increased by at least one.*

Proof Note that any arc (v, u) with $u \in S$ and $v \in T$ must belong to M and hence is tight. We have only the following three cases for every arc:

(a) For a tight arc (u, v) with $u \in S$ and $v \in T$, if $u \in Z \cap S$, then we must have $v \in Z \cap T$. Therefore, after update, (u, v) is still tight. If $v \in Z \cap T$, then we may not have $u \in Z$, but on (u, v), the valid condition $d(u) + d(v) \leq c(u, v)$ is still held.
(b) For a tight arc (v, u) with $u \in S$ and $v \in T$, if $u \in Z \cap S$, then $u \notin R_S$ and hence u can be reached from R_S only from v, which implies $v \in Z$. Therefore, (v, u) is still tight. If $v \in Z \cap T$, then we must have $u \in Z \cap S$ and hence (v, u) keeps tight.
(c) For a loose arc (u, v) with $u \in S$ and $v \in T$, if u and v are both in Z or both not in Z, then the valid condition is clearly held. If $u \in Z$ and $v \notin Z$, then by definition of Δ, the valid condition is still held for (u, v). If $u \notin Z$ and $v \in Z$, then $d(u) + d(v)$ is decreased by Δ and hence the valid condition is held.

Above argument showed that after update, the label is still valid. Moreover, (b) also implies that G_d still contains M. (a) and (b) together imply that every node in Z is still in Z. The definition of Δ implies that one more node gets in Z. □

By Lemmas 7.1.3 and 7.1.4, if M is not maximum, then the algorithm can increase either M or Z. Hence, it can finally obtain a maximum M contained in G_d, i.e., with minimum cost. A pseudocode of the algorithm is included in Algorithm 22.

Algorithm 22 Hungarian algorithm

Input: A complete bipartite graph $G = (S, T, E)$ with $n = |S| = |T|$ and edge cost $c : E \to R_+$.
Output: A minimum cost perfect matching M.
1: $d(u, v) \leftarrow 0$ for $(u, v) \in E$.
2: $M \leftarrow \emptyset$. G_d is the empty graph. $Z \leftarrow \emptyset$.
3: Assign orientation from S to T to every edge in E.
4: **while** $|M| < n$ **do**
5:     Update G_d.
6:     Update Z.
7:     **if** G_d contains an augmenting path P for M **then**
8:         reverse orientation of all arcs in P.
9:         Update M.
10:    **else**
11:        Compute $\Delta = \min\{c(u, v) - d(u) - d(v) \mid u \in Z \cap S, v \in T \setminus Z\}$.
12:        Update label d by
13:        $d(u) \leftarrow d(u) + \Delta$ for $u \in Z \cap S$ and
14:        $d(v) \leftarrow d(v) - \Delta$ for $v \in Z \cap T$.
15:    **end if**
16: **end while**
17: **return** M

Theorem 7.1.5 *Hungarian algorithm produces a minimum cost perfect matching within $O(n^4)$ time.*

Proof In each iteration of while-loop, either M is increased by one or Z is increased by at least one. Between two consecutive ones of increasing M, Z can be increased at most $O(n)$ times. For each time of increasing Z, computing Δ may need $O(n^2)$ time. Note that M can be increased at most n times. Therefore, Hungarian algorithm runs in $O(n^4)$ time. □

Now, we study an example.

Example Consider four workers w_1, w_2, w_3, w_4 and four tasks t_1, t_2, t_3, t_4. The cost $c(i, j)$ of worker w_i finishing task t_j is as shown in the following table:

	t_1	t_2	t_3	t_4
w_1	1	3	2	4
w_2	2	1	4	2
w_3	1	4	3	3
w_4	4	2	1	3

Let us start with a little different initiation. For each worker w_i, assign $d(w_i) = \min\{c(i, j) \mid j = 1, 2, 3, 4\}$. Then for each task t_j, assign $d(t_j) = \min\{c(i, j) - d(w_i) \mid i = 1, 2, 3, 4\}$. Then we obtain

$$
\begin{array}{c|cccc}
 & 0 & 0 & 0 & 1 \\
\hline
1 & 0 & 2 & 1 & 2 \\
1 & 1 & 0 & 3 & 0 \\
1 & 0 & 3 & 2 & 1 \\
1 & 3 & 1 & 0 & 1 \\
\end{array}
$$

where the most left column consists of $d(w_i)$, the top row consists of $d(t_j)$, and the matrix consists of entries $c(i, j) - d(w_i) - d(t_j)$. Therefore, the entry 0 in the cost matrix represents the tight edge. Let M be a maximum matching in G_d of tight edges, consisting of edges represented by 0^*. Mark rows not covered by M, which form R_S. Continue to mark rows and columns according to the following rules:

- If a row is marked, then mark all columns each of which intersects the marked row with an entry 0.
- If a column is marked, then mark all rows each of which intersects the marked column with entry 0.

All marked rows and columns form the set Z. Now, consider all entries located in the intersection of marked rows and unmarked columns. (Each of them is put in a parenthesis.) Let Δ be the minimum of them, i.e., $\Delta = 1$.

$$
\begin{array}{c|cccc|}
 & 0 & 0 & 0 & 1 \\
\hline
1 & 0^* & (2) & (1) & (2) \\
1 & 1 & 0^* & 3 & 0 \\
1 & 0 & (3) & (2) & (1) \\
1 & 3 & 1 & 0^* & 1 \\
\hline
 & * & & & \\
\end{array}
\begin{array}{l}
\\
* \\
\\
* \\
\\
\end{array}
$$

Subtract Δ from marked rows and all Δ to marked columns. We obtain the following matrix with a minimum cost assignment marked 0^*:

$$
\begin{array}{c|cccc|}
 & -1 & 0 & 0 & 1 \\
\hline
2 & 0^* & (1) & (0) & (1) \\
1 & 2 & 0^* & 3 & 0 \\
2 & 0 & (2) & (1) & (0^*) \\
1 & 4 & 1 & 0^* & 1 \\
\hline
 & * & & & \\
\end{array}
\begin{array}{l}
\\
* \\
\\
* \\
\\
\end{array}
$$

In above computation, we see that assigning a value to $d(w_i)$ or $d(t_j)$ is equivalent to adding/subtracting a value on a row or column in cost table. Actually, it is clear that such an operation on cost table will not change optimality of assignment. Therefore, we may use such operations without mentioning label d at all. This will induce a local ratio algorithm. Actually, the primal-dual method and the local ratio method have a very close relationship. A lot of primal-dual algorithms can have their equivalent local ratio companions and vice versa.

Hungarian algorithm can also be used for solving the following problem:

Problem 7.1.6 (Chinese Postman Problem) *Given a graph $G = (V, E)$ with edge cost a, find a postman tour to minimize total edge cost where a postman tour is a cycle passing through every edge at least once.*

A cycle is called as a *Euler tour* if it passes through every edge exactly once. In graph theory, it has been proved that a connected graph has an Euler tour if and only if every node has even degree. A node is called as an *odd node* if its degree is odd. Note that the number of odd degree for any graph is even. Therefore, we may solve the Chinese postman problem in the following way:

- Construct a complete graph H on all old nodes and assign the distance between any two nodes u and v with the shortest distance between u and v in G.
- Find the minimum cost perfect matching M in H.
- Add M to input graph G and the Euler tour in $G \cup M$ is the optimal solution.

7.2   Label-Correcting

In this section, we introduce the label-correcting algorithm for the shortest path problem. This is another look of Bellman-Ford algorithm, which allows negative arc cost and only restriction is no negative cost cycle. The disadvantage is that the running time is slow. However, it induces a faster algorithm for all-pairs shortest paths.

Consider a directed network $G = (V, E)$ with arc cost $c : E \rightarrow R$, an origin node s and a destination node t. The aim of the problem is to find a shortest path from s to t.

Let us start to introduce the label-correcting algorithm by defining the node label. A node label $d : V \rightarrow R$ is said to be *valid* if it satisfies the following conditions:

(a1) $d(s) = 0$.
(a2) $d(v) \leq d(u) + c(u, v)$ for any $(u, v) \in E$ and $v \neq s$.

Let $d^*(v)$ denote the cost of the shortest path from s to v. Then the valid node label has the following properties:

Lemma 7.2.1 *For any valid node label $d(v)$, $d(v) \leq d^*(v)$.*

Proof The proof is by induction on the number of arcs on the shortest path from s to v, denoted by $\hat{d}(v)$. For $\hat{d}(v) = 0$, we have $v = s$ and hence $d(s) = 0 = d^*(s)$.

For $\hat{d}(v) = k > 0$, consider a shortest path P from s to v. Suppose arc (u, v) on P. Then $\hat{d}(v) = \hat{d}(u) + 1$. By induction hypothesis, $d(u) \leq d^*(u)$. Therefore, $d(v) \leq d(u) + c(u, v) \leq d^*(u) + c(u, v) = d^*(v)$. □

Lemma 7.2.2 *Let $d(\cdot)$ be a valid node label. If $d(v)$ is the cost of some path from s to v, then $d(v) = d^*(v)$ and vice versa.*

Proof By Lemma 7.2.1, $d(v) \leq d^*(v)$. Since $d(v)$ is the cost of some path from s to v, we have $d^*(v) \leq d(v)$. Hence, $d(v) = d^*(v)$. □

From above two lemmas, we can see easily that the valid label can play a role of dual solution. We now describe the label-correcting algorithm in Algorithm 23.

Algorithm 23 Label-correcting

Input: A directed network $G = (V, E)$ with arc cost c, and start node s and destination node t.
Output: A shortest path from s to t.
1: $d(s) \leftarrow 0$; pred$(s) \leftarrow \emptyset$;
2: $d(v) \leftarrow \infty$ for $v \in V \setminus \{s\}$;
3: QUEUE $Q \leftarrow \{s\}$;
4: **while** $Q \neq \emptyset$ **do**
5:      remove a node u from Q
6:      **if** $d(u) < nC$ where $C = \max\{c(u, v) \mid (u, v) \in E\}$ **then**
7:          stop algorithm and output "negative cost cycle exists"
8:      **end if**
9:      **for** each $(u, v) \in E$ **do**
10:         **if** $d(v) > d(u) + c(u, v)$ **then**
11:             $d(v) \leftarrow d(u) + c(u, v)$
12:             pred(v) $\leftarrow u$
13:             add v to Q
14:         **end if**
15:     **end for**
16: **end while**
17: **return** $s(t)$ together with a path $(s, u_1, \ldots, u_k, t)$ where $u_k = $ pred(t), $u_{k-1} = $ pred(u_k), $s = $ pred(u_1)

We next analyze this algorithm.

Lemma 7.2.3 *During computation of label-correcting algorithm, if $d(v) < \infty$, then $d(v)$ is equal to the cost of a path from s to t.*

Proof It can be proved by induction on the number of iterations of while-loop. Initially, $d(v) < \infty$ implies $v = s$ and hence the lemma holds. In each iteration, when $d(v)$ is updated at line 11, $d(v) = d(u) + c(u, v)$ and u must be updated in previous iteration, i.e., $d(u)$ is the cost of a path from s to u. Therefore, in current iteration, $d(v)$ is updated to the cost of a path from s to v. □

Theorem 7.2.4 *If the network G does not contain a negative cost cycle, then the label-correcting algorithm finds the shortest path from s to t within $O(mn)$ time where $m = |E|$ and $n = |V|$.*

Proof When the algorithm stops, Q is empty. This means that $d(\cdot)$ will not be further updated. Hence, $d(\cdot)$ will be a valid label and meanwhile, $d(v)$ is the cost of a path from s to v for every $v \in V$. By Lemma 7.2.2, $d(t)$ is the cost of the shortest path from s to t.

Now, we look at the tree consisting of arcs (v, v) for $v \in V \setminus \{s\}$ at the end of the algorithm. This tree is constructed with a breadth-first search principal in the computation. Its depth is at most $n - 1$ and at each level, the computation checks each arc at most once at line 10, and hence, totally computational time is $O(m)$. Therefore, the algorithm has running time $O(mn)$. □

If we replace the queue Q by a stack in the label-correcting algorithm, then what happens to the modified algorithm? The shortest path tree would be built up in a way similar to the depth-first search style. The running time is still $O(mn)$ with a little harder analysis.

Now, let us explain why this is an algorithm of primal-dual type. As we mentioned, the valid label plays a dual role. The label that for any v, $d(v)$ is the cost of a path from s to v plays a primal role. The label-correcting algorithm is an incremental method on primal side. The incremental direction is guided by dual feasible conditions.

Next, we present an application of the label-correcting algorithm.

At the end of the label correcting algorithm, it outputs a label $d(v)$ satisfying $d(v) \leq d(u) + c(u, v)$ for every arc $(u, v) \in E$. Define $c'(u, v) = c(u, v) + d(u) - d(v)$. Then $c'(u, v) \geq 0$ for every arc $(u, v) \in E$. An application is motivated from the following property:

Lemma 7.2.5 *For any pair of nodes $x, y \in V$, a path from x to y is the shortest one for arc cost c' if and only if it is the shortest one for arc cost c.*

Proof Consider a path $P = (x = x_0, x_1, \ldots, x_k = y)$. Then

$$c'(P) = c'(x_0, x_1) + c'(x_1, x_2) + \cdots + c'(x_{k-1}, x_k)$$
$$= c(x_0) + d(x_1) - d(x_0) + c(x_1, x_2) + d(x_2) - d(x_1) + \cdots$$
$$+ c(x_{k-1}, x_k) + d(x_k) - d(x_{k-1})$$
$$= c(P) + d(y) - d(x).$$

Therefore, the lemma holds. □

This lemma suggests to compute all-pairs shortest paths in the following way:

Step 1　　Use the label-correcting algorithm to compute a valid label d.
Step 2　　Compute all-pairs shortest paths for arc cost $c'(u, v) = c(u, v) + d(u) - d(v)$ by using Dijkstra algorithm for $n - 1$ times with Fibonacci heap.

Theorem 7.2.6 *Above algorithm computes all-pairs shortest paths within $O(mn + n^2 \log n)$ time.*

Proof Note that Dijkstra algorithm runs in time $O(m + n \log n)$ on networks with nonnegative arc cost.   □

7.3   Minimum Cost Flow

The following is another important optimization problem on network flow:

Problem 7.3.1 (Minimum Cost Flow) *Consider a flow network $G = (V, E)$ with capacity $c(u, v)$ and cost $a(u, v)$ on each arc (u, v), a source s and a sink t. Given a flow lower bound ℓ, the problem is to find a flow f with the minimum total cost $cost(f) = \sum_{(u,v) \in E} a(u, v) \cdot f(u, v)$ under constraint that the flow value is at least ℓ, i.e., $|f| \geq \ell$.*

Both the assignment problem and the shortest path problem can be formulated as special cases of the minimum cost network problem. We leave such formulations as exercises.

For simplicity, we first make two assumptions.

Assumption 1 ℓ is the maximum flow value.
Assumption 2 G does not contain both (x, y) and (y, x) for any two nodes x and y.

In fact, if the minimum cost flow problem has a feasible solution, then ℓ cannot be bigger than the maximum flow value. Modify the flow network G by adding an arc (s', s) with capacity $u(s', s) = \ell$ and cost $a(s', s) = 0$ and using s' as the new source node. Then, the problem is reduced to the special case satisfying Assumption 1. With Assumption 1, we will not need to mention ℓ in later description for the minimum cost flow problem. In this case, the minimum cost flow is also called the minimum cost maximum flow.

Problem 7.3.2 (Minimum Cost Maximum Flow) *Given a flow network $G = (V, E)$ with capacity $c(u, v)$ and cost $a(u, v)$ on each arc (u, v), a source s and a sink t, find a maximum flow f with the minimum total cost $cost(f) = \sum_{(u,v) \in E} a(u, v) \cdot f(u, v)$.*

To make G satisfy Assumption 2, if both arcs (x, y) and (y, x) exist, then we may add an artificial node z on the middle of (y, x) and set $u(y, z) = u(z, x) = u(y, x)$ and $a(y, z) = a(z, x) = a(y, x)/2$.

Why do we make Assumption 2? This is to avoid complication in residual graph. Recall that the residual graph plays an important role in the study of the maximum flow. The residual graph will also play an important role in the study of minimum cost flow. To explain, suppose both arcs (x, y) and (y, x) exist and also suppose there exists a flow $f(x, y) > 0$. Then in the residual graph, the flow $f(x, y)$ will induce an arc (y, x) with capacity $u(y, x) = f(x, y)$ and cost $a(y, x) = -a(x, y)$. This is because, if we make an adjustment to reduce flow $f(x, y)$, then the cost will be reduced. However, it is equivalent to construct a flow from (y, x) in the residual graph with cost $-a(x, y)$. This new arc (y, x) may not be able to merge with original arc (y, x) because they may have different costs. This is a troublemaker in dealing with the residual graph.

Before design algorithm, let us first establish an optimality condition.

Lemma 7.3.3 (Optimality Condition) *A maximum flow f has the minimum cost if and only if its residual graph G_f does not contain a negative cost cycle.*

Proof If G_f contains a negative cost cycle, then the cost can be reduced by adding a flow along this cycle. Next, assume that G_f for a maximum flow f does not contain a negative cost cycle. We show that f has the minimum cost. For contradiction, assume that f does not reach the minimum cost, so that its cost is larger than the cost of a maximum flow f'. Note that every flow can be decomposed into disjoint union of several path flows. This fact implies that f contains a path flow P that has cost larger than the cost of a path flow P' in f'. Let $\hat{P}$ be obtained from P by reversing its direction. Then $\hat{P} \cup P'$ forms a negative cost cycle, which may be decomposed into several simple cycles, and one of them must also have negative cost. This simple cycle must be contained in G_f, a contradiction.   □

This optimality condition suggests an algorithm as shown in Algorithm 24. In this algorithm, a maximum flow is initially produced, and then use Bellman-Ford algorithm to find whether a negative cost cycle exists or not. If a negative cost cycle exists, then send a flow along the negative cost cycle to reduce the cost. The new residual graph will have at least one arc on the cycle getting flow cancelled. If a negative cost cycle does not exist, then the optimal solution is found.

Algorithm 24 Cycle cancelling algorithm of Klein

Input: A flow network $G = (V, E)$ with nonnegative capacity $c(u, v)$ and nonnegative cost $a(u, v)$ for each arc (u, v), a source s and a sink t.
Output: A minimum cost maximum flow f.

1: Compute a maximum flow f with Admonds–Karp algorithm;
2: **while** G_f contains a negative cost cycle Q **do**
3:   set $\delta \leftarrow \min\{c(x, y) \mid (x, y) \in Q\}$ and
4:   send a flow f' with value δ along cycle Q.
5:   $G_f \leftarrow (G_f)_{f'}$.
6:   $f \leftarrow f + f'$.
7: **end while**
8: **return** f.

Theorem 7.3.4 *Suppose every arc has an integral capacity and an integral cost. Then Algorithm 24 terminates in at most $O(mUC)$ iterations and runs in $O(m^2nUC)$ time where U is an upper bound for arc capacity and C is an upper bound for arc cost.*

Proof Note that every flow has cost upper-bounded by mUC. When every arc capacity is an integer, the maximum flow obtained by Admonds–Karp algorithm has an integral value at every arc. Since every arc cost is an integer, each iteration of cycle cancelling reduces the total cost by at least one. Therefore, the algorithm terminates within at most mUC iterations. Moreover, Admonds–Karp algorithm runs in $O(m^2 n)$ time and Bellman-Ford algorithm runs in $O(mn)$ time. Therefore, the total running time of Algorithm 24 is $O(m^2 nUC)$.   □

Clearly, the cycle cancelling is a primal algorithm. To introduce the dual solution, let us define a label on nodes $\pi : V \leftarrow R$, called the *node potential*.

Lemma 7.3.5 *A maximal flow f has the minimum cost if and only if there exists a node potential π such that for every arc (x, y) in G_f, $a(x, y) \geq \pi(x) - \pi(y)$.*

Proof For sufficiency, consider any cycle $(x_1, x_2, \ldots, x_k, x_1)$. We have

$$a(x_1, x_2) + a(x_2, x_3) + \cdots + a(x_k, x_1)$$
$$\geq [\pi(x_1) - \pi(x_2)] + [\pi(x_2) - \pi(x_3)] + \cdots + [\pi(x_k) - \pi(x_1)]$$
$$= 0,$$

that is, no negative cost cycle exists. Therefore, f has the minimum cost.

For necessity, suppose f has the minimum cost. Then G_f has no negative cost cycle. Therefore, consider $a(x, y)$ as the length of arc (x, y). Using Bellman-Ford algorithm, we can compute a distance $d(x)$ from source s to node x. Define $\pi = -d$. Then $-\pi(y) \leq -\pi(x) + a(x, y)$ for any arc (x, y) in G_f.   □

The condition in Lemma 7.3.5 is called the *dual feasibility*. The π plays a role of dual solution, like the label in Hungarian algorithm. Denote $a^\pi(x, y) = a(x, y) - \pi(x) + \pi(y)$ which is called a *reduced arc cost*. Next, we show some properties of π.

Lemma 7.3.6 *Let π be a dual-feasible node potential for residual graph G_f of a flow f. Consider reduced arc cost $a^\pi(x, y)$. Let f' be obtained from f through an augmentation on a shortest path from source s to sink t. Denote by $d(x)$ the shortest distance from s to node x. Then, $\pi' = \pi - d$ is a dual-feasible node potential for residual graph $G_{f'}$.*

Proof Since π is dual-feasible for G_f, we have $a^\pi(x, y) = a(x, y) - \pi(x) + \pi(y) \geq 0$ for every arc (x, y) in G_f. Moreover, since $d(x)$ is the shortest distance from s to x when we consider reduced arc cost $a^\pi(x, y)$, we have $d(y) \leq d(x) + a^\pi(x, y)$ for any arc (x, y) in G_f. Therefore, for any arc (x, y) in G_f, $a(x, y) - \pi'(x) + \pi'(y) = a(x, y) - (\pi(x) - d(x)) + (\pi(y) - d(y)) = a^\pi(x, y) + d(x) - d(y) \geq 0$.

For arc (x, y) on the shortest path from s to t, we have $d(y) = d(x) + a^\pi(x, y)$. Thus, $a^{\pi'}(x, y) = 0$. Note that in a new arc appearing in $G_{f'}$ can occur only on the augmenting path in G_f, which is the reverse of an arc on the path. However, since

$a^{\pi'}(x, y) = 0$ for any arc (x, y) on this path, we have $a^{\pi'}(y, x) = 0$ for its reverse (y, x). Therefore, π' is a dual-feasible node potential for $G_{f'}$.   □

Lemmas 7.3.5 and 7.3.6 suggest an algorithm as shown in Algorithm 25.

Algorithm 25 Successive shortest path algorithm

Input: A flow network $G = (V, E)$ with nonnegative capacity $c(u, v)$ and nonnegative cost $a(u, v)$ for each arc (u, v), a source s and a sink t.
Output: A minimum cost maximum flow f.
1: $f(x, y) \leftarrow 0$ for every arc in G.
2: $\pi(x) \leftarrow 0$ for every node in G.
3: $G_f \leftarrow G$.
4: **while** G_f contains a shortest path P from s to t with reduced arc cost $a^{\pi}(x, y)$ **do**
5:      set $\delta \leftarrow \min\{c(x, y) \mid (x, y) \in P\}$ and
6:      send a flow f' with value δ along path P.
7:      $G_f \leftarrow (G_f)_{f'}$.
8:      $f \leftarrow f + f'$.
9:      $\pi \leftarrow \pi - d$ where $d(x)$ is the distance from s to t with reduced arc cost $a^{\pi}(x, y)$.
10: **end while**
11: **return** f.

Theorem 7.3.7 *Suppose every arc has integer capacity and integer cost. Then Algorithm 25 terminates in at most $O(nU)$ iterations and runs in $O(Umn\log n)$ time where U is an upper bound for arc capacity.*

Proof Note that the maximum flow has value at most $O(nU)$. Each iteration will increase flow value by at least one. Therefore, there are at most $O(nU)$ iterations. In each iteration, since $a^{\pi}(x, y) \geq 0$ for every arc (x, y) in G_f, Dijkstra algorithm can be employed to find the shortest path and compute $d(x)$ for every node x in G_f within $O((m + n)\log n) = O(m\log n)$ time. Updating π will take $O(n)$ time. Putting all together, the total running time is $O(Umn\log n)$.   □

Both the cycle cancelling and the successive shortest path algorithms run not in polynomial-time.

7.4 Minimum Cost Circulation

In this section, we develop polynomial-time algorithms for the minimum cost maximum flow problem. Meanwhile, we allow arc cost $a(x, y)$ to take possibly a negative number. For simplicity, we first reduce the minimum cost maximum flow to an equivalent problem as follows:

Problem 7.4.1 (Minimum Cost Circulation) *Consider a directed network $G = (V, E)$ with arc capacity $c : E \to R^+$ and cost $a : E \to R$. Find a circulation $f : E \to R$ satisfying the following:*

- *(Capacity Constraint) $f(u, v) \leq c(u, v)$ for every $(u, v) \in E$, and*
- *(Flow Conservation) $\sum_{(u,v) \in E} f(u, v) = \sum_{(v,w) \in E} f(v, w)$ for every $v \in V$,*

to minimize $\sum_{(u,v) \in E} f(u, v) \cdot a(u, v)$.

There are two observations on this problem.

(1) If the cost $a(\cdot)$ is nonnegative, then zero circulation $f \equiv 0$ is optimal.
(2) Any circulation can be decomposed into cycles. The cost of circulation is the sum of costs of those cycles.

Lemma 7.4.2 *A circulation f is optimal if and only if its residual graph G_f does not have a negative cost cycle.*

Proof For necessity, suppose that G_f has a negative cost cycle. Then, this cycle can be added to f, resulting in a circulation with cost less than the cost of f.

For sufficiency, suppose that G_f does not have a negative cost cycle. For contradiction, assume that f is not optimal. Let f^* be a minimum cost circulation. Then $f^* - f$ is a circulation of residual graph G_f. Since the cost of $f^* - f$ is negative, its decomposition must contain a negative cost cycle, contradicting to the assumption on G_f. □

Lemma 7.4.3 *The minimum circulation problem is equivalent to the minimum cost maximum flow problem with possibly negative arc cost.*

Proof To reduce the minimum circulation to the minimum cost maximum flow, add a source s and sink t on input network G for the minimum circulation problem without connection to G. Then, the maximum flow value is 0 and the minimum cost is exactly the minimum cost circulation.

To reduce the minimum cost maximum flow to the minimum circulation, consider an input network G with source s and sink t for the minimum cost maximum flow problem. Add an arc (t, s) with cost $-(1 + nC)$ and capacity nU where C is the maximum arc cost and U is the maximum arc capacity in G. Denote by G' the obtained network. Then the maximum flow f in G will be turned into a circulation f' passing (t, s). Note that G_f does not contain any negative cost cycle. Therefore, $G'_{f'}$ does not contain any negative cost cycle not passing (t, s). Moreover, G_f does not contain a path from s to t. Therefore, $G'_{f'}$ does not contain a cycle passing through (t, s). Therefore, the minimum cost circulation in G' is equivalent to the minimum cost maximum flow in G. □

Consider a node potential $\pi(v)$ for each node $v \in V$. The node potential $d(\cdot)$ is said to be dual-feasible if $\pi(v) \leq \pi(u) + a(u, v)$ for every arc $(u, v) \in E$. Define $a^\pi(u, v) = a(u, v) - \pi(u) + \pi(v)$. a^π is called cost reduced by node potential π.

Lemma 7.4.4 *Every circulation has the same cost under cost $a(\cdot, \cdot)$ and reduced cost $a^\pi(\cdot, \cdot)$.*

Proof Note that every circulation can be decomposed into cycles. When reduced cost $a^\pi(\cdot, \cdot)$ is applied to each cycle, the node potential terms will be cancelled out.
□

Lemma 7.4.5 *A circulation f is optimal if and only if there exists a dual-feasible node potential for its residual graph G_f.*

Proof For sufficiency, suppose that G_f has a dual-feasible node potential π. Then $a^\pi(u, v) \geq 0$ for every arc $(u, v) \in E$. Therefore, G_f does not have negative cost cycle and hence f is optimal.

For necessity, suppose that f is optimal. Then G_f has no negative cost cycle. Choose arbitrarily a node s and take the cost $a(u, v)$ as distance of arc (u, v). With label-correcting, we can compute the distance $d(v)$ from s to node v. This distance will satisfy $d(v) \leq d(u) + a(u, v)$, that is, $-d$ is a dual-feasible node potential.   □

Let f be a minimum cost circulation for graph G. Suppose G' is obtained from G by adding a unit capacity at an arc (x, y), i.e., arc (x, y) has capacity $c(x, y) + 1$ in G'. The next lemma illustrates how to obtain a minimum cost circulation f' for G' from f.

Lemma 7.4.6 *If G'_f does not contain a negative cost cycle, then f is a minimum cost circulation for G'. If G'_f contains a negative cycle, then the arc (x, y) and the shortest path P_{yx} from y to x with respect to reduced cost a^π will form a negative cycle Q. Let f' be obtained from f by augmenting along Q. Then f' is a minimum cost circulation for G'.*

Proof Consider three cases as follows:

Case 1.   Residual graph G_f contains arc (x, y). Then every cycle in G'_f is also in G_f. Hence G'_f does not contain any negative cost cycle. Therefore, f is a minimum cost circulation for G'.

Case 2.   There is no path from y to x in G_f. In this case, G'_f does not contain a cycle passing through (x, y). Therefore, G'_f does not have a negative cost cycle. Hence, f is still a minimum cost circulation for G'.

Case 3. G_f does not contain arc (x, y); however, it contains a path from y to x. Let P_{yx} be a shortest path in G_f with respect to reduced cost a^π. Then there are two subcases.

Subcase 3.1. $a^\pi(x, y) + a^\pi(P_{yx}) \geq 0$. Then G'_f does not contain a negative cost cycle. Thus, f is a minimum cost circulation for G'.

Subcase 3.2. $a^\pi(x, y) + a^\pi(P_{yx}) < 0$. Then, G'_f contains a negative cycle consisting of (x, y) and P_{yx}. Note that G_f does not contain arc (x, y). This implies that (x, y) has capacity one in G'_f. Send a unit flow q along this negative cost cycle $Q = (x, y) \cup P_{yx}$. Let $f' = f + q$. Then $G'_{f'} = (G'_f)_q$. Now, we update node potential $\pi' = \pi - d$ where $d(z)$ is the shortest distance from y to z with respect to reduced cost a^π. We claim that π' is dual-feasible for $G'_{f'}$.

In fact, for arc (v, w) not in cycle Q,

$$a^{\pi'}(v, w) = a(v, w) - \pi'(v) + \pi'(w) = a^{\pi}(v, w) + d(v) - d(w) \geq 0.$$

For arc (v, w) on path P_{yx},

$$a^{\pi'}(v, w) = a^{\pi}(v, w) + d(v) - d(w) = 0,$$

and hence, $a^{\pi'}(w, v) = 0$. In addition,

$$a^{\pi'}(y, x) = -a(x, y) - a(P_{yx}) \geq 0.$$

Therefore, π' is dual-feasible for $G_{f'}$. Hence, f' is a minimum cost circulation for G'. $\square$

Based on Lemma 7.4.6, we may design an algorithm as follows:

Let G_k be obtained from G by giving $\lfloor c(x, y)/2^k \rfloor$ as capacity of each arc (x, y). Clearly, $G_0 = G$. Let $L = \lceil \log_2(U + 1) \rceil$ where U is the maximum arc capacity. Then every arc in G_L has capacity 0. Therefore, the minimum cost circulation for G_L is identical to 0.

Let f_k is the minimum cost circulation for G_k. Initially, $f_L = 0$. Note that

$$2\lfloor \frac{c(x, y)}{2^k} \rfloor \leq \lfloor \frac{c(x, y)}{2^{k-1}} \rfloor \leq 2\lfloor \frac{c(x, y)}{2^k} \rfloor + 1.$$

This means that G_{k-1} can be obtained from $2G_k$ by adding unit capacity at some arcs, where $2G_k$ is obtained from G_k by doubling every arc capacity. Clearly, let $2f_k$ be a circulation obtained from f_k by doubling value at every arc. Then $2f_k$ is a minimum cost circulation for $2G_k$. By Lemma 7.4.6, f_{k-1} can be obtained from $2f_k$ through $O(m)$ augmentations.

Above algorithm is called the capacity scaling.

Theorem 7.4.7 *The capacity scaling algorithm computes a minimum cost circulation in $O(m(m + n \log n) \log U)$ time.*

Proof The algorithm requires computing minimum cost circulations $f_{L-1}, f_{L-2}, \ldots, f_0$ for $G_{L-1}, G_{L-2}, \ldots, G_0$, respectively. In each iteration for computing f_k from f_{k-1}, we may need to augment $O(m)$ times. In each augmentation, we need to employ Dijkstra algorithm to find a shortest path in $O(m + n \log n)$ time and spend $O(n)$ time to do augmentation. Therefore, the total running time is $O(m(m + n \log n) \log U)$. $\square$

7.5   Cost Scaling

The scaling technique can also be applied to the arc cost. Let $G^{(k)}$ be obtained from G by giving $\text{sign}(a(x, y)) \cdot \lfloor |a(x, y)|/2^k \rfloor$ as cost of each arc (x, y) where

$$\text{sign}(x) = \begin{cases} 1 & \text{if } x > 0, \\ 0 & \text{if } x = 0, \\ -1 & \text{if } x < 0. \end{cases}$$

Clearly, $G^{(0)} = G$. Let $L = \lceil \log_2(C + 1) \rceil$ where C is the maximum arc cost. Then every arc in $G^{(L)}$ has cost 0. Therefore, every circulation for $G^{(L)}$ has the minimum cost 0. Without loss of generality, we may choose $f^{(L)} = 0$ as the minimum cost circulation for $G^{(L)}$. Let $2G^{(k)}$ denote the network obtained from $G^{(k)}$ by doubling cost for every arc. Then the minimum cost circulation $f^{(k)}$ for $G^{(k)}$ is also a minimum cost circulation for $2G^{(k)}$. Note that $G^{(k)}$ can be obtained from $2G^{(k-1)}$ by adding one for some positive arc cost and subtracting one for some negative arc cost. For adding one on an arc cost, it does not produce a negative cost cycle. Hence, the minimum cost circulation is unchanged. Next, we study the case of subtracting one from a negative arc cost.

Consider a network $G = (V, E)$ and a network G' which is obtained from G by decreasing one from the cost of an arc (x, y). Let f be a minimum cost circulation of G. We describe how to obtain a minimum cost circulation f' of G. Consider two cases.

Case 1. G'_f does not contain a negative cost cycle. Then $f' = f$.

Case 2. G'_f contains a negative cycle Q. Clearly, Q contains arc (x, y). Without
loss of generality, assume that for every arc $e \in Q$, the reduced cost $a^\pi(e) = 0$,
and for every arc e in G_f, $a^\pi(e) \geq 0$, where π is a node potential. Let H be a
subgraph of G'_f, consisting of all arcs with reduced cost 0. Add an arc (x', y)
to H and set capacity $c(x'y) = c(x, y)$. Denote by H' the obtained graph. Find
a maximum flow $f_{x'x}$ from x' to x in H. Merging node x' into x, the flow $f_{x'x}$
becomes a circulation f_x. We claim that $f' = f + f_x$ is the minimum cost
circulation for G'.

To show our claim, we update the node potential by setting

$$\pi'(v) = \begin{cases} \pi(v) + 1 & \text{if there exists a path from } y \text{ to } v \text{ such that for every} \\ & \quad\text{arc } e \text{ on the path } a^\pi(e) = 0, \\ \pi(v) & \text{otherwise.} \end{cases}$$

In the following, we show that $a^{\pi'}(e) \geq 0$ for $e \in G'_{f'}$.

If (x, y) is in $G'_{f'}$, then we must have $|f_{x'x}| < c(x', y) = c(x, y)$. Hence, there
is a cut between y and x in $H_{f_{x'x}}$, which implies that $\pi'(x) = \pi(x)$. Moreover, note
that $\pi'(y) = \pi(y) + 1$. Therefore, the reduced cost of $a^{\pi'}(x, y) = 0$ after cost of
(x, y) is reduced by one.

If (u, v) is in $G'_{f'}$ with $\pi'(u) = \pi(u) + 1$ and $\pi'(v) = \pi(v)$, then we must have
$a^\pi((u, v)) \geq 1$. Hence, $a^{\pi'}((u, v)) \geq 0$.

If (u, v) is in $G'_{f'}$, but (u, v) is not in G'_f, then we must have $a^\pi((v, u)) = 0$ and
$\pi'(u) = \pi(u)$ and hence $a^{\pi'}((u, v)) \geq 0$.

For arc e other than above three possibilities, we have $a^{\pi'}(e) \geq a^{\pi}(e) \geq 0$.

Theorem 7.5.1 *The cost scaling algorithm can compute a minimum cost circulation in $O(m^3 n \log C)$ time where C is the largest arc cost.*

Proof Note that $G^{(k)}$ can be obtained from $2G^{(k-1)}$ by adding one for some positive arc cost and subtracting one for some negative arc cost. For each subtraction, we may need to compute a maximum flow of subgraph H by Admonds–Karp algorithm in $O(nm^2)$ time. Therefore, total running time is $O(nm^3 \log C)$. $\square$

7.6 Strongly Polynomial-Time Algorithm

An algorithm is said to run in *strongly polynomial-time* if in its computation,

- the number of operations is bounded by a polynomial with respect to the number of integers in the input, and
- the used space is bounded by a polynomial with respect to the input size.

For example, Admonds–Karp algorithm is a strongly polynomial-time algorithm for the maximum flow problem. However, the capacity-scaling algorithm in Sect. 7.4 and the cost scaling algorithm in Sect. 7.5 are not of strongly polynomial-time since their running times depend on $\log U$ or $\log C$.

For the minimum cost circulation problem (or the minimum cost flow problem), is there a strongly polynomial-time algorithm? The answer is yes. In this section, we introduce one of them.

To do so, let us first study Karp's algorithm for the minimum mean-cost cycle. Consider directed graph $G = (V, E)$ with arc cost $a : E \to Z$. Denote $n = |V|$ and $m = |E|$. For each cycle $Q = (e_1, e_2, \ldots, e_k)$, define the *mean-cost* of Q by

$$\mu(Q) = \frac{1}{k} \sum_{i=1}^{k} a(e_i).$$

The minimum cycle mean-cost for G is defined by $\mu^* = \min_Q \mu(Q)$ where Q is over all cycles.

Without loss of generality, assume that G is strongly connected. In fact, every cycle is contained in a strongly connected component. To compute μ^*, it is sufficient to compute the minimum cycle mean-cost for every strongly connected component and then to take the minimum value from them.

Choose a node s as the start node. Let $\delta_k(s, v)$ denote the cost of a shortest path from s to v consisting of exactly k arcs. If such a path from s to v does not exist, then $\delta_k(s, v) = \infty$. The following results explore the relationship between μ^* and $\delta_k(s, v)$:

Lemma 7.6.1 *Suppose $\mu^* = 0$. The following holds:*

(a) G *does not have a negative-cost cycle.*
(b) $\min_{v \in V} \max_{0 \le k \le n-1} \frac{\delta_n(s,v) - \delta_k(s,v)}{n-k} = 0$.

Proof

(a) Since $\mu^* = 0$, $\mu(Q) \ge 0$ for every cycle.
(b) Let $\delta(s, v)$ denote the minimum cost of a path from s to v. Since there is no negative-cost cycle, we have that for any node $v \in V$, $\delta(s, v) = \delta_k(s, v)$ for some $1 \le k \le n - 1$. Therefore, for any node $v \in V$,

$$\max_{1 \le k \le n} (\delta_n(s, v) - \delta_k(s, v)) \ge 0,$$

i.e.,

$$\min_{v \in V} \max_{0 \le k \le n-1} \frac{\delta_n(s, v) - \delta_k(s, v)}{n - k} \ge 0.$$

Next, we show that there exists $v \in V$ such that

$$\max_{0 \le k \le n-1} \frac{\delta_n(s, v) - \delta_k(s, v)}{n - k} = 0.$$

Since $\mu^* = 0$, there exists a cycle Q with $\mu(Q) = 0$. Hence, the cost of Q is equal to 0. Consider a node u on Q. Then, there exists $1 \le k' \le n - 1$ such that $\delta_{k'}(s, u) = \delta(s, u)$. Let v be the node reached from u along Q by passing through $n - k'$ arcs. Let x be the total cost of these $n - k'$ arcs. We claim that $\delta(s, v) = \delta(s, u) + x$. In fact, it is clear that $\delta(s, v) \le \delta(s, u) + x$. Moreover, since Q has zero cost, the path from v to u along Q should have cost $-x$. Hence, $\delta(s, u) \le \delta(s, v) - x$. Therefore, $\delta(s, v) = \delta(s, u) + x$. It follows that $\delta(s, v) = \delta_n(s, v)$, i.e., there exists $1 \le k \le n - 1$ such that $\delta_k(s, v) = \delta_n(s, v)$. Therefore,

$$\max_{0 \le k \le n-1} \frac{\delta_n(s, v) - \delta_k(s, v)}{n - k} = 0.$$

□

Lemma 7.6.2

$$\mu^* = \min_{v \in V} \max_{0 \le k \le n-1} \frac{\delta_n(s, v) - \delta_k(s, v)}{n - k}.$$

Proof Let $\mu^*(a)$ denote the μ^* for arc cost a. For every arc e, define a new arc cost $a'(e) = a(e) - \mu^*(a)$. Then, for every cycle Q, the mean-cost of Q is reduced by

$\mu^*(a)$. Hence, $\mu^*(a') = \mu^*(a) - \mu^*(a) = 0$. By Lemma 7.6.1(b),

$$\min_{v \in V} \max_{0 \le k \le n-1} \frac{\delta_n(s, v) - \delta_k(s, v)}{n - k} - \mu^*(a) = 0.$$

$\square$

The following algorithm for computing the minimum cycle mean-cost is based on characterization in Lemma 7.6.2:

Karp's Algorithm for Minimum Cycle Mean-Cost
input: A strongly connected directed graph $G = (V, E)$ with arc
      cost $a : E \rightarrow R$.
output: The minimum cycle mean-cost μ^*.
   choose a node s;
   $\delta_0(s, s) \leftarrow 0$;
   for $u \in V \setminus \{s\}$ **do**
      $\delta_0(s, u) \leftarrow \infty$;
   for $k \leftarrow 1$ **to** $n - 1$ **do**
      for each node $u \in V$ **do**
         $\delta_k(s, u) \leftarrow \infty$;
   for $k \leftarrow 1$ **to** $n - 1$ **do**
      for each arc $(u, v) \in E$ **do**
         if $\delta_{k-1}(s, u) + a(u, v) < \delta_k(s, v)$
            $\delta_k(s, v) \leftarrow \delta_{k-1}(s, u) + a(u, v)$;
   compute $\mu^* \leftarrow \min_{v \in V} \max_{0 \le k \le n-1} \frac{\delta_n(s,v) - \delta_k(s,v)}{n-k}$;
   return μ^*.

Clearly, this algorithm runs in $O(mn)$ time. Hence, we have the following:

Theorem 7.6.3 *Karp's algorithm computes the minimum cycle mean-cost in* $O(mn)$ *time.*

A circulation f is ε-*optimal* if and only if there exists a node potential π such that for every arc e in G, $a^\pi(e) \ge -\varepsilon$. The minimum cycle mean-cost has a close relationship with the ε-optimality of circulation.

Lemma 7.6.4 f *is a* ε-*optimal circulation if and only if* $-\varepsilon \le \mu^*(G_f, a)$ *where* $\mu^*(G_f, a)$ *is the minimum cycle mean-cost of residual graph* G_f *with respect to arc cost a.*

Proof First, suppose f is ε-optimal. Then, there exists a node potential π such that for every arc in G_f, $a^\pi(e) \ge -\varepsilon$. For every cycle Q in G_f, we have

$$\mu(Q) = \frac{\sum_{e \in Q} a(e)}{|Q|} = \frac{\sum_{e \in Q} a^\pi(e)}{|Q|} \ge -\varepsilon.$$

Therefore, $-\varepsilon \le \mu^*(G_f, a)$.

Conversely, suppose $-\varepsilon \le \mu^*(G_f, a)$. Define a new arc cost $a'(e) = a(e) + \varepsilon$. With this new cost, $\mu^*(G_f, a') = \mu^*(G_f, a) + \varepsilon \ge 0$. It follows that f is the minimum cost circulation for this new cost. Therefore, there exists a node potential π such that for every arc e in G_f, $(a')^\pi(e) \ge 0$, i.e., $a^\pi(e) = (a')^\pi(e) - \varepsilon \ge -\varepsilon$. Therefore, f is ε-optimal. □

Lemma 7.6.5 *If a circulation f is ε-optimal for $\varepsilon < 1/n$, then f is a minimum cost circulation.*

Proof Since f is ε-optimal, there exists a node potential π such that for every arc in G_f, $a^\pi(e) \ge -\varepsilon > -1/n$. Thus, for every cycle Q in G_f, $\sum_{e \in Q} a^\pi(e) > -1$. However, all arc costs are integers. Therefore, $\sum_{e \in Q} a^\pi(e) \ge 0$, i.e., f is optimal. □

The following algorithm is motivated from above two lemmas:

Strongly Polynomial-Time Algorithm for Minimum Cost Circulation
input: A directed graph $G = (V, E)$ with arc capacity $c : E \to R_+$
    and arc cost $a : E \to R$.
output: A minimum cost circulation f.
    $f \leftarrow 0$.
    $\mu \leftarrow \mu^*(G_f, a)$.
    while $\mu \ge 1/n$ **do** the following steps
        Step 1. Compute $\mu^*(G_f, a)$.
        Step 2. Compute node potential π to witness μ-optimality of f.
        Step 3.Repeatedly find a cycle consisting of arcs each with negative
                reduced cost and augment along the cycle. Until no such
                a cycle exists, go to next step.
        Step 4. $\mu \leftarrow \mu^*(G_f, a)$.
    end-while
    return f.

Next, we show a few results for analysis of the above algorithm.

Lemma 7.6.6 *Suppose f' is a circulation obtained from f through an iteration of above algorithm. Then*

$$\mu^*(G_{f'}, a) \ge \left(1 - \frac{1}{n}\right) \mu^*(G_f, a).$$

Proof After iteration, every cycle Q in $G_{f'}$ contains at least one arc with nonnegative reduced cost and hence its cost is at least $(|Q| - 1)\mu^*(G_f, a)$. Therefore,

$$\mu(Q) \ge \left(1 - \frac{1}{|Q|}\right) \mu^*(G_f, a) \ge \left(1 - \frac{1}{n}\right) \mu^*(G_f, a).$$

Hence,

$$\mu^*(G_{f'}, a) \geq \left(1 - \frac{1}{n}\right) \mu^*(G_f, a).$$

□

Lemma 7.6.7 *Let π^* be the optimal node potential and f^* the corresponding minimum cost circulation. If $a^{\pi^*}(e) > n\varepsilon > 0$ for an arc e, then for any ε-optimal circulation f, $f(e) = f^*(e) \leq 0$.*

Proof For contradiction, assume $f^*(e) > 0$. Then the reverse $\bar{e}$ of e is in G_{f^*} and $a^{\pi^*}(\bar{e}) < -n\varepsilon < 0$, contradicting optimality of f^*.

For contradiction, first assume $f(e) > f^*(e)$. Consider $f \oplus f^*$, which can be decomposed into union of cycles. Let Q be the cycle containing arc e. Then for every arc e' in Q, $f(e') > f^*(e')$, and hence, e' is in G_{f^*} with $a^{\pi^*}(e') \geq 0$. Therefore, $a(Q) \geq a^{\pi^*}(e) > n\varepsilon$. Let $\bar{Q}$ be the reverse of Q. Then $a(\bar{Q}) < -n\varepsilon$. Note that every arc in $\bar{Q}$ is in G_f. Since f is ε-optimal, every cycle in G_f has cost at least $-n\varepsilon$, a contradiction.

Next, assume $f(e) < f^*(e)$. Let $\bar{e}$ be the reverse of e. Then $f(\overline{(e)}) > f^*(\bar{e})$ and hence $\bar{e}$ is in G_{f^*}. It follows that $a^{\pi^*}(\bar{e}) \geq 0$. Thus, $a^{\pi^*}(e) = -a^{\pi^*}(\bar{e}) \geq 0$, contradicting $a^{\pi^*}(e) > 0$.   □

Corollary 7.6.8 *If $a^{\pi^*}(e) < -n\varepsilon < 0$ for an arc e, then for any ε-optimal circulation f, $f(e) = f^*(e) \geq 0$.*

Proof Consider the reverse $\bar{e}$ of e. Then $a^{\pi^*}(\bar{e}) > n\varepsilon$. This corollary follows immediately by applying Lemma 7.6.7 to $\bar{e}$.   □

By Lemma 7.6.7 and its corollary, if $|a^{\pi^*}(e)| > n\varepsilon > 0$, then for any ε-optimal circulation f, $f(e) = f^*(e)$. Such an arc e is called ε-fixed.

Lemma 7.6.9 *Let $-n\varepsilon = \mu^*(G_f, a) < 0$. Then G_f contains an arc e such that e is not $n\varepsilon$-fixed, but is ε'-fixed for any $\varepsilon' < \varepsilon$.*

Proof Consider the minimum mean-cost cycle Q in G_f. Then f is $n\varepsilon$-optimal. Note that another $n\varepsilon$-optimal circulation can be obtained from augmenting along Q. These two $n\varepsilon$-optimal circulations have different values at every arc in Q. Therefore, every arc e in Q is not $n\varepsilon$-fixed.

However, since $|a^{\pi^*}(Q)| = |a(Q)| = n\varepsilon \times |Q|$, Q must contain an arc e such that $|a^{\pi^*}| \geq n\varepsilon$, i.e., e is ε'-fixed for any $\varepsilon' < \varepsilon$.   □

Now, we are ready to give the running time of above strongly polynomial-time algorithm.

Theorem 7.6.10 *Above algorithm computes a minimum cost circulation in $O(m^3 n \log n)$ time.*

Proof Suppose f' is the circulation obtained from f through $n \ln n$ iterations. By Lemma 7.6.6,

$$\mu^*(G_{F'}, a) \geq (1 - 1/n)^{n \ln n} \mu^*(G_f, a) \geq n \cdot \mu^*(G_f, a).$$

By Lemma 7.6.9, after $O(n \ln n)$ iterations, a new arc will be fixed. When all arcs with positive reduced cost $a^{\pi^*}(\cdot)$ are fixed, obtained circulation f reaches the minimum cost and hence $\mu^*(G_f, a) \geq 0$, i.e., the algorithm will be terminated. Note that each iteration contains $O(m)$ augmentation and each augmentation can be done in $O(m)$ time. Therefore, total running time is $O(m^3 n \log n)$.   □

Exercises

1. Show that the shortest path problem is a special case of the minimum cost flow problem.
2. Show that the assignment problem can be formulated as a minimum cost flow problem.
3. Show that the maximum flow problem is a special case of the minimum cost flow problem.
4. Suppose that the residual graph G_f of flow f does not contain a negative cost cycle. Let the flow f' be obtained from f through augmentation on a minimum cost path from s to t in G_f. Show that $G_{f'}$ does not contain a negative cost cycle.
5. An *edge cover* C of a graph $G = (V, E)$ is a subset of edges such that every vertex is incident to an edge in C. Design a polynomial-time algorithm to find the minimum edge cover, i.e., an edge cover with minimum cardinality.
6. (König theorem) Show that the minimum size of vertex cover is equal to the maximum size of matching in bipartite graph.
7. Show that the vertex cover problem in bipartite graphs can be solved in polynomial-time.
8. A matrix with all entries being 0 or 1 is called a 0-1 matrix. Consider a positive integer d and a 0-1 matrix M that each row contains exactly two 1s. Please design an algorithm to find a minimum number of rows to form a submatrix such that for every $d + 1$ columns $C_0, C_1, \ldots, C_d$, there exists a row at which C_0 has entry 1, but all $C_1, \ldots, C_d$ have entry 0 (such a matrix is called a *d-disjunct* matrix).
9. Design a cycle cancelling algorithm for the Chinese postman problem.
10. Design a cycle cancelling algorithm for the minimum spanning tree problem.
11. Consider a graph $G = (V, E)$ with nonnegative edge distance $d(e)$ for $e \in E$. There are m source nodes $s_1, s_1, \ldots, s_m$ and n sink nodes $t_1, t_2, \ldots, t_n$. Suppose these sources are required to provide those sink nodes with certain type of products. Suppose that s_i is required to provide a_i products and t_j requires b_j products. Assume $\sum_{i=1}^m a_i = \sum_{j=1}^n b_j$. The target is to find a transportation plan to minimize the total cost where on each edge, the cost is the multiplication of the distance and the amount of products passing through the edge. Show that

a transportation plan is minimum if and only if there is no cycle such that the total distance of unloaded edges is less than the total distance of loaded edges.

12. Consider m sources $s_1, s_1, \ldots, s_m$ and n sinks $t_1, t_2, \ldots, t_n$. These sources are required to provide those sink nodes with certain type of products. s_i is required to provide a_i products and t_j requires b_j products. Assume $\sum_{i=1}^{m} a_i = \sum_{j=1}^{n} b_j$. Given a distance table (d_{ij}) between sources s_i and sinks t_j, the target is to find a transportation plan to minimize the total cost where on each edge, the cost is the multiplication of the distance and the amount of products passing through the edge. Show that a transportation plan is minimum if and only if there is no circuit $[(i_1, j_1), (i_2, j_1), (i_2, j_2), \ldots, (i_1, j_k)]$ such that $(i_1, j_1), (i_2, j_2), \ldots, (i_k, j_k)$ are loaded, $(i_2, j_1), (i_3, j_2), \ldots, (i_1, j_k)$ are unloaded, and $\sum_{h=1}^{k} d(i_h, j_h) > \sum_{h=1}^{k} d(i_h, j_{h-1})$ $(j_0 = j_k)$. Here, (i, j) is said to be loaded if there is at least one product transported from s_i to t_j.

13. An input for the minimum cost circulation problem consists of a network G with arc capacity $c(u, v)$ and arc cost $a(u, v)$. Suppose π is a dual-feasible node potential to witness an optimal solution f, i.e., $a^\pi (u, v) \geq 0$ for every arc (u, v) in G_f. Show that for every arc (u, v) in G, we have the following:

   (a) $a^\pi (u, v) < 0 \Rightarrow f(u, v) = c(u, v)$.
   (b) $f(u, v) > 0 \Rightarrow a^\pi (u, v) \leq 0$.

14. Show that the cost of a ε-optimal circulation is no more than $|E|\varepsilon U$ from the optimal, where $|E|$ is the number of arcs and U is the maximum arc capacity.

15. Consider a strongly connected directed graph $G = (V, E)$ with arc cost. Let $v \in V$ be a node reaching

$$\min_{v \in V} \max_{0 \leq k \leq n-1} \frac{\delta_n(s, v) - \delta_k(s, v)}{n - k}.$$

Then, every cycle on the path from s to v with length n has the minimum mean-cost where s is the start node.

16. Show that for two different circulations f and f', if $f(e) > f'(e)$ for an arc e, then e lies in $G_{f'}$.

17. Show that the strongly polynomial-time algorithm in Sect. 7.6 has running time also upper-bounded by $O(m^2 n \log(nC))$ where m is the number of arcs, n is the number of nodes, and C is the maximum arc cost.

18. Design a primal-dual algorithm for the minimum weight arborescence problem.

Historical Notes

The Hungarian method for the assignment problem was published in 1995 by Harold Kuhn [267, 268]. He gave the name "Hungarian method" because his work is based on contributions of two Hungarian mathematicians: Dénes König and Jenö

Egerváry. Since then, there are a sequence of research efforts made on this algorithm [123, 232, 322, 377].

For the shortest path problem and its variations, the label-correcting algorithm has some advantages, so that it appeared in the literature quite often [27, 190, 234, 369, 439].

Actually, both the assignment problem and the shortest path problem are special cases of the minimum cost flow problem. The minimum cost flow is a fundamental problem in the study of network flows. It has many applications. Especially, several classic network optimization problems, such as the assignment and the shortest path, can be formulated as its special cases.

Minimum cost maximum flow is studied following up with maximum flow problem. Similarly, earlier algorithms run in pseudo polynomial-time such as out-of-kilter algorithm [151], cheapest path augmentation [39], cycle cancelling [257], cut cancelling [129, 205], minimum mean cancelling [176], and successive shortest path [125]. Polynomial-time algorithms were found later such as speed-up successive shortest path, linear programming approach [332], capacity scaling [123], and cost scaling [177]. Strongly polynomial-time algorithm is obtained much later by Tardos [373], Goldberg and Tarjin [176, 177], etc. In the study of strongly polynomial-time algorithm, Karp's algorithm [238] for the minimum cycle mean problem [48] plays an important role. Currently, the fastest strong polynomial-time algorithm has running time is $O(|E|^2 \log^2 |V|)$. This record is also kept by an algorithm of Orlin [332].

Chapter 8
NP-Hard Problems and Approximation Algorithms

The biggest difference between time and space is that you can't reuse time.

—Merrick Furst

8.1 What Is the Class NP?

The class P consists of all polynomial-time solvable decision problems. What is the class NP? There are two popular misunderstandings:

(1) NP is the class of problems which are not polynomial-time solvable.
(2) A decision problem belongs to the class NP if its answer can be checked in polynomial-time.

The misunderstanding (1) comes from incorrect explanation of NP as the brief name for "not polynomial-time solvable." Actually, it is polynomial-time solvable, but in a wide sense of computation, nondeterministic computation, that is, **NP is the class of all nondeterministic polynomial-time solvable decision problems.** Thus, NP is the brief name of "nondeterministic polynomial-time."

What is nondeterministic computation? Let us explain it starting from computation model, Turing machine (TM). A TM consists of three parts, a tape, a head, and a finite control (Fig. 8.1).

The tape has the left end and infinite long in the right direction, which is divided infinitely into many cells. Each cell can hold a symbol. All symbols possibly on the tape form an alphabet Γ, called the *alphabet of tape symbols*. In Γ, there is a special symbol B, called the *blank symbol*, which means the cell is actually empty. Initially, an input string is written on the tape. All symbols possibly in the input string form another alphabet Σ, called the *alphabet of input symbols*. Assume that both Γ and Σ are finite and $B \in \Gamma \setminus \Sigma$.

Fig. 8.1 One-tape Turing machine

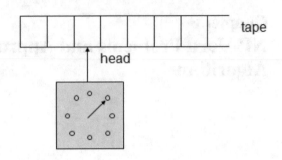

The head can read, erase, and write symbols on the tape. Moreover, it can move to the left and right. In each move, the head can shift a distance of one cell. Please note that in classical one-tape TM, the head is not allowed to stay in the place without moving before the TM halts.

The finite control contains a finite number of states, forming a set Q. The TM's computation depends on function $\delta : Q \times \Gamma \rightarrow Q \times \Gamma \times D$ where $D = \{R, L\}$ is the set of possible moving directions and R means moving to the right, while L means moving to the left. This function δ is called the *transaction function*. For example, $\delta(q, a) = (p, b, L)$ means that when TM in state q reads symbol a, it will change state to p, change symbol a to b, and then move to the left (the uppercase in Fig. 8.2); $\delta(q, a) = (p, b, R)$ means that when TM in state q reads symbol a, it will change state to p, change symbol a to b, and then move to the right (the lowercase in Fig. 8.2); Initially, on an input x, the TM is in a special state s, called the *initial state*, and its head is located at the leftmost cell, which contains the first symbol of x if x is not empty. The TM stops moving if and only if it enters another special state h, called the *final state*. An input x is said to be *accepted* if on x, the TM will finally stop. All accepted inputs form a language, which is called the language accepted by the TM. The language accepted by a TM M is denoted by $L(M)$.

From above description, we see that each TM can be described by the following parameters: an alphabet Σ of input symbols, an alphabet Γ of tape symbols, a finite set Q of states in finite control, a transition function δ, and an initial state s.

The computation time of an TM M on an input x is the number of moves from initial state to final state, denoted by $Time_M(x)$. A TM M is said to be *polynomial-time bounded* if there exists a polynomial p such that for every input $x \in L(M)$, $Time_m(x) \leq p(|x|)$. So far, what TM we have described is the deterministic TM (DTM), that is, for each move, there exists at most one transition determined by the transition function. All languages accepted by polynomial-time bounded DTM form a class, denoted by P.

There are many variations of the TM, in which the TM has more freedom. For example, the head is allowed to stay at the same cell during a move, the tape may have no left end, and multiple tapes exist (Fig. 8.3). However, in terms of

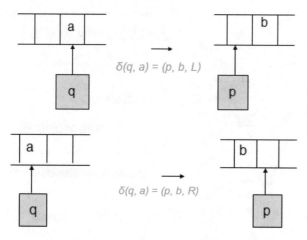

Fig. 8.2   One move to the right

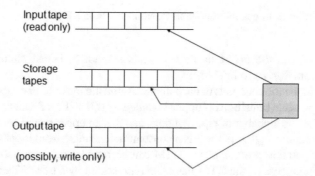

Fig. 8.3   A multi-tape TM

polynomial-time computability, all of them have been proved to have the same power. Based on such experiences, one made the following conclusion:

Extended Church-Turing Thesis   *A function computable in polynomial-time in any reasonable computational model using a reasonable time complexity measure is computable by a deterministic TM in polynomial-time.*

Extended Church-Turing thesis is a natural law of computation. It is similar to physics laws, which cannot have a mathematical proof, but is obeyed by the natural world. By extended Church-Turing thesis, the class P is independent from computational models. In the statement, "reasonable" is an important word. Are there unreasonable computational models? The answer is yes. For example, the nondeterministic Turing machine (NTM) is an important one among them. In an NTM, for each move, there may exist many possible transitions (Fig. 8.4) and the NTM can use any one of them. Therefore, transition function δ in an NTM is a mapping from $Q \times \Gamma$ to $2^{Q \times \Gamma \times \{R,L\}}$, that is, $\delta(q,a)$ is the set of all possible

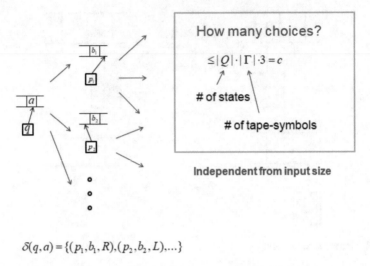

$$\delta(q,a) = \{(p_1,b_1,R),(p_2,b_2,L),\dots\}$$

Fig. 8.4 There are many possible transitions for each move in an NTM

transitions. When the NTM in state q reads symbol a, it can choose any one transition from $\delta(q, a)$ to implement.

It is worth mentioning that for each nondeterministic move of one-tape NTM, the number of possible transitions is upper-bounded by $|Q| \times |\Gamma| \times 3$ where Q is the set of states, Γ is the alphabet of tape symbols, and 3 is an upper bound for the number of moving choices. $|Q| \times |\Gamma| \times 3$ is a constant independent from input size $|x|$.

The computation process of the DTM can be represented by a path, while the computation process of the NTM has to be represented by a tree. When is an input x accepted by an NTM? The definition is that as long as there is a path in the computation tree, leading to the final state, then x is accepted. Suppose that at each move, we make a guess for choice of possible transitions. This definition means that if there exists a correct guess which leads to the final state, we will accept the input. Let us look at an example. Consider the following problem:

Problem 8.1.1 (Hamiltonian Cycle) Given a graph $G = (V, E)$, does G contain a Hamiltonian cycle? Here, a Hamiltonian cycle is a cycle passing through each vertex exactly once.

The following is a nondeterministic algorithm for the Hamiltonian cycle problem:

input a graph $G = (V, E)$.
step 1 guess a permutation of all vertices.
step 2 check if guessed permutation gives a Hamiltonian cycle.
       if yes, **then** accept input.

In step 1, the guess corresponds to nondeterministic moves in the NTM. Note that in step 2, if the outcome of checking is no, then we cannot give any conclusion, and hence nondeterministic computation gets stuck. However, a nondeterministic algorithm is considered to solve a decision problem correctly if there exists a guessed result leading to correct yes-answer. For example, in above algorithm, if input graph contains a Hamiltonian cycle, then there exists a guessed permutation which gives a Hamiltonian cycle and hence gives yes-answer. Therefore, it is a nondeterministic algorithm which solves the HAMILTONIAN CYCLE problem.

Now, let us recall the second popular misunderstanding of NP mentioned at the beginning of this section:

(2) A decision problem belongs to the class NP if its answer can be checked in polynomial-time.

Why (2) is wrong? This is because not only checking step is required to be polynomial-time computable, but also guessing step is required to be polynomial-time computable. How do we estimate guessing time? Let us explain this starting from what is a legal guess. Note that in an NTM, each nondeterministic move can select a choice of transition from a pool with size upper-bound independent from input size. Therefore, **a legal guess is a guess from a pool with size independent from input size.** For example, in above algorithm, guessing in step 1 is not legal because the number of permutation of n vertices is $n!$ which depends on input size.

What is the running time of step 1? It is the number of legal guesses spent in implementation of the guess in step 1. To implement the guess in step 1, we may encode each vertex into a binary code of length $\lceil \log_2 n \rceil$. Then each permutation of n vertices is encoded into a binary code of length $O(n \log n)$. Now, guessing a permutation can be implemented by $O(n \log n)$ legal guesses each of which chooses either 0 or 1. Therefore, the running time of step 1 is $O(n \log n)$.

In many cases, the guessing step is easily implemented by a polynomial number of legal guesses. However, there are some exceptions; one of them is the following:

Problem 8.1.2 Given an $m \times n$ integer matrix A and an n-dimensional integer vector b, determine whether there exists an m-dimensional integer vector x such that $Ax \geq b$.

In order to prove that Problem 8.1.2 is in NP, we may guess an n-dimensional integer vector x and check whether x satisfies $Ax \geq b$. However, we need to make sure that guessing can be done in nondeterministic polynomial-time. That is, we need to show that if the problem has a solution, then there is a solution of polynomial size. Otherwise, our guess cannot find it. This is not an easy job. We include the proof into the following three lemmas:

Let α denote the maximum absolute value of elements in A and b. Denote $q = \max(m, n)$.

Lemma 8.1.3 *If B is a square submatrix of A, then $|\det B| \leq (\alpha q)^q$.*

Proof Let k be the order of B. Then $|\det B| \leq k! \alpha^k \leq k^k \alpha^k \leq q^q \alpha^q = (q\alpha)^q$. $\quad\square$

Lemma 8.1.4 *If $rank(A) = r < n$, then there exists a nonzero vector z such that $Az = 0$ and every component of z is at most $(\alpha q)^q$.*

Proof Without loss of generality, assume that the left-upper $r \times r$ submatrix B is nonsingular. Set $x_{r+1} = \cdots = x_{n-1} = 0$ and $x_n = -1$. Apply Cramer's rule to system of equations

$$B(x_1, \cdots, x_r)^T = (a_{1n}, \cdots, a_{rn})^T$$

where a_{ij} is the element of A on the ith row and the jth column. Then we can obtain $x_i = \det B_i / \det B$ where B_i is a submatrix of A. By Lemma 3.1, $|\det B_i| \le (\alpha q)^q$. Now, set $z_1 = \det B_1, \cdots, z_r = \det B_r, z_{r+1} = \cdots = z_{n-1} = 0$, and $z_n = \det B$. Then $Az = 0$.   □

Lemma 8.1.5 *If $Ax \ge b$ has an integer solution, then it must have an integer solution whose components of absolute value not exceed $2(\alpha q)^{2q+1}$.*

Proof Let a_i denote the ith row of A and b_i the ith component of b. Suppose that $Ax \ge b$ has an integer solution. Then we choose a solution x such that the following set gets the maximum number of elements:

$$\mathcal{A}_x = \{a_i \mid b_i \le a_i x \le b_i + (\alpha q)^{q+1}\} \cup \{e_i \mid |x_i| \le (\alpha q)^q\},$$

where $e_i = (0, \underbrace{\cdots, 0, 1, 0, \cdots}_{i}, 0)$. We first prove that the rank of $\mathcal{A}_x$ is n. For otherwise, suppose that the rank of $\mathcal{A}_x$ is less than n. Then we can find nonzero integer vector z such that for any $d \in \mathcal{A}_x$, $dz = 0$ and each component of z does not exceed $(\alpha q)^q$. Note that $e_k \in \mathcal{A}_x$ implies that kth component z_k of z is zero since $0 = e_k z = z_k$. If $z_k \ne 0$, then $e_k \notin \mathcal{A}_x$, so $|x_k| > (\alpha q)^q$. Set $y = x + z$ or $x - z$ such that $|y_k| < |x_k|$. Then for every $e_i \in \mathcal{A}_x$, $y_i = x_i$, so $e_i \in \mathcal{A}_y$, and for $a_i \in \mathcal{A}_x$, $a_i y = a_i x$, so $a_i \in \mathcal{A}_y$. Thus, $\mathcal{A}_y$ contains $\mathcal{A}_x$. Moreover, for $a_i \notin \mathcal{A}_x$, $a_i y \ge a_i x - |a_i z| \ge b_i + (\alpha q)^{q+1} - n\alpha(\alpha q)^q \ge b_i$. Thus, y is an integer solution of $Ax \ge b$. By the maximality of $\mathcal{A}_x$, $\mathcal{A}_y = \mathcal{A}_x$. This means that we can decrease the value of the kth component again. However, it cannot be decreased forever. Finally, a contradiction would appear. Thus, $\mathcal{A}_x$ must have rank n.

Now, choose n linearly independent vectors $d_1, \cdots, d_n$ from $\mathcal{A}_x$. Denote $c_i = d_i x$. Then $|c_i| \le \alpha + (\alpha q)^{q+1}$. Applying Cramer's rule to the system of equations $d_i x = c_i$, $i = 1, 2, \cdots, n$, we obtain a representation of x through c_i's: $x_i = \det D_i / \det D$ where D is a square submatrix of $(A^T, I)^T$ and D_i is a square matrix obtained from D by replacing the ith column by vector $(c_1, \cdots, c_n)^T$. Note that the determinant of any submatrix of $(A^T, I)^T$ equals to the determinant of a submatrix of A. By Laplace expansion, we obtain that

$$|x_i| \le |\det D_i|$$
$$\le (\alpha q)^q (|c_1| + \cdots + |c_n|)$$
$$\le (\alpha q)^q n(\alpha + (\alpha q)^{q+1})$$
$$\le 2(\alpha q)^{2q+1}.$$

$\square$

Theorem 8.1.6 *Problem 8.1.2 is in NP.*

Proof By Lemma 8.1.5, it is enough to guess a solution x whose total size is at most $n \log_2(2(\alpha q)^{2q+1}) = O(q^2(\log_2 q + \log_2 \alpha))$. Note that the inputs A and b have total length at least

$$\beta = \sum_{i=1}^{m} \sum_{j=1}^{n} \log_2 |a_{ij}| + \sum_{j=1}^{n} \log_2 |b_j| \ge mn + \log_2 \alpha \ge q + \log_2 \alpha.$$

Hence, Problem 8.1.2 is in NP. $\square$

The definition of the class NP involves three concepts, nondeterministic computation, polynomial-time, and decision problems. The first two concepts have been explained as above. Next, we explain what is the decision problem.

A problem is called a *decision problem* if its answer is "Yes" or "No." Each decision problem corresponds to the set of all inputs which receive yes-answer, which is a language when each input is encoded into a string. For example, the Hamiltonian cycle problem and Problem 8.1.2 are decision problems. In case of no confusion, we sometimes use the same notation to denote a decision problem and its corresponding language. For example, we may say that a decision problem A has its characteristic function

$$\chi_A(x) = \begin{cases} 1 \text{ if } x \in A, \\ 0 \text{ otherwise.} \end{cases}$$

Actually, we mean that the corresponding language of decision problem A has characteristic function χ_A.

Usually, combinatorial optimization problems are not decision problems. However, every combinatorial optimization problem can be transformed into a decision version. For example, consider the following:

Problem 8.1.7 (Traveling Salesman) Given n cities and a distance table between n cities, find the shortest Hamiltonian tour where a Hamiltonian tour is a Hamiltonian cycle in the complete graph on the n cities.

Its decision version is as follows:

Problem 8.1.8 (Decision Version of Traveling Salesman) Given n cities, a distance table between n cities, and an integer $K > 0$, is there a Hamiltonian tour with total distance at most K?

Clearly, if the traveling salesman problem can be solved in polynomial-time, so is its decision version. Conversely, if its decision version can be solved in polynomial-time, then we may solve the traveling salesman problem in the following way within polynomial-time.

Let us assume that all distances between cities are integers.[1] Let d_{min} and d_{max} be the smallest distance and the maximum distance between two cities. Let $a = nd_{min}$ and $b = nd_{max}$. Set $K = \lceil (a + b)/2 \rceil$. Determine whether there is a tour with total distance at most K by solving the decision version of the traveling salesman problem. If the answer is yes, then set $b \leftarrow K$; else set $a \leftarrow K$. Repeat this process until $|b - a| \leq 1$. Then, compute the exact optimal objective function value of the traveling salesman problem by solving its decision version twice with $K = a$ and $K = b$, respectively. In this way, suppose the decision version of the traveling salesman problem can be solved in polynomial-time $p(n)$. Then the traveling salesman problem can be solved in polynomial-time $O(\log(nd_{max})p(n))$.

Now, we may find that actually, Problem 8.1.2 is closely related to the decision version of the following integer program:

Problem 8.1.9 (0-1 Integer Program)

$$\max\ cx$$
$$\text{subject to}\ \ Ax \geq b$$
$$x \in \{0, 1\}^n,$$

where A is an $m \times n$ integer matrix, c is an n-dimensional integer row vector, and b is an m-dimensional integer column vector.

8.2   What Is NP-Completeness?

In 1965, J. Admonds conjectured the following:

Conjecture 8.2.1 *The traveling salesman problem does not have a polynomial-time solution.*

In the study of this conjecture, S. Cook introduced the class NP and showed the first NP-complete problem in 1971.

---

[1] If they are rational numbers, then we can transform them into integers. If some of them are irrational numbers, then we have to touch the complexity theory of real number computation, which is out of scope of this book.

A problem is NP-hard if the existence of polynomial-time solution for it implies the existence of polynomial-time solution for every problem in NP. An NP-hard problem is NP-complete if it also belongs to the class NP.

To introduce Cook's result, let us recall some knowledge on Boolean algebra.

A *Boolean function* is a function whose variable values and function value all are in $\{true, false\}$. Here, we would like to denote true by 1 and false by 0. In the following table, there are two Boolean functions of two variables, *conjunction* $\wedge$ and *disjunction* $\vee$, and a Boolean function of a variable, *negation* $\neg$.

x	y	$x \wedge y$	$x \vee y$	$\neg x$
0	0	0	0	1
0	1	0	1	1
1	0	0	1	0
1	1	1	1	0

For simplicity, we also write $x \wedge y = xy$, $x \vee y = x + y$ and $\neg x = \bar{x}$. The conjunction and disjunction follow the commutative, associative, and distributive laws. An interesting and important law about negation is De Morgan's law, i.e.,

$$\overline{xy} = \bar{x} + \bar{y} \text{ and } \overline{x + y} = \bar{x}\bar{y}.$$

The SAT problem is defined as follows:

Problem 8.2.2 (Satisfiability (SAT)) Given a Boolean formula F, is there a satisfied assignment for F?

Here, an assignment to variables of F is *satisfied* if the assignment makes F equal to 1. A Boolean formula F is *satisfiable* if there exists a satisfied assignment for F.

The SAT problem has many applications. For example, the following puzzle can be formulated into an instance of SAT:

Example 8.2.3 After three men were interviewed, the department Chair said: "We need Brown and if we need John, then we need David, if and only if we need either Brown or John and don't need David." If this department actually needs more than one new faculties, which ones were they?

Solution. Let B, J, and D denote respectively Brown, John, and David. What the Chair said can be written as a Boolean formula:

$$[B(\bar{J} + D)][\overline{(B + J)\bar{D}}] + \overline{B(\bar{J} + D)} \cdot (B + J)\bar{D}$$
$$= B\bar{D}\bar{J} + (\bar{B} + J\bar{D})(\bar{B}\bar{J} + D)$$
$$= B\bar{D}\bar{J} + \bar{B}\bar{J} + \bar{B}D$$

Fig. 8.5 Polynomial-time
many-one reduction

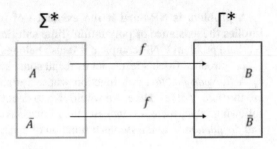

Since this department actually needs more than one new faculty, there is only one way to satisfy this Boolean formula, that is, $B = 0$, $D = J = 1$. Thus, John and David will be hired. $\square$

Now, we are ready to state Cook's result.

Theorem 8.2.4 (Cook Theorem) *The* SAT *problem is NP-complete.*

After the first NP-complete problem is discovered, a large number of problems have been found to be NP-hard or NP-complete. Indeed, there are many tools passing the NP-hardness from one problem to another problem. We introduce one of them as follows:

Consider two decision problems A and B. A is said to be polynomial-time many-one reducible to B, denoted by $A \leq^p_m B$, if there exists a polynomial-time computable function f mapping from all inputs of A to inputs of B such that A receives yes-answer on input x if and only if B receives yes-answer on input $f(x)$ (Fig. 8.5).

For example, we have

Example 8.2.5 The Hamiltonian cycle problem is polynomial-time many-one reducible to the decision version of the traveling salesman problem.

Proof To construct this reduction, for each input graph $G = (V, E)$ of the Hamiltonian cycle problem, we consider V as the set of cities and define a distance table D by setting

$$d(u, v) = \begin{cases} 1 & \text{if } (u, v) \in E \\ |V| + 1 & \text{otherwise.} \end{cases}$$

Moreover, set $K = |V|$. If G contains a Hamiltonian cycle, this Hamiltonian cycle would give a tour with total distance $|V| = K$ for the traveling salesman problem on defined instance. Conversely, if the traveling salesman problem on defined instance has a Hamiltonian tour with total distance at most K, then this tour cannot contain an edge $(u, v) \notin E$ and hence it induces a Hamiltonian cycle in G. Since the reduction can be constructed in polynomial-time, it is a polynomial-time many-one reduction from the Hamiltonian cycle problem to the decision version of the traveling salesman problem. $\square$

Fig. 8.6 The proof of
Proposition 8.2.6

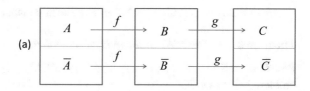

(a)

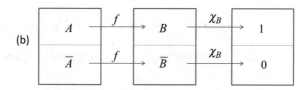

(b)

There are two important properties of the polynomial-time many-one reduction
(Fig. 8.6).

Proposition 8.2.6

(a) If $A \leq_m^p B$ and $B \leq_m^p C$, then $A \leq_m^p C$.
(b) If $A \leq_m^p B$ and $B \in P$, then $A \in P$.

Proof

(a) Let $A \leq_m^p B$ via f and $B \leq_m^p C$ via g. Then $A \leq_m^p C$ via h where
 $h(x) = g(f(x))$. Let f and g be computable in polynomial-times $p(n)$ and
 $q(n)$, respectively. Then for any x with $|x| = n$, $|f(x)| \leq p(n)$. Hence, h can
 be computed in time $p(n) + q(p(n))$
(b) Let $A \leq_m^p B$ via f. f is computable in polynomial-time $p(n)$ and B can
 be solved in polynomial-time $q(n)$. Then A can be solved in polynomial-time
 $p(n) + q(p(n))$.   □

Property (a) indicates that $\leq_m^p$ is a partial ordering. Property (b) gives us a simple
way to establish the NP-hardness of a decision problem. To show the NP-hardness
of a decision problem B, it suffices to find an NP-complete problem A and prove
$A \leq_m^p B$. In fact, if $B \in P$, then $A \in P$. Since A is NP-complete, every problem in
NP is polynomial-time solvable. Therefore, B is NP-hard.

The SAT problem is the root to establish the NP-hardness of almost all other
problems. However, it is hard to use the SAT problem directly to construct reduction.
Often, we use an NP-complete special case of the SAT problem. To introduce this
special case, let us first explain a special type of Boolean formulas, 3CNF.

A *literal* is either a Boolean variable or the negation of a Boolean variable. An
elementary sum is a sum of several literals. Consider an elementary sum c and a
Boolean function f. If $c = 0$ implies $f = 0$, then c is called a *clause* of f. A CNF
(conjunctive normal form) is a product of its clauses. A CNF is called a 3CNF if
each clause of the CNF contains exactly three distinct literals about three variables.

Problem 8.2.7 3SAT: Given a 3CNF F, determine whether the F is satisfiable.

Theorem 8.2.8 *The 3SAT problem is NP-complete.*

Proof Since SAT is in NP, as a special case of SAT, it is easy to see that the 3SAT problem belongs to NP. Next, we show $SAT \leq_m^p 3SAT$.

First, we show two facts.

(a) $w = x + y$ if and only if $p(w, x, y)$ is satisfiable where

$$p(w, x, y) = (\bar{w} + x + y)(w + \bar{x} + y)(w + x + \bar{y})(w + \bar{x} + \bar{y}).$$

(b) $w = xy$ if and only if $q(w, x, y)$ is satisfiable where

$$q(w, x, y) = p(\bar{w}, \bar{x}, \bar{y}).$$

To show (a), we note that $w = x + y$ if and only if $\bar{w}\bar{x}\bar{y} + w(x + y) = 1$. Moreover, we have

$$
\begin{aligned}
&\bar{w}\bar{x}\bar{y} + w(x + y) \\
&= (\bar{w} + x + y)(\bar{x}\bar{y} + w) \\
&= (\bar{w} + x + y)(\bar{x} + w)(\bar{y} + w) \\
&= (\bar{w} + x + y)(w + \bar{x} + y)(w + x + \bar{y})(w + \bar{x} + \bar{y}) \\
&= q(w, x, y).
\end{aligned}
$$

Therefore, (a) holds.

(b) can be derived from (a) by noting that $w = xy$ if and only if $\bar{w} = \bar{x} + \bar{y}$.

Now, consider a Boolean formula F. F must contain a term xy or $x + y$ where x and y are two literals. In the former case, replace xy by a new variable w in F and set $F \leftarrow q(w, x, y)F$. In the latter case, replace $x + y$ by a new variable w in F and set $F \leftarrow p(w, x, y)F$. Repeat this operation until F becomes a literal z. Let u and v be two new variables. Finally, set $F \leftarrow F(z + u + v)(z + \bar{u} + v)(z + u + \bar{v})(z + \bar{u} + \bar{v})$. Then the original F is satisfiable if and only if the new F is satisfiable. $\qquad\square$

Starting from the 3SAT problem through polynomial-time many-one reduction, there are a very large number of combinatorial optimization problems; their decision versions have been proved to be NP-complete. Moreover, none of them have been found to have a polynomial-time solution. If one of them has a polynomial-time solution, so do others. This fact makes one confidently say: An NP-hard problem is **unlikely** to have a polynomial-time solution. This "unlikely" can be removed only if P$\neq$NP is proved, which is a big open problem in the literature.

Since for NP-hard combinatorial optimization problems, they are unlikely to have polynomial-time exact solution, we have to move our attention from exact solutions to approximation solutions. How do we design and analyze approximation solution?

Fig. 8.7 Intersection of NP and co-NP

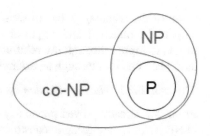

Those techniques will be studied systematically in the next few chapters. Before doing so, we would touch a few fundamental NP-complete problems and their related combinatorial optimization problems with their approximation solutions in later sections of this chapter.

To end this section, let us mention a rough way to judge whether a problem has a possible polynomial-time solution or not. Note that in many cases, it is easy to judge whether a problem belongs to NP or not. For a decision problem A in NP, if it is hard to find a polynomial-time solution, then we may study its complement $\bar{A} = \{x \mid x \notin A\}$. If $\bar{A} \in$ NP, then we may need to try hard to find a polynomial-time solution. If it is hard to show $\bar{A} \in NP$, then we may try to show NP-hardness of problem A.

Actually, let co-NP denote the class consisting of all complements of decision problems in NP. Then class P is contained in the intersection of NP and co-NP (Fig. 8.7). So far, no natural problem has been found to exist in (NP∩co-NP)\P.

In the history, there are two well-known open problems existing in NP∩co-NP, and they were unknown to have polynomial-time solutions for many years. They are the primality test and the decision version of linear program. Finally, they both have been found to have polynomial-time solutions.

8.3 Hamiltonian Cycle

Consider an NP-complete decision problem A and a possible NP-hard decision problem B. How do we construct a polynomial-time many-one reduction? Every reader who has no experience would like to know the answer of this question. Of course, we would not have an efficient method to produce such a reduction. Indeed, we do not know if such a method exists. However, we may give some idea to follow.

Let us recall how to show a polynomial-time many-one reduction from A to B.

(1) Construct a polynomial-time computable mapping f from all inputs of problem A to inputs of problem B.
(2) Prove that problem A on input x receives yes-answer if and only if problem B on input $f(x)$ receives yes-answer.

Since the mapping f has to satisfy (2), the idea is to find the relationship of output of problem A and output of problem B, that is, **find the mapping from inputs to inputs through the relationship between outputs of two problems.** Let us explain this idea through an example.

Theorem 8.3.1 *The Hamiltonian cycle problem is NP-complete.*

Proof We already proved previously that the Hamiltonian cycle problem belongs to NP. Next, we are going to construct a polynomial-time many-one reduction from the NP-complete 3SAT problem to the Hamiltonian cycle problem.

The input of the 3SAT problem is a 3CNF F and the input of the Hamiltonian cycle problem is a graph G. We need to find a mapping f such that for any 3CNF F, $f(F)$ is a graph such that F is satisfiable if and only if $f(F)$ contains a Hamiltonian cycle. What can make F satisfiable? It is a satisfied assignment. Therefore, our construction should give a relationship between assignments and Hamiltonian cycles. Suppose F contains n variables $x_1, x_2, \ldots, x_n$ and m clauses $C_1, C_2, \ldots, C_m$. To do so, we first build a ladder H_i with $4m + 2$ levels, corresponding to a variable x_i as shown in Fig. 8.8. In this ladder, there are exactly two Hamiltonian paths corresponding to two values 0 and 1 for x_i. Connect n ladders into a cycle as shown in Fig. 8.9. Then we obtain a graph H with exactly 2^n Hamiltonian cycles corresponding to 2^n assignments of F.

Now, we need to find a way to involve clauses. An idea is to represent each clause C_j by a point and represent the fact "clause C_j is satisfied under an assignment" by the fact "point C_j is included in the Hamiltonian cycle corresponding to the assignment." To realize this idea, for each literal x_i in clause C_j, we connected point C_j to two endpoints of an edge, between the $(4j - 1)$th level and the $(4j)$th level, on the path corresponding to $x_i = 1$ (Fig. 8.10), and for each $\bar{x}_i$ in clause C_j, we connected point C_j to two endpoints of an edge on the path corresponding to $x_i = 0$. This completes our construction for graph $f(F) = G$.

Fig. 8.8 A ladder H_i contains exactly two Hamiltonian paths between two ends

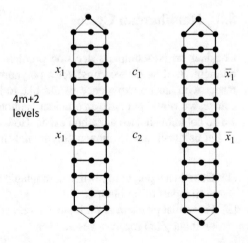

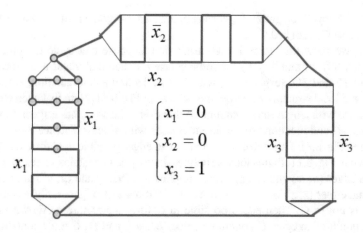

$$\begin{cases} x_1 = 0 \\ x_2 = 0 \\ x_3 = 1 \end{cases}$$

Fig. 8.9 Each Hamiltonian cycle of graph H represents an assignment

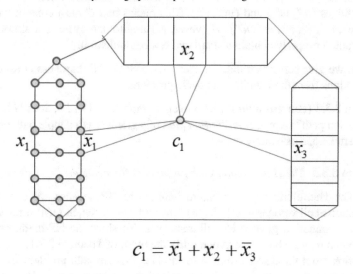

$$c_1 = \overline{x}_1 + x_2 + \overline{x}_3$$

Fig. 8.10 A point C_1 is added

To see this construction meeting our requirement, we first assume F has a satisfied assignment σ and show that G has a Hamiltonian cycle. To this end, we find the Hamiltonian cycle C in H corresponding to the satisfied assignment. Note that each clause C_j contains a literal $y = 1$ under assignment σ. Thus, C_j is connected to endpoints of an edge (u, v) on the path corresponding to $y = 1$. Replacing this edge (u, v) by two edges (C_j, u) and (C_j, v) would include point C_j into the cycle, which will become a Hamiltonian cycle of G when all points C_j are included.

Conversely, suppose G has a Hamiltonian cycle C. We claim that in C, each point C_j must connect to two endpoints of an edge (u, v) in H. If our claim holds, then replace two edges (C_j, u) and (C_j, v) by edge (u, v). We would obtain a

Hamiltonian cycle of graph H, corresponding an assignment of F, which makes every clause C_j satisfied.

Now, we show the claim. For contradiction, suppose that cycle C contains its two edges (C_j, u) and (C_j, v) for some clause C_j such that u and v are located in different H_i and $H_{i'}$, respectively, with $i \neq i'$. To find a contradiction, we look at closely the local structure of vertex u as shown in Fig. 8.11. Note that each ladder is constructed with length longer enough so that every clause C_j has a special location in ladder H_i and locations for different clauses with at least distance 3 away each other (see Fig. 8.8). This makes that at vertex u, edges possible in C form a structure as shown in Fig. 8.11. In this local structure, since cycle C contains vertex w, C must contain edges (u, w) and (w, z), which implies that (u, u') and (u, u'') are not in C. Note that either (z, z') or (z, z'') is not in C. Without loss of generality, assume that (z, z'') is not in C. Then edges possible in C form a structure as shown in 8.11. Since Hamiltonian cycle C contains vertices u'', w'', and z'', C must contain edges (u'', u'''), (u'', w''), (w'', z''), and (z'', z'''). Since C contains vertex w''', C must contain edges (u''', w''') and (w''', z'''). This means that C must contain the small cycle $(u'', w'', z'', z''', w''', u''')$. However, a Hamiltonian cycle is a simple cycle which cannot properly contain a small cycle, a contradiction.   □

Next, we give some examples in each of which the NP-hardness is established by reductions from the Hamiltonian cycle problem.

Problem 8.3.2 (Hamiltonian Path) Given a graph $G = (V, E)$, does G contain a Hamiltonian path? Here, a *Hamiltonian path* of a graph G is a simple path on which every vertex appears exactly once.

Theorem 8.3.3 *The Hamiltonian path problem is NP-complete.*

Proof The Hamiltonian path problem belongs to NP because we can guess a permutation of all vertices in $O(n \log n)$ time and then check, in $O(n)$ time, whether guessed permutation gives a Hamiltonian path. To show the NP-hardness of the Hamiltonian path problem, we may modify the proof of Theorem 8.3.1, to construct a reduction from the 3SAT problem to the Hamiltonian path problem by making a little change on graph H, which is obtained from connecting all H_i into a path

Fig. 8.11 Local structure near vertex u

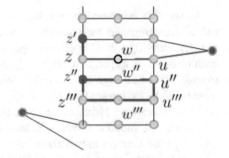

instead of a cycle. However, in the following, we would like to give a simple proof by reducing the Hamiltonian cycle problem to the Hamiltonian path problem.

We are going to find a polynomial-time computable mapping f from graphs to graphs such that G contains a Hamiltonian cycle if and only if $f(G)$ contains a Hamiltonian path. Our analysis starts from how to build a relationship between a Hamiltonian cycle of G and a Hamiltonian path of $f(G)$. If $f(G) = G$, then from a Hamiltonian cycle of G, we can find a Hamiltonian path of $f(G)$ by deleting an edge; however, from a Hamiltonian path of $f(G)$, we may not be able to find a Hamiltonian cycle of G. To have "if and only if" relation, we first consider a simple case that there is an edge (u, v) such that if G contains a Hamiltonian cycle C, then C must contain edge (u, v). In this special case, we may put two new edges (u, u') and (v, v') at u and v, respectively.

For simplicity of speaking, we may call these two edges as two *horns*. Now, if G has the Hamiltonian cycle C, then $f(G)$ has a Hamiltonian path between endpoints of two horns, u' and v'. Conversely, if $f(G)$ has a Hamiltonian path, then this Hamiltonian path must have two endpoints u' and v'; hence, we can get back C by deleting two horns and putting back edge (u, v).

Now, we consider the general case that such an edge (u, v) may not exist. Note that for any vertex u of G, suppose u have k neighbors $v_1, v_2, \ldots, v_k$. Then a Hamiltonian cycle of G must contain one of edges (u, v_1), (u, v_2), $\ldots$, (u, v_k). Thus, we may first connect all $v_1, v_2, \ldots, v_k$ to a vertex u' and put two horns (u, w) and (u', w') (Fig. 8.12). This construction would work similarly as above.   □

As a corollary of Theorem 8.3.1, we have

Corollary 8.3.4 *The traveling salesman problem is NP-hard.*

Proof In Example 8.2.5, a polynomial-time many-one reduction has been constructed from the Hamiltonian cycle problem to the traveling salesman problem.   □

The longest path problem is a maximization problem as follows:

Problem 8.3.5 (Longest Path) Given a graph $G = (V, E)$ with positive edge length $c : E \rightarrow R^+$, and two vertices s and t, find a longest simple path between s and t.

Fig. 8.12 Install two horns at u and its copy u'

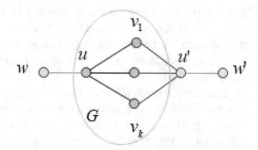

As another corollary of Theorem 8.3.1, we have

Corollary 8.3.6 *The longest path problem is NP-hard.*

Proof We will construct a polynomial-time many-one reduction from the Hamiltonian cycle problem to the decision version of the longest path problem as follows: *Given a graph $G = (V, E)$ with positive edge length $c : E \rightarrow R^+$, two vertices s and t, and an integer $K > 0$, is there a simple path between s and t with length at least K?.*

Let graph $G = (V, E)$ be an input of the Hamiltonian cycle problem. Choose a vertex $u \in V$. We make a copy of u by adding a new vertex u' and connecting u' to all neighbors of u. Add two new edges (u, s) and (u', t). Obtained graph is denoted by $f(G)$. Let $K = |V| + 2$. We show that G contains a Hamiltonian cycle if and only if $f(G)$ contains a simple path between s and t with length at most K.

First, assume that G contains a Hamiltonian cycle C. Break C at vertex u by replacing an edge (u, v) with (u', v). We would obtain a simple path between u and u' with length $|V|$. Extend this path to s and t. We would obtain a simple path between s and t with length $|V| + 2 = K$.

Conversely, assume that $f(G)$ contains a simple path between s and t with length at most K. Then this path contains a simple subpath between u and u' with length $|V|$. Merge u and u' by replacing edge (u', v), on the subpath, with edge (u, v). Then we would obtain a Hamiltonian cycle of G.   □

For NP-hard optimization problems like the traveling salesman problem and the longest path problem, it is unlikely to have an efficient algorithm to compute their exact optimal solution. Therefore, one usually study algorithms which produce approximation solutions for them. Such algorithms are simply called *approximations*.

For example, let us study the traveling salesman problem. When the given distance table satisfies the triangular inequality, that is,

$$d(a, b) + d(b, c) \geq d(a, c)$$

for any three vertices a, b, and c where $d(a, b)$ is the distance between a and b, there is an easy way to obtain a tour (i.e., a Hamiltonian cycle) with total distance within twice from the optimal.

To do so, at the first compute a minimum spanning tree in the input graph and then travel around the minimum spanning tree (see Fig. 8.13). During this trip, a vertex which appears at the second time can be skipped without increasing the total distance of the trip due to the triangular inequality. Note that the length of a minimum spanning tree is smaller than the minimum length of a tour. Moreover, this trip uses each edge of the minimum spanning tree exactly twice. Thus, the length of the Hamiltonian cycle obtained from this trip is within twice from the optimal.

Christofides in 1976 introduced an idea to improve above approximation. After computing the minimum spanning tree, he considers all vertices of odd degree (called *odd vertices*) in the tree and computes a minimum perfect matching among

Fig. 8.13 Travel around the
minimum spanning tree

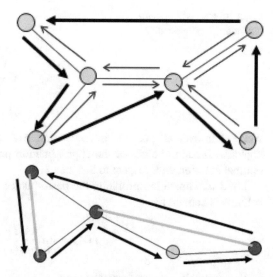

Fig. 8.14 Christofides
approximation

these odd vertices. Because in the union of the minimum spanning tree and the
minimum perfect matching, every vertex has even degree, one can travel along
edges in this union using each edge exactly once. This trip, called *Euler tour*, can be
modified into a traveling salesman tour (Fig. 8.14), without increasing the length by
the triangular inequality. Thus, an approximation is produced with length bounded
by the length of minimum spanning tree plus the length of the minimum perfect
matching on the set of vertices with odd degree. We claim that each Hamiltonian
cycle (namely, a traveling salesman tour) can be decomposed into a disjoint union of
two parts that each is not smaller than the minimum perfect matchings for vertices
with odd degree. To see this, we first note that the number of vertices with odd
degree is even since the sum of degrees over all vertices in a graph is even. Now,
let $x_1, x_2, \cdots, x_{2k}$ denote all vertices with odd degree in clockwise ordering of
the considered Hamiltonian cycle. Then $(x_1, x_2), (x_3, x_4), \cdots, (x_{2k-1}, x_{2k})$ form
a perfect matching for vertices with odd degree and $(x_2, x_3), (x_4, x_5), \cdots, (x_{2k}, x_1)$
form the other perfect matching. The claim then follows immediately from the
triangular inequality. Thus, the length of the minimum matching is at most half
of the length of the minimum Hamiltonian cycle. Therefore, Christofides gave an
approximation within a factor of 1.5 from the optimal.

From the above example, we see that the ratio of objective function values
between approximation solution and optimal solution is a measure for the performance of an approximation.

For a minimization problem, the *performance ratio* of an approximation algorithm A is defined as follows:

$$r(A) = \sup_I \frac{A(I)}{opt(I)}$$

Fig. 8.15 Extremal case for
Christofides approximation

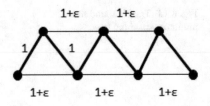

where I is over all possible instances and $A(I)$ and $opt(I)$ are respectively the
objective function values of the approximation produced by algorithm A and the
optimal solution with respect to instance I.

For a maximization problem, the performance ratio of an approximation algorithm A is defined by

$$r(A) = \inf_{I} \frac{A(I)}{opt(I)}.$$

For example, the performance ratio of Christofides approximation is at most
$3/2$ as we showed in the above. Actually, the performance ratio of Christofides
approximation is exactly $3/2$. To see this, we consider $2n + 1$ points (vertices) with
distances as shown in Fig. 8.15. The minimum spanning tree of these $2n + 1$ points
has distance $2n$. It has only two odd vertices with distance $n(1 + \varepsilon)$. Hence, the
length of Christofides approximation is $2n + n(1 + \varepsilon)$. Moreover, the minimum
tour has length $(2n - 1)(1 + \varepsilon) + 2$. Thus, in this example, $A(I)/opt(I) =
(3n + n\varepsilon)/(2n + 1 + (2n - 1)\varepsilon)$, which is approach to $3/2$ as ε goes to 0 and n
goes to infinity.

Theorem 8.3.7 *For the traveling salesman problem in metric space, the
Christofides approximation A has the performance ratio $r(A) = 3/2$.*

For simplicity, an approximation A is said to be α-approximation if $r(A) \leq \alpha$
for minimization and $r(A) \geq \alpha$ for maximization, that is, for every input I,

$$opt(I) \leq A(I) \leq \alpha \cdot opt(I)$$

for minimization, and

$$opt(I) \geq A(I) \geq \alpha \cdot opt(I)$$

for maximization. For example, Christofides approximation is a 1.5-approximation,
but not α-approximation of the traveling salesman problem in metric space for any
constant $\alpha < 1.5$.

Not every problem has a polynomial-time approximation with constant performance ratio. An example is the traveling salesman problem without triangular
inequality condition on distance table. In fact, for contradiction, suppose that its
performance ratio $r(A) \leq K$ for a constant K. Then we can show that the

Hamiltonian cycle problem can be solved in polynomial-time. For any graph $G = (V, E)$, define that for any pair of vertices u and v,

$$d(u, v) = \begin{cases} 1 & \text{if } \{u, v\} \in E \\ K \cdot |V| & \text{otherwise} \end{cases}$$

This gives an instance I for the traveling salesman problem. Then, G has a Hamiltonian cycle if and only if for I, the travel salesman has a tour with length at most $K|V|$. The optimal tour has length $|V|$. Applying approximation algorithm A to I, we will obtain a tour of length at most $K|V|$. Thus, G has a Hamiltonian cycle if and only if approximation algorithm A produces a tour of length at most $K|V|$. This means that the Hamiltonian cycle problem can be solved in polynomial-time. Because the Hamiltonian cycle problem is NP-complete, we obtain a contradiction. The above argument proved the following:

Theorem 8.3.8 *If $P \neq NP$, then no polynomial-time approximation algorithm for the traveling salesman problem in general case has a constant performance ratio.*

For the longest path problem, there exists also a negative result.

Theorem 8.3.9 *For any $\varepsilon > 0$, the longest path problem has no polynomial-time $n^{1-\varepsilon}$-approximation unless $P = NP$.*

8.4 Vertex Cover

A vertex subset C is called a *vertex cover* if every edge has at least one endpoint in C. Consider the following problem:

Problem 8.4.1 (Vertex Cover) Given a graph $G = (V, E)$ and a positive integer K, is there a vertex cover of size at most K?

The vertex cover problem is the decision version of the minimum vertex cover problem as follows:

Problem 8.4.2 (Minimum Vertex Cover) Given a graph $G = (V, E)$, compute a vertex cover with minimum cardinality.

Theorem 8.4.3 *The vertex cover problem is NP-complete.*

Proof To show that the vertex cover problem is in NP, we can guess a vertex subset within $O(n \log n)$ time and check whether obtained vertex subset is a vertex cover or not. Next, we show that the vertex cover problem is NP-hard.

Let F be a 3CNF with m clauses $C_1, \ldots, C_m$ and n variables $x_1, \ldots, x_n$. We construct a graph $G(F)$ of $2n + 3m$ vertices as follows: For each variable x_i, we give an edge with two endpoints labeled by two literals x_i and $\bar{x}_i$. For each clause

Fig. 8.16 $G(F)$

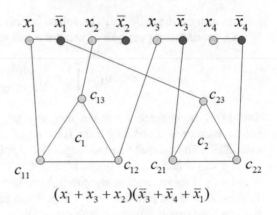

$$(x_1 + x_3 + x_2)(\bar{x}_3 + \bar{x}_4 + \bar{x}_1)$$

$C_j = x + y + z$, we give a triangle $j_1 j_2 j_3$ and connect j_1 to literal x, j_2 to literal y, and j_3 to literal z (Fig. 8.16). Now, we prove that F is satisfiable if and only if $G(F)$ has a vertex cover of size at most $n + 2m$.

First, suppose that F is satisfiable. Consider an assignment satisfying F. Let us construct a vertex cover S as follows: (1) S contains all truth literals; (2) for each triangle $j_1 j_2 j_3$, put two vertices into S such that the remainder j_k is adjacent to a truth literal. Then S is a vertex cover of size exactly $n + 2m$.

Conversely, suppose that $G(F)$ has a vertex cover S of size at most $n + 2m$. Since each triangle $j_1 j_2 j_3$ must have at least two vertices in S and each edge $(x_i, \bar{x}_i)$ has at least one vertex in S, S must contain exactly two vertices in each triangle $j_1 j_2 j_3$ and exactly one vertex for each edge $(x_i, \bar{x}_i)$. Set

$$x_i = \begin{cases} 1 \text{ if } x_i \in S, \\ 0 \text{ if } \bar{x}_i \in S. \end{cases}$$

Then each clause C_j must have a truth literal which is the one adjacent to the j_k not in S. Thus, F is satisfiable.

The above construction is clearly polynomial-time computable. Hence, the 3SAT problem is polynomial-time many-one reducible to the vertex cover problem. $\square$

Corollary 8.4.4 *The minimum vertex cover problem is NP-hard.*

Proof It is NP-hard since its decision version is NP-complete. $\square$

There are two combinatorial optimization problems closely related to the minimum vertex cover problem.

Problem 8.4.5 (Maximum Independent Set) Given a graph $G = (V, E)$, find an independent set with maximum cardinality.

Here, an *independent set* is a subset of vertices such that no edge exists between any two vertices in the subset. A subset of vertices is an independent set if and only if its complement is a vertex cover. In fact, from the definition, every edge has to have at least one endpoint in the complement of an independent set, which means that the complement of an independent set must be a vertex cover. Conversely, if the complement of a vertex subset I is a vertex cover, then every edge has an endpoint not in I and hence I is independent. Furthermore, we see that a vertex subset I is the maximum independent set if and only if the complement of I is the minimum vertex cover.

Problem 8.4.6 (Maximum Clique) Given a graph $G = (V, E)$, find a clique with maximum size.

Here, a *clique* is a complete subgraph of input graph G and its size is the number of vertices in the clique. Let $\bar{G}$ be the complementary graph of G, that is, an edge e is in $\bar{G}$ if and only if e is not in G. Then a vertex subset I is induced a clique in G if and only if I is an independent set in $\bar{G}$. Thus, a subgraph on a vertex subset I is a maximum clique in G if and only if I is a maximum independent set in $\bar{G}$.

From their relationship, we see clearly the following:

Corollary 8.4.7 *Both the maximum independent set problem and the maximum clique problem are NP-hard.*

Next, we study the approximation of the minimum vertex cover problem.

Theorem 8.4.8 *The minimum vertex cover problem has a polynomial-time 2-approximation.*

Proof Compute a maximal matching. The set of all endpoints of edges in this maximal matching form a vertex cover, which is a 2-approximation for the minimum vertex cover problem since each edge in the matching must have an endpoint in the minimum vertex cover. □

The minimum vertex cover problem can be generalized to hypergraphs. This generalization is called the hitting set problem as follows:

Problem 8.4.9 (Hitting Set) Given a collection $\mathcal{C}$ of subsets of a finite set X, find a minimum subset S of X such that every subset in $\mathcal{C}$ contains an element in S. Such a set S is called a *hitting set*.

For the maximum independent set problem and the maximum clique problem, there are negative results on their approximation.

Theorem 8.4.10 *For any $\varepsilon > 0$, the maximum independent set problem has no polynomial-time $n^{1-\varepsilon}$-approximation unless $NP = P$.*

Theorem 8.4.11 *For any $\varepsilon > 0$, the maximum clique problem has no polynomial-time $n^{1-\varepsilon}$-approximation unless $NP = P$.*

8.5   Three-Dimensional Matching

Consider another well-known NP-complete problem.

Problem 8.5.1 (Three-Dimensional Matching (3DM)) Consider three disjoint sets X, Y, Z each with n elements and 3-sets each consisting of three elements belonging to X, Y, and Z, respectively. Given a collection $\mathcal{C}$ of 3-sets, determine whether $\mathcal{C}$ contains a three-dimensional matching, where a subcollection $\mathcal{M}$ of $\mathcal{C}$ is called a *three-dimensional matching* if $\mathcal{M}$ consists of n 3-sets such that each element of $X \cup Y \cup Z$ appears exactly once in 3-sets of $\mathcal{M}$.

Theorem 8.5.2 *The 3DM problem is NP-complete.*

Proof First, the 3DM problem belongs to NP because we can guess a collection of n 3-sets within $O(n \log n)$ time and check, in $O(n + m)$ time, whether obtained collection is a three-dimensional matching in given collection $\mathcal{C}$.

Next, we show the NP-hardness of the 3DM problem by constructing a polynomial-time many-one reduction from the 3SAT problem to the 3DM problem. Consider an input 3CNF F of the 3SAT problem. Suppose that F contains n variables $x_1, \ldots, x_n$ and m clauses $C_1, \ldots, C_m$. Construct a collection $\mathcal{C}$ of 3-sets as follows:

- For each variable x_i, construct $2m$ 3-sets $\{x_{i1}, y_{i1}, z_{i1}\}$, $\{x_{i2}, y_{i1}, z_{i2}\}$, $\{x_{i2}, y_{i2}, z_{i3}\}$, $\ldots$, $\{x_{i1}, y_{im}, z_{2m}\}$. They form a cycle as shown in Fig. 8.17.
- For each clause C_j consisting of variables $x_{i_1}, x_{i_2}, x_{i_3}$, construct three 3-sets, $\{x_{0j}, y_{0j}, z_{i_1k_1}\}$, $\{x_{0j}, y_{0j}, z_{i_2k_2}\}$, and $\{x_{0j}, y_{0j}, z_{i_3k_3}\}$ where for $h = 1, 2, 3$,

$$k_h = \begin{cases} 2j - 1 & \text{if } C_j \text{ contains } x_{i_h}, \\ 2j & \text{if } C_j \text{ contains } \bar{x}_{i_h}. \end{cases}$$

Fig. 8.17 Proof of
Theorem 8.5.2

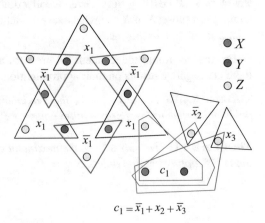

$$c_1 = \bar{x}_1 + x_2 + \bar{x}_3$$

- For each $1 \leq h \leq m(n-1)$, $1 \leq i \leq n$, $1 \leq k \leq 2m$, construct 3-set $\{x_{n+1,h}, y_{n+1,h}, z_{ik}\}$.
- Collect all above x_{pq}, y_{pq}, and z_{pq} to form sets X, Y, and Z, respectively.

Now, suppose C_F has a three-dimensional matching $\mathcal{M}$. Note that each element appears in $\mathcal{M}$ exactly once. For each variable x_i, $\mathcal{M}$ contains either

$$P_i = \{\{x_{i1}, y_{i1}, z_{i1}\}, \{x_{i2}, y_{i2}, z_{i3}\}, \ldots, \{x_{im}, y_{im}, z_{i,2m-1}\}\}$$

or

$$Q_i = \{\{x_{i1}, y_{i2}, z_{i2}\}, \{x_{i2}, y_{i3}, z_{i4}\}, \ldots, \{x_{im}, y_{i1}, z_{i,2m}\}\}$$

Define

$$x_i = \begin{cases} 1 \text{ if } P_i \subseteq \mathcal{M}, \\ 0 \text{ if } Q_i \subseteq \mathcal{M}. \end{cases}$$

Then this assignment will satisfy F. In fact, for any clause C_j, in order to have elements x_{0j} and y_{0j} appear in $\mathcal{M}$, $\mathcal{M}$ must contain 3-set $\{x_{0j}, y_{0j}, z_{i_h k_h}\}$ for some $h \in \{1, 2, 3\}$. This assignment will assign 1 to the hth literal of C_j according to the construction. Conversely, suppose F has a satisfied assignment. We can construct a three-dimensional matching $\mathcal{M}$ as follows:

- If $x_i = 1$, then put P_i into $\mathcal{M}$. If $x_i = 0$, then put Q_i into $\mathcal{M}$.
- If hth literal of clause C_j is equal to 1, then put 3-set $\{x_{0j}, y_{0j}, z_{i_h k_h}\}$ into $\mathcal{M}$.
- So far, all elements in $X \cup Y$ have been covered by 3-sets put in $\mathcal{M}$. However, there are $m(n-1)$ elements of Z that are left outside. We now use 3-sets $\{x_{n+1,h}, y_{n+1,h}, z_{ik}\}$ to play a role of garbage collector. For each z_{ik} not appearing in 3-sets in $\mathcal{M}$, select a pair of $x_{n+1,h}$ and $y_{n+1,h}$, and then put 3-set $\{x_{n+1,h}, y_{n+1,h}, z_{ik}\}$ into $\mathcal{M}$.

□

There is a combinatorial optimization problem closely related to the three-dimensional matching problem.

Problem 8.5.3 (Set Cover) Given a collection $\mathcal{C}$ of subsets of a finite set X, find a minimum set cover $\mathcal{A}$ where a set cover $\mathcal{A}$ is a subcollection of $\mathcal{C}$ such that every element of X is contained in a subset in $\mathcal{A}$.

Theorem 8.5.4 *The set cover problem is NP-hard.*

Proof Note that the decision version of the set cover problem is as follows: Given a collection $\mathcal{C}$ of subsets of a finite set X and a positive integer $k \leq |X|$, determine whether there exists a set cover of size at most k.

We construct a polynomial-time many-one reduction from the three-dimensional matching problem to the decision version of the set cover problem. Let $(X, Y, Z, \mathcal{C})$

be an instance of the three-dimensional matching problem. Construct an instance (X, C, k) of the decision version of the set cover problem by setting

$$X \leftarrow X \cup Y \cup Z,$$

$$C \leftarrow C,$$

$$k \leftarrow |X \cup Y \cup Z|.$$

Clearly, for instance (X, Y, Z, C), a three-dimensional matching exists if and only if for instance (X, C, k), a set cover of size k exists.   □

For any subcollection $\mathcal{A} \subseteq C$, define

$$f(\mathcal{A}) = |\cup_{A \in \mathcal{A}} A|.$$

The set cover problem has a greedy approximation as follows:

Algorithm 26 Greedy algorithm SC

Input: A finite set X and a collection of subsets of X.
Output: A subcollection $\mathcal{A}$ of C.
1: $\mathcal{A} \leftarrow \emptyset$.
2: **while** $f(\mathcal{A}) < |X|$ **do**
3:    choose $A \in C$ to maximize $f(\mathcal{A} \cup \{A\})$
4:    and set $\mathcal{A} \leftarrow \mathcal{A} \cup \{A\}$
5: **end while**
6: **return** $\mathcal{A}$.

This approximation can be analyzed as follows:

Lemma 8.5.5 *For any two subcollections $\mathcal{A} \subset \mathcal{B}$ and any subset $A \subseteq X$,*

$$\Delta_A f(\mathcal{A}) \geq \Delta_A f(\mathcal{B}), \tag{8.1}$$

where $\Delta_A f(\mathcal{A}) = f(\mathcal{A} \cup \{A\}) - f(\mathcal{A})$.

Proof Since $\mathcal{A} \subset \mathcal{B}$, we have

$$\Delta_A f(\mathcal{A}) = |A \setminus \cup_{S \in \mathcal{A}} S| \geq |A \setminus \cup_{S \in \mathcal{B}} S| = \Delta_A f(\mathcal{B}).$$

□

Theorem 8.5.6 *Greedy algorithm SC is a polynomial-time $(1 + \ln \gamma)$-approximation for the set cover problem, where γ is the maximum cardinality of a subset in input collection C.*

Proof Let $A_1, \ldots, A_g$ be subsets selected in turn by greedy algorithm SC. Denote $\mathcal{A}_i = \{A_1, \ldots, A_i\}$. Let *opt* be the number of subsets in a minimum set cover.

Let $\{C_1, \ldots, C_{opt}\}$ be a minimum set cover. Denote $\mathcal{C}_j = \{C_1, \ldots, C_j\}$.
By the greedy rule,

$$f(\mathcal{A}_{i+1}) - f(\mathcal{A}_i) = \Delta_{A_{i+1}} f(\mathcal{A}_i) \geq \Delta_{C_j} f(\mathcal{A}_i)$$

for $1 \leq j \leq opt$. Therefore,

$$f(\mathcal{A}_{i+1}) - f(\mathcal{A}_i) \geq \frac{\sum_{j=1}^{opt} \Delta_{C_j} f(\mathcal{A}_i)}{opt}.$$

On the other hand,

$$\frac{|X| - f(\mathcal{A}_i)}{opt} = \frac{f(\mathcal{A}_i \cup \mathcal{C}_{opt}) - f(\mathcal{A}_i)}{opt}$$

$$= \frac{\sum_{j=1}^{opt} \Delta_{C_j} f(\mathcal{A}_i \cup \mathcal{C}_{j-1})}{opt}.$$

By Lemma 8.5.5,

$$\Delta_{C_j} f(\mathcal{A}_i) \geq \Delta_{C_j} f(\mathcal{A}_i \cup \mathcal{C}_{j-1}).$$

Therefore,

$$f(\mathcal{A}_{i+1}) - f(\mathcal{A}_i) \geq \frac{|X| - f(\mathcal{A}_i)}{opt}, \tag{8.2}$$

that is,

$$|X| - f(\mathcal{A}_{i+1}) \leq (|X| - f(\mathcal{A}_i)) \left(1 - \frac{1}{opt}\right)$$

$$\leq |X|(1 - \frac{1}{opt})^{i+1}$$

$$\leq |X| e^{-(i+1)/opt}.$$

Choose i such that $|X| - f(\mathcal{A}_{i+1}) < opt \leq |X| - f(\mathcal{A}_i)$. Then

$$g \leq i + opt$$

and

$$opt \leq |X| e^{-i/opt}.$$

Therefore,

$$g \leq opt \left(1 + \ln \frac{|X|}{opt}\right) \leq opt(1 + \ln \gamma).$$

$\square$

The following theorem indicates that above greedy approximation has the best possible performance ratio for the set cover problem:

Theorem 8.5.7 *For $\rho < 1$, there is no polynomial-time $(\rho \ln n)$-approximation for the set cover problem unless $NP = P$ where $n = |X|$.*

In the worst case, we may have $\gamma = n$. Therefore, this theorem indicates that the performance of Greedy algorithm is tight in some sense.

The hitting set problem is equivalent to the set cover problem. To see this equivalence, for each element $x \in X$, define $\mathcal{S}_x = \{C \in \mathcal{C} \mid x \in C\}$. Then the set cover problem on input $(X, \mathcal{C})$ is equivalent to the hitting set problem on input $(\mathcal{C}, \{\mathcal{S}_x \mid x \in X\})$. In fact, $\mathcal{A} \subseteq \mathcal{C}$ covers X if and only if $\mathcal{A}$ hits every $\mathcal{S}_x$. From this equivalence, the following is obtained immediately:

Corollary 8.5.8 *The hitting set problem is NP-hard and has a greedy $(1 + \ln \gamma)$-approximation. Moreover, for any $\rho < 1$, it has no polynomial-time $\rho \ln \gamma$-approximation unless $NP = P$.*

8.6   Partition

The partition problem is defined as follows:

Problem 8.6.1 (Partition) Given n positive integers $a_1, a_2, \ldots, a_n$, is there a partition (N_1, N_2) of $[n]$ such that $\sum_{i \in N_1} a_i = \sum_{i \in N_2} a_i$?

To show NP-completeness of this problem, we first study another problem.

Problem 8.6.2 (Subsum) Given $n+1$ positive integers $a_1, a_2, \ldots, a_n$ and L where $1 \leq L \leq S = \sum_{i=1}^{n} a_i$, is there a subset N_1 of $[n]$ such that $\sum_{i \in N_1} a_i = L$?

Theorem 8.6.3 *The subsum problem is NP-complete.*

Proof The subsum problem belongs to NP because we can guess a subset N_1 of $[n]$ in $O(n)$ time and check, in polynomial-time, whether $\sum_{i \in N_1} a_i = L$.

Next, we show $3SAT \leq_m^p$ SUBSUM. Let F be a 3CNF with n variables $x_1, x_2, \ldots, x_n$ and m clauses $C_1, C_2, \ldots, C_m$. For each variable x_i, we construct two positive decimal integers b_{x_i} and $b_{\bar{x}_i}$, representing two literals x_i and $\bar{x}_i$, respectively. Each b_{x_i} ($b_{\bar{x}_i}$) contains $m + n$ digits. Let $b_{x_i}[k]$ ($b_{\bar{x}_i}[k]$) be the kth rightmost digit of b_{x_i} ($b_{\bar{x}_i}$). Set

$$b_{x_i}[k] = b_{\bar{x}_i}[k] = \begin{cases} 1 \text{ if } k = i, \\ 0 \text{ otherwise} \end{cases}$$

for recording the ID of variable x_i. To record information on relationship between literals and clauses, set

$$b_{x_i}[n+j] = \begin{cases} 1 \text{ if } x_i \text{ appears in clause } C_j, \\ 0 \text{ otherwise,} \end{cases}$$

and

$$b_{\bar{x}_i}[n+j] = \begin{cases} 1 \text{ if } \bar{x}_i \text{ appears in clause } C_j, \\ 0 \text{ otherwise.} \end{cases}$$

Finally, define $2m + 1$ additional positive integers c_j, c_j' for $1 \le j \le m$ and L as follows:

$$c_j[k] = c_j'[k] = \begin{cases} 1 \text{ if } k = n + j, \\ 0 \text{ otherwise.} \end{cases}$$

$$L = \overbrace{3 \ldots 3}^{m} \overbrace{1 \ldots 1}^{n}.$$

For example, if $F = (x_1 + x_2 + \bar{x}_3)(\bar{x}_2 + \bar{x}_3 + x_4)$, then we would construct the following $2(m + n) + 1 = 13$ positive integers:

$$b_{x_1} = 010001, \quad b_{\bar{x}_1} = 000001,$$
$$b_{x_2} = 010010, \quad b_{\bar{x}_2} = 100010,$$
$$b_{x_3} = 000100, \quad b_{\bar{x}_3} = 110100,$$
$$b_{x_4} = 101000, \quad b_{\bar{x}_4} = 001000,$$
$$c_1 = c_1' = 010000, \quad c_2 = c_2' = 100000,$$
$$L = 331111.$$

Now, we show that F has a satisfied assignment if and only if $A = \{b_{i,j} \mid 1 \le n, j = 0, 1\} \cup \{c_j, c_j' \mid 1 \le j \le m\}$ has a subset A' such that the sum of all integers in A' is equal to L.

First, suppose F has a satisfied assignment σ. For each variable x_i, put b_{x_i} into A' if $x_i = 1$ under assignment σ and put $b_{\bar{x}_i}$ into A' if $x_i = 0$ under assignment σ. For each clause C_j, put both c_j and c_j' into A' if C_j contains exactly one satisfied literal under assignment σ, put only c_j into A' if C_j contains exactly two satisfied literal under assignment σ, and put neither c_j nor c_j' into A' if all three literals in C_j

are satisfied under assignment σ. Clearly, obtained A' meets the condition that the sum of all numbers in A' is equal to L.

Conversely, suppose that there exists a subset A' of A such that the sum of all numbers in A is equal to L. Since $L[i] = 1$ for $1 \leq i \leq n$, A' contains exactly one of b_{x_i} and $b_{\bar{x}_i}$. Define an assignment σ by setting

$$x_i = \begin{cases} 1 \text{ if } b_{x_i} \in A', \\ 0 \text{ if } b_{\bar{x}_i} \in A'. \end{cases}$$

We claim that σ is a satisfied assignment for F. In fact, for any clause C_j, since $L[n + j] = 3$, there must be a b_{x_i} or $b_{\bar{x}_i}$ in A' whose the $(n + j)$th leftmost digit is 1. This means that there is a literal with assignment 1, appearing in C_j, i.e., making C_j satisfied.   □

Now, we show the NP-completeness of the partition problem.

Theorem 8.6.4 *The partition problem is NP-complete.*

Proof The partition problem can be seen as the subsum problem in the special case that $L = S/2$ where $S = a_1 + a_2 + \cdots + a_n$. Therefore, it is in NP. Next, we show subsum $\leq_m^p$ partition.

Consider an instance of the subsum problem, consisting of $n+1$ positive integers $a_1, a_2, \ldots, a_n$ and L where $0 < L \leq S$. Since the partition problem is equivalent to the subsum problem with $2L = S$, we may assume without of generality that $2L \neq S$. Now, consider an input for the partition problem, consisting of $n + 1$ positive integers $a_1, a_2, \ldots, a_n$ and $|2L - S|$. We will show that there exists a subset N_1 of $[n]$ such that $\sum_{i \in N_1} a_i = L$ if and only if $A = \{a_1, a_2, \ldots, a_n, |2L - S|\}$ has a partition (A_1, A_2) such that the sum of all numbers in A_1 equals the sum of all numbers in A_2. Consider two cases as follows:

Case 1 $2L > S$. First, suppose there exists a subset N_1 of $[n]$ such that $\sum_{i \in N_1} a_i = L$. Let $A_1 = \{a_i \mid i \in N_1\}$ and $A_2 = A - A_1$. Then, the sum of all numbers in A_2 is equal to

$$\sum_{i \in [n] - N_1} a_i + 2L - S = S - L + 2L - S = L = \sum_{i \in N_1} a_i.$$

Conversely, suppose A has a partition (A_1, A_2) such that the sum of all numbers in A_1 equals the sum of all numbers in A_2. Without loss of generality, assume $2L - S \in A_2$. Note that the sum of all numbers in A equals $S + 2L - S = 2L$. Therefore, the sum of all numbers in A_1 equals L.

Case 2 $2L < S$. Let $L' = S - L$ and $N_2 = [n] - N_1$. Then $2L' - S > 0$ and $\sum_{i \in N_1} a_i = L$ if and only if $\sum_{i \in N_2} a_i = L'$. Therefore, this case can be done in a way similar to Case 1 by replacing L and N_1 with L' and N_2, respectively.

□

We next study an optimization problem.

Problem 8.6.5 (Knapsack) Suppose you get in a cave and find n items. However, you have only a knapsack to carry them and this knapsack cannot carry all of them. The knapsack has a space limit S and the ith item takes space a_i and has value c_i. Therefore, you would face a problem of choosing a subset of items, which can be put in the knapsack, to maximize the total value of chosen items. This problem can be formulated into the following linear 0-1 programming:

$$\max \quad c_1 x_1 + c_2 x_2 + \cdots + c_n x_n$$
$$\text{subject to} \quad a_1 x_1 + a_2 x_2 + \cdots + a_n x_n \leq S$$
$$x_1, x_2, \ldots, x_n \in \{0, 1\}$$

In this 0-1 linear programming, variable x_i is an indicator that $x_i = 1$ if the ith item is chosen, and $x_i = 0$ if the ith item is not chosen.

Theorem 8.6.6 *The knapsack problem is NP-hard.*

Proof The decision version of the knapsack problem is as follows: Given positive integers $a_1, a_2, \ldots, a_n$, $c_1, c_2, \ldots, c_n$, S and k, does the following system of inequalities have 0-1 solution?

$$c_1 x_1 + c_2 x_2 + \cdots + c_n x_n \geq k,$$
$$a_1 x_1 + a_2 x_2 + \cdots + a_n x_n \leq S.$$

We construct a polynomial-time many-one reduction from the partition problem to the decision version of the knapsack problem. Consider an instance of the partition problem, consisting of positive integers $a_1, a_2, \ldots, a_n$. Define an instance of the decision version of the knapsack problem by setting

$$c_i = a_i \text{ for } 1 \leq i \leq n$$
$$k = S = \lfloor (a_1 + a_2 + \cdots + a_n)/2 \rfloor.$$

Then the partition problem receives yes-answer if and only if the decision version of knapsack problem receives yes-answer.   □

The knapsack problem has a simple 1/2-approximation (Algorithm 27).

Without loss of generality, assume $a_i \leq S$ for every $1 \leq i \leq n$. Otherwise, item i can be removed from our consideration because it cannot be put in the knapsack. First, sort all items into ordering $\frac{c_1}{a_1} \geq \frac{c_2}{a_2} \geq \cdots \geq \frac{c_n}{a_n}$. Then put items one by one into knapsack according to this ordering, until no more items can be put in. Suppose that above process stops at the kth item, that is, either $k = n$ or first k items have been placed into the knapsack and the $(k+1)$th item cannot be put in. In the former case, all n items can be put in the knapsack. In the latter case, if $\sum_{i=1}^{k} c_i > c_{k+1}$,

---

Algorithm 27 1/2-approximation for knapsack

Input: n items $1, 2, \ldots, n$ and a knapsack with volume S. Each item i is associated with a positive volume a_i and a positive value c_i. Assume $a_i \leq S$ for all $i \in [n]$.
Output: A subset A of items with total value c_G.
1: sort all items into ordering $c_1/a_1 \geq c_2/a_2 \geq \cdots \geq c_n/a_n$;
2: $A \leftarrow \emptyset, k \leftarrow 1$;
3: **if** $\sum_{i=1}^{n} a_i \leq S$ **then**
4:      $A \leftarrow [n]$
5: **else**
6:      **while** $\sum_{i \in A} \leq S$ and $k < n$ **do**
7:          $k \leftarrow k + 1$
8:      **end while**
9:      **if** $\sum_{i=1}^{k_1} c_i > c_k$ **then**
10:          $A \leftarrow [k - 1]$
11:      **else**
12:          $A \leftarrow \{k\}$
13:      **end if**
14: **end if**
15: $c_G \leftarrow \sum_{i \in A} c_i$;
16: **return** A and c_G.

---

then take the first k items to form a solution; otherwise, take the $(k + 1)$th item as a solution.

Theorem 8.6.7 *Algorithm 27 produces a 1/2-approximation for the knapsack problem.*

Proof If all items can be put in the knapsack, then this will give a simple optimal solution. If not, then $\sum_{i=1}^{k} c_i + c_{k+1} > opt$ where opt is the objective function value of an optimal solution. Hence, $\max(\sum_{i=1}^{k} c_i, c_{k+1}) \geq 1/2 \cdot opt$. $\qquad \square$

From above 1/2-approximation, we may have the following observation: For an item selected into the knapsack, two facts are considered:

- The first fact is the ratio c_i/a_i. The larger ratio means that volume is used for higher value.
- The second fact is the c_i. When putting an item with small c_i and bigger c_i/a_i into the knapsack may affect the possibility of putting items with bigger c_i and smaller c_i/a_i, we may select the one with bigger c_i.

By properly balancing consideration on these two facts, we can obtain a construction for $(1 + \varepsilon)$-approximation for any $\varepsilon > 0$ (Algorithm 28).

Denote $\alpha = c_G \cdot \frac{2\varepsilon}{1+\varepsilon}$ where c_G is the total value of a 1/2-approximation solution obtained by Algorithm 27. Classify all items into two sets A and B. Let A be the set of all items each with value $c_i < \alpha$ and B the set of all items each with value $c_i \geq \alpha$. Suppose $|A| = m$. Sort all items in A in ordering $c_1/a_1 \geq c_2/a_2 \geq \cdots \geq c_m/a_m$.

Algorithm 28 $(1+\varepsilon)$ approximation for knapsack

Input: n items $1, 2, \ldots, n$, a knapsack with volume S and a positive number ε. Each item i is associated with a positive volume a_i and a positive value c_i. Assume $a_i \leq S$ for all $i \in [n]$.
Output: A subset A_ε of items with total value c_ε.

1: run Algorithm 27 to obtain c_G;
2: $\alpha \leftarrow c_G \cdot \frac{2\varepsilon}{1+\varepsilon}$;
3: classify all items into A and B where
4: $A \leftarrow \{i \in [n] \mid c_i < \alpha\}$, $B \leftarrow \{i \in [n] \mid c_i \geq \alpha\}$;
5: sort all items of A into ordering $c_1/a_1 \geq c_2/a_2 \geq \cdots \geq c_m/a_m$;
6: $A_\varepsilon \leftarrow \emptyset, k \leftarrow 1$;
7: $\mathcal{B} \leftarrow \{I \subseteq B \mid |I| \leq 1 + 1/\varepsilon\}$;
8: **for** each $I \in \mathcal{B}$ **do**
9:     **if** $\sum_{i \in I} > S$ **then**
10:         $\mathcal{B} \leftarrow \mathcal{B} \setminus \{I\}$
11:     **else**
12:         $S \leftarrow S - \sum_{i \in I} c_i$;
13:         **if** $\sum_{i \in A} a_i > S$ **then**
14:             **while** $\sum_{i=1}^{k} a_i \leq S$ and $k < m$ **do**
15:                 $k \leftarrow k + 1$
16:             **end while**
17:             **if** $\sum_{i=1}^{k_1} c_i > c_k$ **then**
18:                 $A(I) \leftarrow [k-1]$
19:             **else**
20:                 $A(I) \leftarrow \{k\}$
21:             **end if**
22:         **else**
23:             $A(I) \leftarrow A$
24:         **end if**
25:     **end if**
26:     $c(I) \leftarrow \sum_{i \in I \cup A(I)} c_i$;
27: **end for**
28: $I \leftarrow \operatorname{argmax}_{I \in \mathcal{B}} c(I)$;
29: $A_\varepsilon \leftarrow I \cup A(I)$;
30: $c_\varepsilon \leftarrow c(I)$;
31: **return** A_ε and c_ε.

For any subset I of B, with $|I| \leq 1 + 1/\varepsilon$, if $\sum_{i \in I} a_i > S$, then define $c(I) = 0$; otherwise, select the largest $k \leq m$ satisfying $\sum_{i=1}^{k} a_i \leq S - \sum_{i \in I} c_i$ and define $c(I) = \sum_{i \in I} c_i + \sum_{i=1}^{k} c_i$.

Lemma 8.6.8 *Let* $c_\varepsilon = \max_I c(I)$. *Then*

$$c_\varepsilon \geq \frac{1}{1+\varepsilon} \cdot opt$$

where opt *is the objective function value of an optimal solution.*

Proof Let $I_b = B \cap OPT$ and $I_a = A \cap OPT$ where OPT is an optimal solution. Note that for $I \subseteq B$ with $|I| > 1 + 1/\varepsilon$, we have

$$\sum_{i \in I} a_i > \alpha \cdot (1 + 1/\varepsilon)$$

$$= c_G \cdot \frac{2\varepsilon}{1+\varepsilon} \cdot (1 + 1/\varepsilon)$$

$$\geq opt.$$

Thus, we must have $|I_b| \leq 1 + 1/\varepsilon$ and hence $c_\varepsilon \geq c(I_b)$. Moreover, we have

$$c(I_b) = \sum_{i \in I_b} c_i + \sum_{i \in I_a} c_i$$

$$\geq opt - \alpha$$

$$= opt - c_G \cdot \frac{2\varepsilon}{1+\varepsilon}$$

$$\geq opt - \frac{opt}{2} \cdot \frac{2\varepsilon}{1+\varepsilon}$$

$$= opt \cdot \frac{1}{1+\varepsilon}.$$

Therefore, $c_\varepsilon \geq opt \cdot \frac{1}{1+\varepsilon}$.   □

Lemma 8.6.9 *Algorithm 28 runs in $O(n^{2+1/\varepsilon})$ time.*

Proof Note that there are at most $n^{1+1/\varepsilon}$ subsets I of B with $|I| \leq 1 + 1/\varepsilon$. For each such I, the algorithm runs in $O(n)$ time. Hence, the total time is $O(n^{2+1/\varepsilon})$.
                                                                                                □

An optimization problem is said to have PTAS (polynomial-time approximation scheme) if for any $\varepsilon > 0$, there is a polynomial-time $(1 + \varepsilon)$-approximation for the problem. By Lemmas 8.6.8 and 8.6.9, the knapsack problem has a PTAS.

Theorem 8.6.10 *Algorithm 28 provides a PTAS for the knapsack problem.*

A PTAS is called a FPTAS (fully polynomial-time approximation scheme) if for any $\varepsilon > 0$, there exists a $(1 + \varepsilon)$-approximation with running time which is a polynomial with respect to $1/\varepsilon$ and the input size. Actually, the knapsack problem also has a FPTAS. To show it, let us first study exact solutions for the knapsack problem.

Let $opt(k, S)$ be the objective function value of an optimal solution of the following problem:

$$\max \ c_1 x_1 + c_2 x_2 + \cdots + c_k x_k$$

$$\text{subject to } a_1 x_1 + a_2 x_2 + \cdots + a_k x_k \leq S$$

$$x_1, x_2, \ldots, x_k \in \{0, 1\}.$$

Then

$$opt(k, S) = \max(opt(k-1, S), c_k + opt(k-1, S - a_k)).$$

This recursive formula gives a dynamic programming to solve the knapsack problem within $O(nS)$ time. This is a pseudopolynomial-time algorithm, not a polynomial-time algorithm because the input size of S is $\lceil \log_2 S \rceil$, not S.

To construct a PTAS, we need to design another pseudopolynomial-time algorithm for the knapsack problem.

Let $c(i, j)$ denote a subset of index set $\{1, \ldots, i\}$ such that

(a) $\sum_{k \in c(i,j)} c_k = j$ and
(b) $\sum_{k \in c(i,j)} s_k = \min\{\sum_{k \in I} s_k \mid \sum_{k \in I} c_k = j, I \subseteq \{1, \ldots, i\}\}$.

If no index subset satisfies (a), then we say that $c(i, j)$ is undefined, or write $c(i, j) = nil$. Clearly, $opt = \max\{j \mid c(n, j) \neq nil$ and $\sum_{k \in c(i,j)} s_k \leq S\}$. Therefore, it suffices to compute all $c(i, j)$. The following algorithm is designed with this idea.

Initially, compute $c(1, j)$ for $j = 0, \ldots, c_{sum}$ by setting

$$c(1, j) := \begin{cases} \emptyset & \text{if } j = 0, \\ \{1\} & \text{if } j = c_1, \\ nil & \text{otherwise,} \end{cases}$$

where $c_{sum} = \sum_{i=1}^{n} c_i$.

Next, compute $c(i, j)$ for $i \geq 2$ and $j = 0, \ldots, c_{sum}$.

for $i = 2$ **to** n **do**
    for $j = 0$ **to** c_{sum} **do**
        case 1 $[c(i-1, j - c_i) = nil]$
            set $c(i, j) = c(i-1, j)$
        case 2 $[c(i-1, j - c_i) \neq nil]$
            and $[c(i-1, j) = nil]$
            set $c(i, j) = c(i-1, j - c_i) \cup \{i\}$
        case 3 $[c(i-1, j - c_i) \neq nil]$
            and $[c(i-1, j) \neq nil]$
            if $[\sum_{k \in c(i-1,j)} s_k > \sum_{k \in c(i-1,j-c_i)} s_k + s_i]$
            then $c(i, j) := c(i-1, j - c_i) \cup \{i\}$
            else $c(i, j) := c(i-1, j)$;
Finally, set $opt = \max\{j \mid c(n, j) \neq nil$ and $\sum_{k \in c(i,j)} s_k \leq S\}$.

This algorithm computes the exact optimal solution for the knapsack problem with running time $O(n^3 M \log(MS))$ where $M = \max_{1 \leq k \leq n} c_k$, because the algorithm contains two loops, the outside loop runs in $O(n)$ time, the inside loop runs in $O(nM)$ time, and the central part runs in $O(n \log(MS))$ time. This is a pseudopolynomial-time algorithm because the input size of M is $\log_2 M$ and the running time is not a polynomial with respect to input size.

Now, we use the second pseudopolynomial-time algorithm to design a FPTAS.

For any $\varepsilon > 0$, choose integer $h > 1/\varepsilon$. Denote $c'_k = \lfloor c_k n(h+1)/M \rfloor$ for $1 \le k \le n$ and consider a new instance of the knapsack problem as follows:

$$\max \quad c'_1 x_1 + c'_2 x_2 + \cdots + c'_n x_n$$

$$\text{subject to} \quad s_1 x_1 + s_2 x_2 + \cdots + s_n x_n \le S$$

$$x_1, x_2, \ldots, x_n \in \{0, 1\}.$$

Apply the second pseudopolynomial-time algorithm to this new problem. The running time will be $O(n^4 h \log(nhS))$, a polynomial-time with respect to n, h, and $\log S$. Suppose x^h is an optimal solution of this new problem. Set $c^h = c_1 x_1^h + \cdots + c_n x_n^h$. We claim that

$$\frac{c^*}{c^h} \le 1 + \frac{1}{h},$$

that is, x^h is a $(1 + 1/h)$-approximation.

To show our claim, let $I^h = \{k \mid x_k^h = 1\}$ and $c^* = \sum_{k \in I^*} c_k$. Then, we have

$$c^h = \sum_{k \in I^h} \frac{c_k n(h+1)}{M} \cdot \frac{M}{n(h+1)}$$

$$\ge \sum_{k \in I^h} \lfloor \frac{c_k n(h+1)}{M} \rfloor \cdot \frac{M}{n(h+1)}$$

$$= \frac{M}{n(h+1)} \sum_{k \in I^h} c'_k$$

$$\ge \frac{M}{n(h+1)} \sum_{k \in I^*} c'_k$$

$$\ge \frac{M}{n(h+1)} \sum_{k \in I^*} \left(\frac{c_k n(h+1)}{M} - 1 \right)$$

$$\ge opt - \frac{M}{h+1}$$

$$\ge opt \left(1 - \frac{1}{h+1} \right).$$

Theorem 8.6.11 *The knapsack problem has FPTAS.*

For an application of this result, we study a scheduling problem.

Problem 8.6.12 (Scheduling $P\|C_{max}$) Suppose there are m identical machines and n jobs $J_1, \ldots, J_n$. Each job J_i has a processing time a_i, which does not allow preemption, i.e., the processing cannot be cut. All jobs are available at the beginning. The problem is to find a scheduling to minimize the complete time, called *makespan*.

Theorem 8.6.13 *The scheduling $P\|C_{max}$ problem is NP-hard.*

Proof For $m = 2$, this problem is equivalent to find a partition (N_1, N_2) for $[n]$ to minimize $\max(\sum_{i \in N_1} a_i, \sum_{i \in N_2} a_i)$. Thus, we can reduce the partition problem to the decision version of this problem by requiring the makespan not exceed $\lfloor (\sum_{i=1}^{n} a_i)/2 \rfloor$. □

For $m = 2$, we can also obtain a FPTAS from the FPTAS of the knapsack problem.

To this end, we consider the following instance of the knapsack problem:

$$\max \quad a_1 x_1 + a_2 x_2 + \cdots + a_n x_n$$

$$\text{subject to} \quad a_1 x_1 + a_2 + \cdots + a_n x_n \leq S/2$$

$$x_1, x_2, \ldots, x_n \in \{0, 1\}$$

where $S = a_1 + a_2 + \cdots + a_n$. Note that if opt_k is the objective function value of an optimal solution for this knapsack problem, then $opt_s = S - opt_k$ is the objective function value of an optimal solution of above scheduling problem.

Applying the FPTAS to above instance of the knapsack problem, we may obtain a $(1 + \varepsilon)$-approximation solution $\hat{x}$. Let $N_1 = \{i \mid \hat{x}_i = 1\}$ and $N_2 = \{i \mid \bar{x}_i = 0\}$. Then (N_1, N_2) is a partition of $[n]$ and moreover, we have

$$\max \left(\sum_{i \in N_1} a_i, \sum_{i \in N_2} a_i \right) = \sum_{i \in N_2} a_i = S - \sum_{i \in N_1} a_i$$

and

$$\frac{opt_k}{\sum_{i \in N_1} a_i} \leq 1 + \varepsilon.$$

Therefore,

$$\frac{S - opt_s}{S - \sum_{i \in N_2} a_i} \leq 1 + \varepsilon,$$

that is,

$$S - \sum_{i \in N_2} a_i \geq (S - opt_s)/(1 + \varepsilon).$$

Thus,

$$\sum_{i \in N_2} a_i \leq \frac{\varepsilon S + opt_s}{1 + \varepsilon} \leq \frac{\varepsilon \cdot 2opt_s + opt_s}{1 + \varepsilon} \leq opt_s(1 + \varepsilon).$$

Therefore, (N_1, N_2) is a $(1+\varepsilon)$-approximation solution for the scheduling problem.

8.7   Planar 3SAT

A CNF F is *planar* if graph $G^*(F)$, defined as follows, is planar.

- The vertex set consists of all variables $x_1, x_2, \ldots, x_n$ and all clauses $C_1, C_2, \ldots, C_m$.
- The edge set $E(G^*(F)) = \{(x_i, C_j) \mid x_i$ appears in $C_j\}$.

A CNF F is *strongly planar* if graph $G(F)$, defined as follows, is planar.

- The vertex set consists of all literals $x_1, \bar{x}_1, x_2, \bar{x}_2, \ldots, x_n, \bar{x}_n$ and all clauses $C_1, C_2, \ldots, C_m$.
- The edge set $E(G^*(F)) = \{(x_i, \bar{x}_i) \mid i = 1, 2, \ldots, m\} \cup \{(x_i, C_j) \mid x_i \in C_j\} \cup \{(\bar{x}, C_j) \mid \bar{x}_i \in C_j\}$.

Corresponding two types of planar CNF, there are two problems.

Problem 8.7.1 (Planar 3SAT) Given a planar 3CNF F, determine whether F is satisfiable.

Problem 8.7.2 (Strongly Planar 3SAT) Given a strongly planar 3CNF F, determine whether F is satisfiable.

Theorem 8.7.3 *The planar 3SAT problem is NP-complete.*

Proof The problem is a special case of the 3SAT problem and hence longs to NP. We next construct a reduction to witness 3SAT $\leq_m^P$ planar 3SAT. To do so, consider a 3CNF F and $G^*(F)$. $G^*(F)$ may contain many cross-points. For each cross-point, we use a crosser to remove it. As shown in Fig. 8.18, this crosser is constructed with three $\oplus$ operations each defined by

$$x \oplus y = x\bar{y} + \bar{x}y.$$

We next show that for each $\oplus$ operation $x \oplus y = z$, there exists a planar 3CNF $F_{x \oplus y = z}$ such that

$$x \oplus y = z \Leftrightarrow F_{x \oplus y = z} \in SAT$$

Fig. 8.18 A crosser

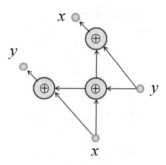

Fig. 8.19 CNF $c(x, y, z)$ is planar

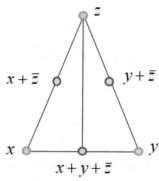

that is, $F_{x \oplus y = z}$ is satisfiable.

Note that a CNF $c(x, y, z) = (x + y + \bar{z})(\bar{x} + z)(\overline{(y)} + z)$ is planar as shown in Fig. 8.19. Moreover, we have

$$x + y = z \Leftrightarrow c(x, y, z) \in SAT,$$

$$x \cdot y = z \Leftrightarrow c(\bar{x}, \bar{y}, \bar{z}) \in SAT.$$

Since

$$x \oplus y = (x + y) \cdot \bar{y} + \bar{x} \cdot (x + y),$$

we have

$$x \oplus y = z \Leftrightarrow F'_{x \oplus y = z} = c(x, y, u)c(\bar{u}, y, \bar{v})c(x, \bar{u}, \bar{w})c(v, w, z) \in SAT.$$

As shown in Fig. 8.20, $F'_{x \oplus y = z}$ is planar. $F'_{x \oplus y = z}$ contains some clauses with two literals. Each such clause $x + y$ can be replaced by two clauses $(x + y + w)(x + y + \bar{w})$ with a new variable w as shown in Fig. 8.21. Then we can obtain a planar 3CNF $F_{x \oplus y = z}$ such that

$$x \oplus y = z \Longleftrightarrow F_{x \oplus y = z} \in SAT.$$

Fig. 8.20 CNF $F'_{x \oplus y = z}$ is
planar

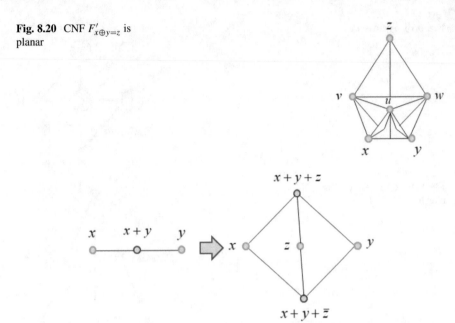

Fig. 8.21 A clause of size 2 can be replaced by two clauses of size 3

Finally, look back at the instance 3CNF F of the 3SAT problem at the beginning.
Let F^* be the product of F and all 3CNFs for all $\oplus$ operations appearing in crossers
used for removing cross-points in $G^*(F)$. Then F^* is planar and

$$F \in SAT \iff F^* \in SAT.$$

This completes our reduction from the 3SAT problem to the planar 3SAT problem.
□

Theorem 8.7.4 *The strongly planar 3SAT problem is NP-complete.*

Proof Clearly, the strongly planar 3SAT problem belongs to NP. Next, we construct
a polynomial-time many-one reduction from the planar 3SAT problem to the
strongly planar 3SAT problem. Consider a planar 3CNF F. In $G^*(F)$, if we replace
each vertex labeled with variable x by an edge with endpoints x and $\bar{x}$, then some
cross-points will be introduced.

To overcome this trouble, we replace the vertex with label x in $G^*(F)$ by a cycle
$G(F_x)$ (Fig. 8.22) where

$$F_x = (x + \bar{w}_1)(w_1 + \bar{w}_2) \cdots (w_k + \bar{x}),$$

and k is selected in the following way: For each edge (x, C_j), label it with x if
C_j contains lateral X, and $\bar{x}$ if C_j contains literal $\bar{x}$. Select k to be the number of

Fig. 8.22 Cycle $G(F_x)$

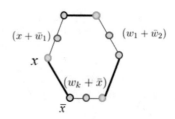

Fig. 8.23 Each vertex x is replaced properly by a cycle $G(F_x)$

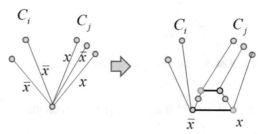

changes from edge x to $\bar{x}$ when travel around vertex x. Note that

$$(x + \bar{w}_1)(w_1 + \bar{w}_2) \cdots (w_k + \bar{x}) = 1 \Rightarrow x = w_1 = \cdots = w_k.$$

Now, each edge (x, C_j) in $G^*(F)$ is properly replaced by an edge (x, C_j) or (w_i, C_j) (Fig. 8.23). We will obtain a planar $G(F')$ where $F' = F \cdot \prod_x F_x$ and F' is a strongly planar CNF. Note that F' contains some clauses of size 2. Finally, we can replace them by clauses of size 3 as shown in Fig. 8.21.   □

As an application, let us consider a problem in planar graphs.

Problem 8.7.5 (Planar Vertex Cover) Given a planar graph G, find a minimum vertex cover of G.

Theorem 8.7.6 *The planar vertex cover problem is NP-hard.*

Proof We construct a polynomial-time many-one reduction from the strongly planar 3SAT problem to the planar vertex cover problem. Let F be a strongly planar 3CNF with n variables $x_1, x_2, \ldots, x_n$ and m clauses $C_1, C_2, \ldots, C_m$. Consider $G(F)$. For each clause C_j, replace vertex C_j properly by a triangle $C_{j1}C_{j2}C_{j3}$ (Fig. 8.24), and we will obtain a planar graph G' such that G' has a vertex cover of size at most $2m + n$ if and only if F is satisfiable.   □

The planar vertex cover problem has PTAS. Actually, a lot of combinatorial optimization problem in planar graph and geometric plan or space have PTAS. We will discuss them in later chapters.

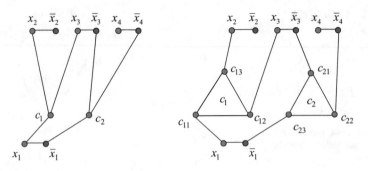

Fig. 8.24 Each vertex C_j is replaced properly by a triangle $C_{j1}C_{j2}C_{j3}$

8.8 Complexity of Approximation

In previous sections, we studied several NP-hard combinatorial optimization problems. Based on their approximation solutions, they may be classified into the following four classes:

1. PTAS, consisting of all combinatorial problems each of which has a PTAS, e.g., the knapsack problem and the planar vertex cover problem.
2. APX, consisting of all combinatorial optimization problems each of which has a polynomial-time $O(1)$-approximation, e.g., the vertex cover problem and the Hamiltonian cycle problem with triangular inequality.
3. Log-APX, consisting of all combinatorial optimization problems each of which has a polynomial-time $O(\ln n)$-approximation for minimization, or $(1/O(\ln n))$-approximation for maximization, e.g., the set cover problem.
4. Poly-APX, consisting of all combinatorial optimization problems each of which has a polynomial-time $p(n)$-approximation for minimization, or $(1/p(n))$-approximation for maximization for some polynomial $p(n)$, e.g., the maximum independent set problem and the longest path problem.

Clearly, PTAS $\subset$ APX $\subset$ Log-APX $\subset$ Poly-APX. Moreover, we have

Theorem 8.8.1 *If NP $\neq$ P, then PTAS $\neq$ APX $\neq$ Log-APX $\neq$ Poly-APX.*

Actually from previous sections, we know that the set cover problem has $(1 + \ln n)$-approximation and if NP $\neq$ P, then it has no polynomial-time $O(1)$-approximation. Hence, the set cover problem separates APX and Log-APX. The maximum independent set problem has a trivial $1/n$-approximation by taking a single vertex as solution and hence it belongs to Poly-APX. Moreover, if NP $\neq$ P, then it has no polynomial-time $n^{\varepsilon-1}$-approximation. Hence, the maximum independent set problem separates Log-APX from Poly-APX. Next, we study a problem which separates PTAS from APX.

Problem 8.8.2 (*k*-**Center**) Given a set C of n cities with a distance table, find a subset S of k cities as centers to minimize

$$\max_{c \in C} \min_{s \in S} d(c, s)$$

where $d(c, s)$ is the distance between c and s.

Theorem 8.8.3 *The k-center problem with triangular inequality has a polynomial-time 2-approximation.*

Proof Consider the following algorithm:

Initially, choose arbitrarily a vertex $s_1 \in C$ and set
$S_1 \leftarrow \{s_1\}$;
for $i = 2$ **to** k **do**
    select $s_i = \text{arcmax}_{c \in C} d(c, S_{i-1})$, and set
    $S_i \leftarrow S_{i-1} \cup \{s_i\}$;
output S_k.

We will show that this algorithm gives a 2-approximation.
    Let S^* be an optimal solution. Denote

$$opt = \max_{c \in C} d(c, S^*).$$

Classify all cities into k clusters such that each cluster contains a center $s^* \in S^*$ and $d(c, s^*) \leq d^*$ for every city c in the cluster. Now, we consider two cases.

Case 1 Every cluster contains a member $s_i \in S_k$. Then for each city c in the cluster with center s^*, $d(c, s_i) \leq d(c, s^*) + d(s^*, s_i) \leq 2 \cdot opt$.

Case 2 There is a cluster containing two members $s_i, s_j \in S_k$ with $i < j$. Suppose the center of this cluster is s^*. Then for any $c \in C$,

$$
\begin{aligned}
d(c, S_k) &\leq d(c, S_{j-1}) \\
&\leq d(s_j, S_{j-1}) \\
&\leq d(s_j, s_i) \\
&\leq d(s_i, s^*) + d(s^*, s_j) \\
&\leq 2 \cdot opt.
\end{aligned}
$$

$\square$

Fig. 8.25 For each edge
(u, v), add a new vertex x_{uv}
and two edges (x_{uv}, u) and
(x_{uv}, v)

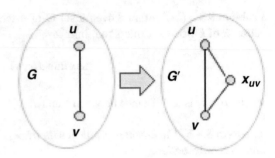

A corollary of Theorem 8.8.3 is that the k-center problem belongs to APX. Before we show that the k-center problem does not belong to PTAS unless NP=P, let us study another problem.

Problem 8.8.4 (Dominating Set) Given a graph G, find the minimum dominating set where a dominating set is a subset of vertices such that every vertex is either in the subset or adjacent to a vertex in the subset.

Lemma 8.8.5 *The decision version of the dominating set problem is NP-complete.*

Proof Consider an input graph $G = (V, E)$ of the vertex cover problem. For each edge (u, v), create a new vertex x_{uv} together with two edges (u, x_{uv}) and (x_{uv}, v) (Fig. 8.25). Then we obtain a modified graph G'. If G has a vertex cover of size $\leq k$, then the same vertex subset must be a dominating set of G', also of size $\leq k$.

Conversely, if G' has a dominating set D of size $\leq k$, then without loss of generality, we may assume $D \subseteq E$. In fact, if $x_{uv} \in D$, then we can replace x_{uv} by either u or v, which results in a dominating set of the same size. Since $D \subseteq E$ dominating all x_{uv} in G', D covers all edges in G.   □

Now, we come back to the k-center problem.

Theorem 8.8.6 *For any $\varepsilon > 0$, the k-center problem with triangular inequality does not have a polynomial-time $(2 - \varepsilon)$-approximation unless NP=P.*

Proof Suppose that the k-center problem has a polynomial-time $(2 - \varepsilon)$-approximation algorithm A. We use algorithm A to construct a polynomial-time algorithm for the decision version of the dominating set problem.

Consider an instance of the decision version of the dominating set problem, consisting of a graph $G = (V, E)$ and a positive integer k. Construct an instance of the k-center problem by choosing all vertices as cities with distance table defined as follows:

$$d(u, v) = \begin{cases} 1 & \text{if } (u, v) \in E, \\ |V| + 1 & \text{otherwise.} \end{cases}$$

If G has a dominating set of size at most k, then the k-center problem will have an optimal solution with $opt = 1$. Therefore, algorithm A produces a solution with objective function value at most $(2 - \varepsilon)$, actually has to be one. If G does not have a dominating set of size at most k, then the k-center problem will have its optimal solution with $opt \geq 2$. Hence, algorithm A produces a solution with objective function value at least two. Therefore, from objective function value of solution produced by algorithm A, we can determine whether G has a dominating set of size $\leq k$ or not. By Lemma 8.8.5, we have NP $=$ P. □

By Theorems 8.8.3 and 8.8.6, the k-center problem with triangular inequality separates PTAS and APX.

Actually, APX is a large class which contains many problems not in PTAS if NP$\neq$P. Those are called APX-complete problems. There are several reductions to establish the APX-completeness. Let us introduce a popular one, the polynomial-time L-reduction.

Consider two combinatorial optimization problems Π and Γ. Π is said to be polynomial-time L-reducible to Γ, written as $\Pi \leq_L^p \Gamma$, if there exist two polynomial-time computable functions h and g, and two positive constants a and b such that

(L1) h maps from instances x of Π to instances $h(x)$ of Γ such that

$$opt_\Gamma(h(x)) \leq a \cdot opt_\Pi(x)$$

where $opt_\Pi(x)$ is the objective function value of an optimal solution for Π on instance x;

(L2) g maps from feasible solutions y of Γ on instance $h(x)$ to feasible solutions $g(y)$ of Π on instance x such that

$$|obj_\Pi(g(y)) - opt_\Pi(x)| \leq b \cdot |obj_\Gamma(y) - opt_\Gamma(h(x))|$$

where $obj_\Gamma(y)$ is the objective function value of feasible solution y for Γ (Fig. 8.26).

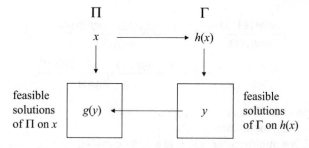

Fig. 8.26 Definition of L-reduction

Fig. 8.27 The proof of
Theorem 8.8.7

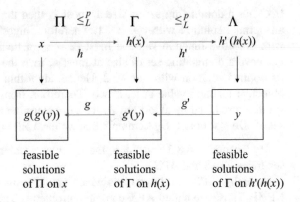

This reduction has the following properties:

Theorem 8.8.7 $\Pi \leq_L^p \Gamma, \Gamma \leq_L^p \Lambda \Leftarrow \Pi \leq_L^p \Lambda.$

Proof As shown in Fig. 8.27, we have

$$opt_\Lambda(h'(h(x))) \leq a' \cdot opt_\Gamma(h(x)) \leq a'a \cdot opt_\Pi(x)$$

and

$$|obj_\Pi(g(g'(y))) - opt_\Pi(x)|$$
$$\leq b \cdot |obj_\Gamma(g'(y)) - opt_\Gamma(h(x))|$$
$$\leq bb' \cdot |obj_\Lambda(y) - opt_\Lambda(h'(h(x)))|.$$

$\square$

Theorem 8.8.8 *If* $\Pi \leq_L^p \Gamma$ *and* $\Gamma \in PTAS,$ *then* $\Pi \in PTAS.$

Proof Consider four cases.

Case 1. Both Π and Γ are minimization problems:

$$\frac{obj_\Pi(g(y))}{opt_\Pi(x)} = 1 + \frac{obj_\Pi(g(y)) - opt_\Pi(x)}{opt_\Pi(x)}$$
$$\leq 1 + \frac{ab(obj_\Gamma(y) - opt_\Gamma(h(x)))}{opt_\Gamma(h(x))}.$$

If y is a polynomial-time $(1 + \varepsilon)$-approximation for Γ, then $g(y)$ is a polynomial-time $(1 + ab\varepsilon)$-approximation for Π.

Case 2. Π is a minimization and Γ is a maximization:

$$\frac{obj_\Pi(g(y))}{opt_\Pi(x)} = 1 + \frac{obj_\Pi(g(y)) - opt_\Pi(x)}{opt_\Pi(x)}$$

$$\leq 1 + \frac{ab(opt_\Gamma(h(x)) - obj_\Gamma(y))}{opt_\Gamma(h(x))}$$

$$\leq 1 + \frac{ab(opt_\Gamma(h(x)) - obj_\Gamma(y))}{obj_\Gamma(y)}.$$

If y is a polynomial-time $(1+\varepsilon)^{-1}$-approximation for Γ, then $g(y)$ is a polynomial-time $(1 + ab\varepsilon)$-approximation for Π.

Case 3 Π is a maximization and Γ is a minimization:

$$\frac{obj_\Pi(g(y))}{opt_\Pi(x)} = 1 - \frac{opt_\Pi(x) - obj_\Pi(g(y))}{opt_\Pi(x)}$$

$$\geq 1 - \frac{ab(obj_\Gamma(y) - opt_\Gamma(h(x)))}{opt_\Gamma(h(x))}.$$

If y is a polynomial-time $(1 + \varepsilon)$-approximation for Γ, then $g(y)$ is a polynomial-time $(1 - ab\varepsilon)$-approximation for Π.

Case 4 Both Π and Γ are maximization:

$$\frac{obj_\Pi(g(y))}{opt_\Pi(x)} = 1 - \frac{opt_\Pi(x) - obj_\Pi(g(y))}{opt_\Pi(x)}$$

$$\geq 1 - \frac{ab(opt_\Gamma(h(x)) - obj_\Gamma(y))}{opt_\Gamma(h(x))}$$

$$\geq 1 - \frac{ab(opt_\Gamma(h(x)) - obj_\Gamma(y))}{obj_\Gamma(y)}.$$

If y is a polynomial-time $(1+\varepsilon)^{-1}$-approximation for Γ, then $g(y)$ is a polynomial-time $(1 - ab\varepsilon)$-approximation for Π. $\quad\square$

Let us look at some examples for APX-complete problems.

Problem 8.8.9 (MAX3SAT-3) Given a 3CNF F that each variable appears in at most three clauses, find an assignment to maximize the number of satisfied clauses.

Theorem 8.8.10 *The MAX3SAT-3 problem is APX-complete.*

Let us use this APX-completeness as a root to derive others.

Problem 8.8.11 (MI-b) Given a graph G with vertex-degree upper-bounded by b, find a maximum independent set.

Fig. 8.28 The proof of
Theorem 8.8.12

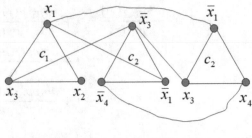

$$(x_1 + x_3 + x_2)(\bar{x}_3 + \bar{x}_4 + \bar{x}_1)(\bar{x}_1 + x_3 + x_4)$$

Theorem 8.8.12 *The MI-4 problem is APX-complete.*

Proof First, we show MI-4 $\in$ APX. Given a graph $G = (V, E)$, construct a maximal independent set by selecting vertices iteratively, and at each iteration, select a vertex and delete it together with its neighbors until no vertex is left. Clearly, this maximal independent set contains at least $|V|/5$ vertices. Therefore, it gives a polynomial-time 1/5-approximation.

Next, we show MAX3SAT $\leq_L^P$ MI-4. Consider an instance 3CNF F of MAX3SAT, with n variables $x_1, x_2, \ldots, x_n$ and m clauses $C_1, C_2, \ldots, C_m$. For each clause C_j, create a triangle with three vertices C_{j1}, C_{j2}, C_{j3} labeled by three literals of C_j, respectively. Then connect every vertex with label x_i to every vertex with label $\bar{x}_i$ as shown in Fig. 8.28. This graph is denoted by $h(F)$. For each independent set y of $h(F)$, we define an assignment $g(y)$ to make every vertex in the independent set with a true literal. Thus, F has at least $|y|$ clauses satisfied.

To show (L1), we claim that

$$opt_{MAX3SAT}(F) = opt_{MI-4}(h(F)).$$

Suppose x^* is an optimal assignment. Construct an independent set y^* by selecting one vertex with true label in each satisfied clause. Then

$$opt_{MAX3SAT}(F) = |y^*| \leq opt_{MI-4}(h(F)).$$

Conversely, suppose y^* is a maximum independent set of $h(F)$. We have F have at least $|y^*|$ satisfied clauses with assignment $g(y^*)$. Therefore,

$$opt_{MAX3SAT} F \geq |y^*| = opt_{MI-4}(h(F)).$$

To see (L2), we note that

$$|opt_{MAX3SAT}(F) - obj_{MAX3SAT}(g(y))| \leq |opt_{MI-4}(h(F)) - obj_{MI-4}(y)|$$

since $obj_{MAX3SAT}(g(y)) \geq obj_{MI-4}(y)$. $\qquad\square$

Fig. 8.29 Vertex u is replaced by a path $(u_1, v_1, u_2, v_2, u_3, v_3, u_4)$

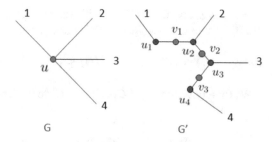

Theorem 8.8.13 *The MI-3 problem is APX-complete.*

Proof Since MI-4 $\in$ APX, so is MI-3. We next show MI-4 $\leq_L^P$ MI-3. Consider a graph $G = (V, E)$ with vertex degree at most 4. For each vertex u with degree 4, we put a path $(u_1, v_1, u_2, v_2, u_3, v_3, u_4)$ as shown in Fig. 8.29. Denote obtained graph by $G' = h(G)$. Then (L1) holds since

$$opt_{MI_3}(G') \leq |V(G')| \leq 7|V(G)| \leq 28 \cdot opt_{MI-4}(G)$$

where $V(G)$ denotes the vertex set of graph G. To see (L2), note that the path $(u_1, v_1, u_2, v_2, u_3, v_3, u_4)$ has unique maximum independent set $I_u = \{u_1, u_2, u_3, u_4\}$. For any independent set I of G', define $g(I)$ to be obtained from I by replacing set I_u by vertex u and removing all other vertices not in G. Then $g(I)$ is an independent set of G. We claim

$$opt_{MI-4}(G) - |g(I)| \leq opt_{MI-3}(G') - |I|.$$

To show it, define I' to be obtained from I by $\{v_1, v_2, v_3\}$ if I contains one of v_1, v_2, v_3. Clearly, I' is still an independent set of G' and $|I| \leq |I'|$ and $g(I) = g(I')$. Then, we have

$$opt_{M-4}(G) - g(I) = opt_{MI-4}(G) - g(I')$$
$$= opt_{MI-3}(G') - |I'|$$
$$\leq opt_{MI-3}(G') - |I|.$$

□

Problem 8.8.14 (VC-b) Given a graph G with vertex degree upper-bounded by b, find the minimum vertex cover.

Theorem 8.8.15 *The VC-3 problem is APX-complete.*

Proof Since the vertex cover problem has a polynomial-time 2-approximation, so does the VC-3 problem. Hence, VC-3 $\in$ APX. Next, we show MI-3 $\leq_L^P$ VC-3.

For any graph $G = (V, E)$ with vertex degree at most 3, define $h(G) = G$. To see (L1), note that $opt_{MI-3} \geq |V|/4$ and $|E| \leq 3|V|/2 = 1.5|V|$. Hence,

$$opt_{VC-3}(G) = |V| - opt_{MI-3}(G) \leq |V| - |V|/4 = (3/4)|V| \leq 3 \cdot opt_{MI-3}(G).$$

Now, for any vertex cover C, define $g(C)$ to be the complement of C. Then we have

$$opt_{MI-3}(G) - |g(C)| = |V| - opt_{VC-3}(G) - (|V| - |C|)$$
$$= |C| - opt_{VC-3}(G).$$

Therefore, (L2) holds.   □

There are also many problems in Log-APX \ APX, such as various optimization problems on covering and dominating. The lower bound for their approximation performance is often established based on Theorem 8.5.7 with a little modification.

Theorem 8.8.16 *Theorem 8.5.7 still holds in special case that each input consisting of a collection C of subsets of a final set X with condition $|C| \leq |X|$, that is, in this case, we still have that for any $0 < \rho < 1$, the set cover problem does not have a polynomial-time $(\rho \ln n)$-approximation unless NP=P where $n = |X|$.*

Here is an example.

Theorem 8.8.17 *For any $0 < \rho < 1$, the dominating set problem does not have a polynomial-time $(\rho \ln n)$-approximation unless NP=P where n is the number of vertices in input graph.*

Proof Suppose there exists a polynomial-time $(\rho \ln n)$-approximation for the dominating set problem. Consider instance (X, C) of the set cover problem with $|C| \leq |X|$. Construct a bipartite graph $G = (C, X, E)$. For $S \in C$ and $x \in X$, there exists an edge $(S, x) \in E$ if and only if $x \in S$. Add two new vertices o and o' together with edges (o, o') and (S, o) for all $S \in C$. Denote this new graph by G' as shown in Fig. 8.30.

First, note that

$$opt_{ds}(G') \leq opt_{sc}(X, C) + 1$$

where $opt_{ds}(G')$ denotes the size of minimum dominating set of G' and $opt_{sc}(G)$ denotes the cardinality of minimum set cover on input (X, C). In fact, suppose that C^* is a minimum set cover on input (X, C). Then $C \cup \{o\}$ is a dominating set of G'.

Next, consider a dominating set D of G', generated by the polynomial-time $(\rho \ln n)$-approximation for the dominating set problem. Then, we have $|D| \leq (\rho \ln(2|X| + 2))opt_{ds}(G')$. Construct a set cover S as follows:

Step 1.   If D does not contain o, then add o. If D contains a vertex with label $x \in X$, then replace x by a vertex with label $C \in C$ such that $x \in C$.

We will obtain a dominating set D' of G' with size $|D'| \leq |D| + 1$ and without vertex labeled by element in X.

Step 2.   Remove o and o' from D'. We will obtain a set cover $\mathcal{S}$ for X with size $|\mathcal{S}| \leq |D|$.

Note that

$$|\mathcal{S}| \leq |D|$$
$$\leq (\rho \ln(2|X| + 2)) opt_{ds}(G')$$
$$\leq (\rho \ln(2|X| + 2))(1 + opt_{sc}(X, \mathcal{C}))$$
$$= \frac{\ln(2|X| + 2)}{\ln|X|} \cdot (1 + \frac{1}{opt_{sc}(X, \mathcal{C})}) \cdot (\rho \ln|X|) opt_{sc}(X, \mathcal{C}).$$

Select two sufficiently large positive constants α and β such that

$$\rho' = \frac{\ln(2\alpha + 2)}{\ln \alpha} \cdot (1 + \frac{1}{\beta}) \cdot \rho < 1.$$

Then for $|X| \geq \alpha$ and $opt_{sc}(X, \mathcal{C}) \geq \beta$,

$$|\mathcal{S}| \leq (\rho' \ln|X|) \cdot opt_{sc}(X, \mathcal{C}).$$

For $|X| < \alpha$ or $opt_{sc}(X, \mathcal{C}) < \beta$, an exactly optimal solution can be computed in polynomial-time. Therefore, there exists a polynomial-time $(\rho' \ln n)$-approximation for the set cover problem and hence NP=P by Theorem 8.8.16.   □

Class Poly-APX may also be further divided into several levels.

Polylog-APX, consisting of all combinatorial optimization problems each of which has a polynomial-time $O(\ln^i n)$-approximation for minimization, or $(1/O(\ln^i n))$-approximation for maximization for some $i \geq 1$.

Fig. 8.30 The proof of Theorem 8.8.17

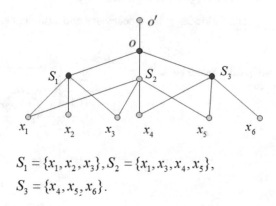

$$S_1 = \{x_1, x_2, x_3\}, S_2 = \{x_1, x_3, x_4, x_5\},$$
$$S_3 = \{x_4, x_5, x_6\}.$$

Sublinear-APX, consisting of all combinatorial optimization problems each of which has a polynomial-time $O(n^a)$-approximation for minimization, or $(1/n^a)$-approximation for maximization for some $0 < a < 1$.

Linear-APX, consisting of all combinatorial optimization problems each of which has a polynomial-time $O(n)$-approximation for minimization, or $(1/n)$-approximation for maximization.

In the literature, we can find the group Steiner tree problem [168, 203] and the connected set cover problem [442] in Polylog-APX \ Log-APX unless some complexity class collapses; the directed Steiner tree problem [46] and the densest k subgraph problem [28] in Sublinear-APX \ Log-APX. However, there are quite a few problems in Linear-APX \ Sublinear-APX. Especially, we may meet such a problem in the real world. Next, we give an example in the study of wireless sensors.

Consider a set of wireless sensors lying in a rectangle which is a piece of boundary area of region of interest. The region is below the rectangle and outside is above the rectangle. The monitoring area of each sensor is a unit disk, i.e., a disk with radius of unit length. A point is said to be covered by a sensor if it lies in the monitoring disk of the sensor. The set of sensors is called a *barrier cover* if they can cover a line (not necessarily straight) connecting two vertical edges (Fig. 8.31) of the rectangle. The barrier cover is used for protecting any intruder coming from outside. Sensors are powered with batteries and hence lifetime is limited. Assume that all sensors have unit lifetime. When several disjoint barrier covers are available, they are often working one by one, so that a security problem is raised.

In Fig. 8.31, a point a lies behind barrier cover B_1 and in front of barrier cover B_2. Suppose B_2 works first and after B_2 stops, B_1 starts to work. Then the intruder can go to point a during the period that B_2 works. After B_2 stops, the intruder enters the area of interest without getting monitored by any sensor. Thus, scheduling (B_2, B_1) is not secure. The existence of point b in Fig. 8.31 indicates that scheduling (B_1, B_2) is not secure neither. Thus, in Fig. 8.31, secure scheduling can contain only one barrier cover. In general, we have the following problem:

Problem 8.8.18 (Secure Scheduling) Given n disjoint barrier covers $B_1, B_2, \ldots,$ B_n, find a longest secure scheduling.

The following gives a necessary and sufficient condition for a secure scheduling:

Fig. 8.31 Sensor barrier covers

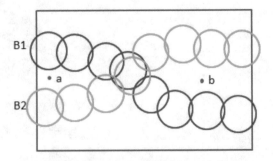

Lemma 8.8.19 *A scheduling* $(B_1, B_2, .., B_k)$ *is secure if and only if for any* $1 \leq i \leq k - 1$, *there is no point* a *lying above* B_i *and below* B_{i+1}.

Proof If such a point a exists, then the scheduling is not secure since the intruder can walk to point a during B_i works and enters into the area of interest during B_{i+1} works. Thus, the condition is necessary.

For sufficiency, suppose the scheduling is not secure. Consider the moment at which the intruder gets the possibility to enter the area of interest and the location a where the intruder lies. Let B_i work before this moment. Then a must lie above B_i and below B_{i+1}. □

This lemma indicates that the secure scheduling can be reduced to the longest path problem in directed graphs in the following way:

- Construct a directed graph G as follows: For each barrier cover B_i, create a node i. For two barrier covers B_i and B_j, if there exists a point a lying above barrier cover B_i and below barrier cover B_j, add an arc (i, j).
- Construct the complement $\bar{G}$ of graph G, that is, $\bar{G}$ and G have the same node set and an arc in $\bar{G}$ if and only if it is not in G.

By Lemma 8.8.19, each secure scheduling of barrier covers a corresponding simple path in $\bar{G}$, and a secure scheduling is maximum if and only if a corresponding simple path is the longest one. Actually, the longest path problem can also be reduced to the secure scheduling problem as shown in the proof of the following theorem:

Theorem 8.8.20 *For any* $\varepsilon > 0$, *the secure scheduling problem has no polynomial-time* $n^{1-\varepsilon}$-*approximation unless* $NP = P$.

Proof Let us reduce the longest path problem in directed graph to the secure scheduling problem. Consider a directed graph $G = (V, E)$. Let $\bar{G} = (V, \bar{E})$ be the complement of G, i.e., $\bar{E} = \{(i, j) \in V \times V \mid (i, j) \notin E\}$. Draw a horizontal line L and for each arc $(i, j) \in \bar{E}$, create a point (i, j) on the line L. All points (i, j) are apart from each other with distance 6 units (Fig. 8.32). At each point (i, j), add a disk S_{ij} with center (i, j) and unit radius. Cut line L into a segment L' to include all disks between two endpoints. Add more unit disks with centers on the segment L' to cover the uncovered part of L' such that point (i, j) is covered only by S_{ij}. Let B_0 denote the set of sensors with constructed disks as their monitoring areas.

Now, let B_i be obtained from B_0 in the following way:

- For any $(i, j) \in \bar{E}$, remove S_{ij} to break B_0 into two parts. Add two unit disks S_{i1}^{ij} and S_{i2}^{ij} to connect the two parts, such that point (i, j) lies above them.
- For any $(j, i) \in \bar{E}$, remove S_{ji} to break B_0 into two parts. Add two unit disks S_{i1}^{ij} and S_{i2}^{ij} to connect the two parts, such that point (i, j) lies below them.
- To make all constructed barrier covers disjoint, unremoved disks in B_0 will be made copies and put those copies into B_i (see Fig. 8.32).

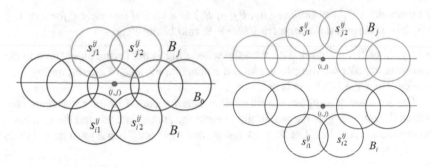

Fig. 8.32 Sensor barrier covers

Clearly, G has a simple path $(i_1, i_2, \ldots, i_k)$ if and only if there exists a secure scheduling $(B_{i_1}, B_{i_2}, \ldots, B_{i_k})$. Therefore, our construction gives a reduction from the longest path problem to the secure scheduling problem. Hence, this theorem can be obtained from Theorem 8.3.9 for the longest path problem.[2]   □

Other than the polynomial-time L-reduction, there exist many reductions preserving or amplifying the gap between approximation solutions and optimal solutions. They are very useful tools to establish the inapproximability of a target problem through transformation from known inapproximability of another problem. The reader may find more in Chapter 10 of [100].

Exercises

1. For any language A, Kleene closure $A^* = A^0 \cup A^1 \cup A^2 \cup \cdots$. Solve the following:

   (a) Design a deterministic Turing machine to accept language $\varnothing^*$.
   (b) Show that if $A \in$ P, then $A^* \in$ P.
   (c) Show that if $A \in$ NP, then $A^* \in$ NP.

2. Given a graph $G = (V, E)$ with edge weight $w : E \rightarrow R^+$, assign each vertex u with a weight x_u to satisfy $x_u + x_v \geq w(u, v)$ for every edge $(u, v) \in E$, and to minimize $\sum_{u \in V} x_u$. Find a polynomial-time solution and a faster 2-approximation.

3. Given a graph $G = (V, E)$ and a positive integer k, find a set C of k vertices to cover the maximum number of edges. Show the following:

   (a) This problem has a polynomial-time $(1/2)$-approximation.

---

[2] Theorem 8.3.9 states for undirected graphs. The same theorem also holds for directed graphs since the graph can be seen as special case of directed graphs.

(b) If this problem has a polynomial-time $1/\gamma$-approximation, then the minimum vertex cover problem has a polynomial-time γ-approximation.

4. Show that the following problems are NP-hard:

   (a) Given a directed graph, find the minimum subset of edges such that every directed cycle contains at least one edge in the subset.
   (b) Given a directed graph, find the minimum subset of vertices such that every directed cycle contains at least one vertex in the subset.

5. Show the NP-completeness of the following problem: Given a sequence of positive integers $a_1, a_2, \ldots, a_n$, determine whether the sequence can be partitioned into three parts with equal sums.

6. Given a Boolean formula F, determine whether F has at least two satisfying assignments. Show that this problem is NP-complete.

7. Show NP-hardness of the following problem: Given a graph G and an integer $k > 0$, determine whether G has a vertex cover C of size at most k, satisfying the following conditions:

   (a) The subgraph $G|_C$ induced by C has no isolated vertex.
   (b) Every vertex in C is adjacent to a vertex not in C.

8. Show that all internal nodes of a depth-first search tree form a vertex cover, which is 2-approximation for the minimum vertex cover problem.

9. Given a directed graph, find an acyclic subgraph containing maximum number of arcs. Design a polynomial-time 1/2-approximation for this problem.

10. A wheel is a cycle with a center (not on the cycle) which is connected to every vertex on the cycle. Prove the NP-completeness of the following problem: Given a graph G, does G have a spanning wheel?

11. Given a 2-connected graph G and a vertex subset A, find the minimum vertex subset B such that $A \cup B$ induces a 2-connected subgraph. Show that this problem is NP-hard.

12. Show that the following problems are NP-hard:

   (a) Given a graph G, find a spanning tree with minimum number of leaves.
   (b) Given a graph G, find a spanning tree with maximum number of leaves.

13. Given two graphs G_1 and G_2, show the following:

   (a) It is NP-complete to determine whether G_1 is isomorphic to a subgraph of G_2 or not.
   (b) It is NP-hard to find a subgraph H_1 of G_1 and a subgraph H_2 of G_1 such that H_1 is isomorphic to H_2 and $|E(H_1)| = |E(H_2)|$ reaches the maximum common.

14. Given a collection $\mathcal{C}$ of subsets of three elements in a finite set X, show the following:

   (a) It is NP-complete to determine whether there exists a set cover consisting of disjoint subsets in $\mathcal{C}$.

(b) It is NP-hard to find a minimum set cover, consisting of subsets in $\mathcal{C}$.

15. Given a graph, find the maximum number of vertex-disjoint paths with length 2. Show the following:

   (a) This problem is NP-hard.
   (b) This problem has a polynomial-time 2-approximation.

16. Design a polynomial-time 2-approximation for the following problem: Given a graph, find a maximal matching with minimum cardinality.

17. (Maximum 3DM) Given three disjoint sets X, Y, Y with $|X| = |Y| = |Z|$ and a collection $\mathcal{C}$ of 3-sets, each 3-set consisting of exactly one element in X, one element in Y, and one element in Z, find the maximum number of disjoint 3-sets in $\mathcal{C}$. Show the following:

   (a) This problem is NP-hard.
   (b) This problem has polynomial-time 3-approximation.
   (c) This problem is APX-complete.

18. There are n students who studied at a late night for an exam. The time has come to order pizzas. Each student has his own list of required toppings (e.g., pepperoni, sausage, mushroom, etc.). Everyone wants to eat at least one third of a pizza, and the topping of the pizza must be in his required list. To save money, every pizza must have only one topping. Find the minimum number of pizzas to order in order to make everybody happy. Please answer the following questions:

   (a) Is it an NP-hard problem?
   (b) Does it belong to APX?
   (c) If everyone wants to eat at least a half of a pizza, is there a change about the answer for above questions?

19. Show that the following is an NP-hard problem: Given two collections $\mathcal{C}$ and $\mathcal{D}$ of subsets of X and a positive integer d, find a subset A with at most d elements of X to minimize the total number of subsets in $\mathcal{C}$ not hit by A and subsets in $\mathcal{D}$ hit by A, i.e., to minimize

$$|\{S \in \mathcal{C} \mid S \cap A = \emptyset\} \cup \{S \in \mathcal{D} \mid S \cap A \neq \emptyset\}|.$$

20. Design a FPTAS for the following problem: Consider n jobs and m identical machine. Assume that m is a constant. Each job j has a processing time p_j and a weight w_j. The processing does not allow preemption. The problem is to find a scheduling to minimize $\sum_j w_j C_j$ where C_j is the completion time of job j.

21. Design a FPTAS for the following problem: Consider a directed graph with a source node s and a sink node t. Each edge e has an associated cost $c(e)$ and length $\ell(e)$. Given a length bound L, find a minimum cost path from s to t of total length at most L.

22. Show the NP-completeness of the following problem: Given n positive integers $a_1, a_2, \ldots, a_n$, is there a partition (I_1, I_2) of $[n]$ such that $|\sum_{i \in I_1} a_i - \sum_{i \in I_2} a_i| \leq 2$?

23. (Ron Graham's Approximation for Scheduling $P||C_{max}$) Show that the following algorithm gives a 2-approximation for the scheduling $P||C_{max}$ problem:

   • List all jobs. Process them according to the ordering in the list.
   • Whenever a machine is available, move the first job from the list to the machine until the list becomes empty.

24. In the proof of Theorem 8.7.4, if letting k be the degree of vertex x, then the proof can also work. Please complete the construction of replacing vertex x by cycle $G(F_x)$.

25. (1-in-3SAT) Given a 3CNF F, is there an assignment such that for each clause of F, exactly one literal gets value 1? This is called the 1-in-3SAT problem. Show the following:

   (a) The 1-in-3SAT problem is NP-complete.
   (b) The planar 1-in-3SAT problem is NP-complete.
   (c) The strongly planar 1-in-3SAT is NP-complete.

26. (NAE3SAT) Given a 3CNF F, determine whether there exists an assignment such that for each clause of F, is there an assignment such that for each clause of F, not all three literals are equal? This is called the NAE3SAT problem. Show the following:

   (a) The NAE3SAT problem is NP-complete.
   (b) The planar NAE3SAT is in P.

27. (Planar 3SAT with Variable Cycle) Given a 3CNF F which has $G^*(F)$ with property that all variables can be connected into a cycle without crossing, is F satisfiable?

   (a) Show that this problem is NP-complete.
   (b) Show that the planar Hamiltonian cycle problem is NP-hard.

28. Show that the planar dominating set problem is NP-hard.

29. Show that the following are APX-complete problems:

   (a) (Maximum 1-in-3SAT) Given a 3CNF F, find an assignment to maximize the number of 1-in-3 clauses, i.e., exactly one literal equal to 1.
   (b) (Maximum NAE3SAT) Given a 3CNF F, find an assignment to maximize the number of NAE clauses, i.e., either one or two literals equal to 1.

30. (Network Steiner Tree) Given a network $G = (V, E)$ with nonnegative edge weight, and a subset of nodes, P, find a tree interconnecting all nodes in P, with minimum total edge weight. Show that this problem is APX-complete.

31. (Rectilinear Steiner Arborescence) Consider a rectilinear plan with origin O. Given a finite set of terminals in the first of this plan, find the shortest arborescence to connect all terminals, that is, the shortest directed tree rooted at

origin O such that for each terminal t, there is a path from O to t and the path
is allowed to go only to the right or upward. Show that this problem is NP-hard.

32. (Connected Vertex Cover) Given a graph $G = (V, E)$, find a minimum vertex
cover which induces a connected subgraph. Show that this problem has a
polynomial-time 3-approximation.

33. (Weighed Connected Vertex Cover) Given a graph $G = (V, E)$ with nonnegative vertex weight, find a minimum total weight vertex cover which induces a
connected subgraph. Show the following:

   (a) This problem has a polynomial-time $O(\ln n)$-approximation where $n = |V|$.
   (b) For any $0 < \rho < 1$, this problem has no polynomial-time $(\rho \ln n)$-
       approximation unless NP=P.

34. (Connected Dominating Set) In a graph G, a subset C is called a *connected
dominating set* if C is a dominating set and induces a connected subgraph.
Given a graph, find a minimum connected dominating set. Show that for any
$0 < \rho < 1$, this problem has no polynomial-time $(\rho \ln n)$-approximation unless
NP=P where n is the number of vertices in input graph.

35. Show that the following problem is APX-complete: Given a graph with vertex
degree upper-bounded by a constant b, find a clique of the maximum size.

36. Show that the traveling salesman problem does not belong to Poly-APX if the
distance table is not required to satisfy the triangular inequality.

Historical Notes

The concept of NP-completeness was established by Cook [71] in 1971, and it is
widespread in many areas, especially its ubiquitous existence in combinatorial optimization which was made by Karp [236]. Garey and Johnson [166] accumulated
many examples of NP-complete and NP-hard problems, which is still a very good
reference book in the study on theory of NP-completeness.

The study of approximation algorithms was initiated by Graham [184] in 1966.
The importance of this work was not fully understood until NP-hardness was
established. For efforts at earlier stage in the study of approximation algorithms,
it is worth mentioning the following: Sahni [353] designed PTAS, and Ibarra and
Kim [223] discovered the first FPTAS for the knapsack problem. Christofides
[66] constructed a polynomial-time 3/2-approximation for the traveling salesman
problem with triangle inequality. Until today, no approximation has been found to
have better performance ratio than Christofides approximation.

In the study of approximation, there are two research directions, algorithm design
and inapproximability proof. In the 1970s, while approximation algorithms received
some efforts, inapproximability proofs also got some progress, such as Garey and
John [164], Sahni and Gonzalez [354], and Ko [259]. This progress was speed
up in the 1990s. Papadimitriou and Yannakakis [336] introduced L-reduction and

MAXSNP class. This class is extended to APX by Khanna et al. [242]. Motivated from the study of MAXSNP-completeness, the PCP theorem and the PCP system were initiated by Arora et al. [10, 11] and Arora and Safra [13, 14]. With the PCP system, many results are generated on inapproximability, such as Hastad [206, 207], Lund and Yannakalis [305], Feige [134], Raz and Safra [347], and Zuckerman [464, 465].

Recently, techniques developed in above classic results on algorithm designs and inapproximability proofs have been widely used in the study of wireless sensor networks and social networks, such as secure scheduling of barrier covers [452] and influence maximization in various models [302–304].

Chapter 9
Restriction and Steiner Tree

The universe has no restriction. You place restrictions on the universe with your expectations.

—Deepak Chopra

Restriction is a major technique in design of approximation algorithms. The Steiner minimum tree is a classic NP-hard combinatorial optimization problem. In the study of the Steiner minimum tree and its variations, restriction plays an important role.

9.1 Idea of Restriction

As shown in Fig. 9.1, consider problem

$$\min \quad f(x)$$
$$\text{subject to} \quad x \in \Omega$$

where Ω is a feasible domain. By restriction, we mean to put some constraints to feasible solutions so that the feasible domain Ω is shrunken to a smaller domain Γ, on which the minimization of objective function $f(x)$ can be solved or approximated easily. Then, the optimal or approximate solution of the restricted problem can be utilized to approximate the original one.

There is a general idea to analyze the performance of approximation which is designed with restriction techniques. To explain it, suppose $\min\{f(x) \mid x \in \Gamma\}$ can be solved in polynomial-time and y^* is the optimal solution. We intend to use y^* as an approximation solution for $\min\{f(x) \mid x \in \Omega\}$. Then, the approximation performance can be obtained in the following way:

- Consider an optimal solution x^* for $\min\{f(x) \mid x \in \Omega\}$.
- Modify x^* to y satisfying the restriction, i.e., $y \in \Gamma$.

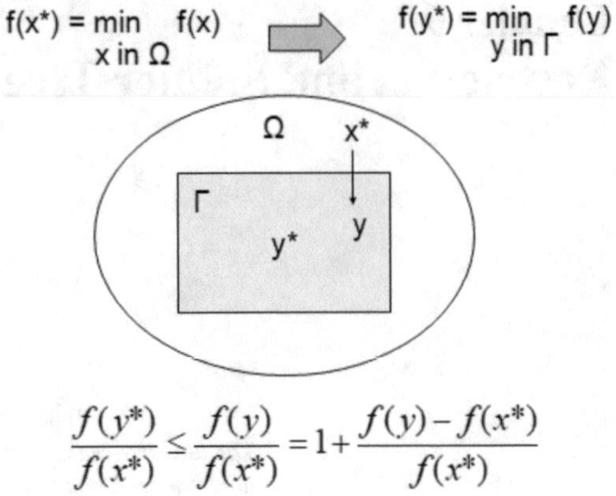

Fig. 9.1 Idea of restriction

$$f(x^*) = \min_{x \text{ in } \Omega} f(x) \implies f(y^*) = \min_{y \text{ in } \Gamma} f(y)$$

$$\frac{f(y^*)}{f(x^*)} \leq \frac{f(y)}{f(x^*)} = 1 + \frac{f(y) - f(x^*)}{f(x^*)}$$

- Meanwhile, estimate the cost of modification. Suppose this cost is within a factor of α from the minimum solution, i.e.,

$$\frac{f(y) - f(x^*)}{f(x^*)} \leq \alpha.$$

Then, the performance ratio is $1 + \alpha$, i.e.,

$$\frac{f(y^*)}{f(x^*)} \leq \frac{f(y)}{f(x^*)} \leq 1 + \frac{f(y) - f(x^*)}{f(x^*)} \leq 1 + \alpha.$$

Let us show an example. Consider the following problem:

Problem 9.1.1 (Minimum Length Rectangular Partition) *Given a rectangle with point-holes inside, partition it into smaller rectangles without hole to minimize the total length of cuts.*

Note that a rectangular partition may not be a guillotine partition (Fig. 9.2). Actually, this is an NP-hard problem, while the minimum length guillotine partition problem is polynomial-time solvable by using dynamic programming (Sect. 3.1).

Let P be the optimal solution for the minimum length guillotine partition. We want to use P as an approximation solution for the minimum length rectangular partition and analyze its performance as follows:

Theorem 9.1.2 *The optimal solution for the minimum length guillotine partition is 2-approximation for the minimum length rectangular partition.*

Proof Consider an optimal solution P^* for the minimum length rectangular partition. Let us modify P^* into a guillotine partition by adding line segments. Note that the guillotine partition is constituted by a sequence of guillotine cuts, each of

Fig. 9.2 Rectangular
partition, but not guillotine
partition

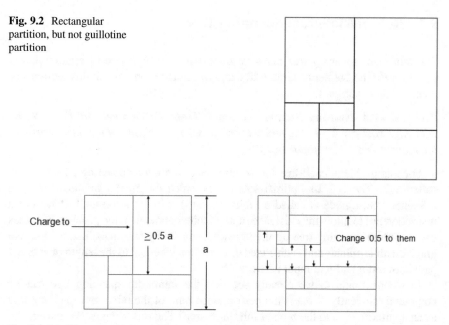

Fig. 9.3 Modify a rectangular partition to a guillotine partition

which is on a rectangle. Therefore, the modification is also made sequentially, and at each step, cut a rectangle into two.

Suppose at current step, we consider rectangle R. If P^* contains a line segment which can be a guillotine cut for R, then apply this cut. Otherwise, we consider the following two cases:

Case 1 P^* contains a vertical segment AB of length at least $0.5a$ where a is the length of vertical edge of R (the left side of Fig. 9.3). In this case, extend segment AB into a guillotine cut. This extension will contain added segments of total length at most $0.5a$, which charges to segment AB.

Case 2 Case 1 does not occur. In this case, we cut R by a horizontal segment CD at the middle of R. CD must lie between horizontal segments in P^*, i.e., from each point on CD, going above and below must meet horizontal segments in P^*. Therefore, the total length of horizontal segments above CD and directly facing CD is equal to the length of CD. So is the total length of horizontal segments below CD and directly facing CD. We charge 0.5 to those horizontal segments in P^*. Note that each horizontal segment in P^* can be charged at most twice, one from above and one from below. Therefore, the total charge is at most one.

From argument in the above two cases, we can see that modifying P^* into a guillotine partition needs to add new cut segments of total length not exceeding the total length of P^*. Therefore, the optimal solution for the minimum length guillotine partition is 2-approximation for the minimum length rectangular partition.   □

9.2   Role of Minimum Spanning Tree

The minimum spanning tree plays an important role in design of approximation solutions for the Steiner minimum tree and its related problems. In this section, we present a few examples.

Problem 9.2.1 (Network Steiner Minimum Tree) *Given a network $G = (V, E)$ with edge cost $c : E \rightarrow R_+$ and a terminal set $P \subseteq V$, find a minimum cost tree interconnecting all terminals in P.*

For this problem, each feasible solution is a tree interconnecting all terminals, called a *Steiner tree*. Its optimal solution is called the *Steiner minimum tree*. In a Steiner tree, a node is called a *Steiner node* if it is not a terminal. If we put a restriction not to allow the existence of any Steiner node, then the problem becomes the minimum spanning tree on the network induced by terminals, i.e., a complete graph on all terminals and each edge (u, v) has length equal to the length of shortest path between u and v in input network G.

In Chaps. 1 and 4, we already see that the minimum spanning tree can be computed efficiently. What is the performance ratio of the minimum spanning tree as an approximation to the Steiner minimum tree? The following is the answer:

Theorem 9.2.2 *In network, the minimum spanning tree is 2-approximation for the Steiner minimum tree.*

Proof Consider a Steiner minimum tree T. Construct a Euler tour using each edge twice. Suppose $p_1, p_2, \ldots p_n$ are all terminals lying one by one along the Euler tour. For every $i = 1, 2, \ldots, n - 1$, connect p_i and p_{i+1} with a shortest path between them. This results in a spanning tree. This means that the length of such a Euler tour must be not smaller than the length of minimum spanning tree. Therefore, the minimum spanning tree is 2-approximation for the Steiner minimum tree.   □

Actually, 2 is tight for the approximation performance ratio of the minimum spanning tree. To see this, consider a star graph with n edges each of unit length. Let terminal set consist of all leaves. Then, the distance between any two terminals is 2 in the star. Therefore, the minimum spanning tree on those terminals has length $2(n - 1)$. However, the Steiner minimum tree can have the center as a Steiner node and hence has length n. Therefore, the ratio of lengths of the minimum spanning tree and the Steiner minimum tree is $2n/(n - 1)$. As n goes to infinity, this ratio goes to 2. This means that the performance ratio cannot be a constant smaller than 2.

An interesting and early improvement got the performance ratio $2(1 - \frac{1}{\ell})$ where ℓ is the number of leaves in a Steiner minimum tree. The detail can be found in an exercise and also in [265].

In many variations of the Steiner minimum tree problems, the minimum (length) spanning tree is often a good candidate for an approximation solution, even if the objective function is not about length. This is due to a special property of the minimum spanning tree as follows:

Lemma 9.2.3 *Let T be a spanning tree and T^* be a minimum spanning tree. Let $E(T)$ and $E(T^*)$ be their edge sets. Then, there is a one-to-one onto mapping f from $E(T^*)$ to $E(T)$ such that*

$$c(e) \le c(f(e)) \text{ for any } e \in E(T^*),$$

where $c(e)$ is the length of edge e.

Proof Let T' be a minimum spanning tree produced by Kruskal algorithm, with edges $e_1', e_2', \ldots, e_t'$ in ordering $c(e_1') \le c(e_2') \le \cdots \le c(e_t')$. Let T be a spanning tree with edges $e_1, e_2, \ldots, e_t$ in ordering $c(e_1) \le c(e_2) \le \cdots \le c(e_t)$. We claim $c(e_i') \le c(e_i)$ for all $i = 1, 2, \ldots, t$.

For contradiction, suppose the claim is not true. Let k be the smallest index such that $c(e_k') > c(e_k)$. Denote $A = \{e_1', \ldots, e_{k-1}'\}$ and $B = \{e_1, \ldots, e_k\}$. Let $G[A]$ denote the subgraph of G, induced by edge set A. By the rule in Kruskal algorithm, for each $e_i \in B$, either e_i is an edge in $T'[A]$ or $T'[A] \cup e_i$ contains a cycle. Therefore, $T'[A]$ is a spanning forest of $G[A \cup B]$. It implies that every acyclic subgraph of $G[A \cup B]$ contains at most $k - 1$ edges, contradicting the fact that $|B| = k$.

Now, let T^* be a minimum spanning tree with edges $e_1^*, e_2^*, \ldots, e_t^*$ in ordering $c(e_1^*), c(e_2^*), \ldots, c(e_t^*)$. From above proved claim, we know that $c(e_i') \le c(e_i^*)$ for all $i = 1, 2, \ldots, t$. Moreover, we have $\sum_{i=1}^t c(e_i') = \sum_{i=1}^t c(e_i^*)$. Hence, $c(e_i') = c(e_i^*)$ for all $i = 1, 2, \ldots, t$. Now, define $f(e_i^*) = e_i$. Then, f meets the requirement.   □

Next, we study two variations of Steiner tree.

Problem 9.2.4 (Steiner Tree with Minimum Number of Steiner Points (ST-MSP)) *Given n terminals in the Euclidean plane and a positive number R, compute a Steiner tree interconnecting all terminals with the minimum number of Steiner points such that every edge has Euclidean length at most R.*

The ST-MSP problem has a strong application background in wavelength division multiplexing (WDM) optical network design. Suppose there are n sites required to be interconnecting with WDM optical network. Due to the limit of transmission power, signals can only travel a limited distance R. Therefore, for distances larger than R, some amplifiers or receivers/transmitters are placed on the intermediate locations to break the distance into pieces of length at most R.

A Steiner tree as a feasible solution for ST-MSP may contain a Steiner point with degree two. If all Steiner points have degree two, then this Steiner tree can be obtained from a spanning tree by putting Steiner points on its edges, which will be called a *steinerized spanning tree* (Fig. 9.4). We reserve the term "minimum spanning tree" for a spanning tree with minimum length. Define the *minimum steinerized spanning tree* to be the steinerized spanning tree with minimum number of Steiner points. The minimum spanning tree and the minimum steinerized spanning tree have a close relationship.

Fig. 9.4 Steinerized
spanning tree

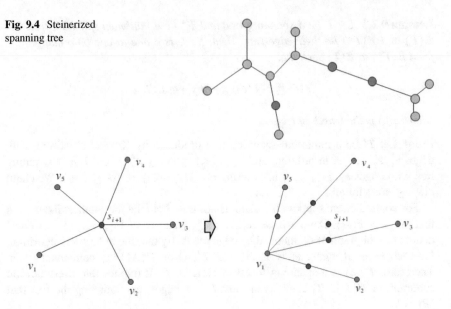

Fig. 9.5 Proof of Theorem 9.2.6

Lemma 9.2.5 *By breaking each edge with length longer than R into smaller pieces of length at most R, the minimum length spanning tree will induce a minimum steinerized spanning tree.*

Proof It follows immediately from Lemma 9.2.3.   □

Theorem 9.2.6 *Suppose that for any set of terminals, there always exists a minimum spanning tree with vertex degree at most D. Then, the minimum steinerized spanning tree is* $(D-1)$*-approximation for ST-MSP.*

Proof Let P be a set of terminals and S^* an optimal tree on input P for ST-MSP. Suppose S^* contains k Steiner points $s_1, s_2, \ldots, s_k$ in the ordering of the breadth-first search starting from a node of S^*. Let $N(P)$ denote the number of Steiner points in a minimum steinerized spanning tree induced from the minimum length spanning tree on P. We first show a claim that

$$N(P \cup \{s_1, \ldots, s_i\}) \le N(P \cup \{s_1, \ldots, s_{i+1}\}) + (D-1). \tag{9.1}$$

To do so, consider a minimum length spanning tree T for $P \cup \{s_1, \ldots, s_{i+1}\}$, with degree at most D. Suppose s_{i+1} has adjacent nodes $v_1, \ldots, v_d$ ($d \le D$). Then, one of the edges $(s_{i+1}, v_1), \ldots, (s_{i+1}, v_d)$ has length not exceeding R because $P \cup \{s_1, \ldots, s_i\}$ has distance at most R from s_{i+1} (Fig. 9.5).

Without loss of generality, assume the length of (s_{i+1}, v_1) is at most R. Delete edges $(s_{i+1}, v_1), \ldots, (s_{i+1}, v_d)$, and add $d-1$ edges $(v_1, v_2), \ldots, (v_1, v_d)$. This results in a spanning tree T' on $P \cup \{s_1, \ldots, s_i\}$. Since $d(v_1, v_j) \le d(v_1, s_{i+1}) +$

$d(s_{i+1}, v_j) \leq R + d(s_{i+1}, v_j)$ where $d(x, y)$ is the Euclidean distance between x and y, we have that breaking edge $d(v_1, v_j)$ into pieces of length at most R needs one more degree-two Steiner point than that breaking (s_{i+1}, v_j) needs. It follows that inducing T' into a steinerized spanning tree needs $d-1$ more degree-two Steiner points than $N(P \cup \{s_1, .., s_i\})$. Thus, this spanning tree T' contains at most $d - 1$ more Steiner points than T. Therefore, by Lemma 9.2.5, (9.1) holds. Hence,

$$N(P) \leq N(P \cup \{s_1, \ldots, s_k\}) + k(D - 1) = k(D - 1).$$

$\square$

Note that for any set of terminals in the Euclidean plane, there is a minimum spanning tree with degree at most 5 (an exercise in Chap. 1). Therefore, we have the following:

Corollary 9.2.7 *The minimum steinerized spanning tree is 4-approximation for ST-MSP in the Euclidean plane.*

Proof It follows immediately from the fact that for any set of terminals in the Euclidean plane, there is a minimum spanning tree with degree at most 5. We leave the proof of this fact as an exercise. $\square$

The following problem is closely related to ST-MSP:

Problem 9.2.8 (Bottleneck Steiner Tree) *Given a set P of terminals in the Euclidean plane and a positive integer k, find a Steiner tree on P with at most k Steiner nodes, to minimize the length of longest edge.*

Consider a spanning tree T on P. The steinerized spanning tree induced by T is defined to be the tree obtained in the following way:

Optimal Steinerization:
input: A spanning tree T.
output: A steinerized spanning tree T.
    for every edge $e \in T$ **do** $n(e) \leftarrow 0$;
    for $i = 1$ **to** k **do**
        choose $e \in T$ to maximize $c(e)/n(e)$
        (remind: $c(e)$ is the length of edge e)
        and set $n(e) \leftarrow n(e) + 1$;
    for every edge $e \in T$ **do**
        cut e evenly with $n(e)$ Steiner points;
    return T.

Next, we show two lemmas:

Lemma 9.2.9 *Among steinerized spanning tree obtained from T by adding k Steiner points, optimal Steinerization gives the one with the minimum of longest edge length.*

Proof We prove it by induction on k. For $k = 0$, it is trivial. Next, consider $k \geq 1$.
Let $e_1, e_2, \ldots, e_t$ be all edges of T. Suppose that after adding k Steiner points,

$$\frac{c(e_1)}{n(e_1)} = \max_{1 \leq i \leq t} \frac{c(e_i)}{n(e_i)}.$$

Denote by $Opt(k; e_1, \ldots, e_t)$ the minimum value of longest edge length after
adding k Steiner points on edges $e_1, \ldots, e_t$. We will show that

$$Opt(k + 1; e_1, \ldots, e_t) = \max \left(\max_{2 \leq i \leq t} \frac{c(e_i)}{n(e_i)}, \frac{c(e_1)}{n(e_1) + 1} \right). \tag{9.2}$$

By induction hypothesis,

$$Opt(k; e_1, \ldots, e_t) = \frac{c(e_1)}{n(e_1)}.$$

Note that in the algorithm on input $e_1, e_2, \ldots, e_t$, if we ignore the step for adding
points on e_1, then the remaining steps are exactly those steps in the algorithm on
input $e_2, \ldots, e_t$. Therefore, by induction hypothesis, we also have

$$Opt(k - n(e_1); e_2, \ldots, e_t) = \max_{2 \leq i \leq t} \frac{c(e_1)}{n(e_1)}.$$

Note that

$$Opt(k + 1; e_1, \ldots, e_t) \leq \max \left(Opt(k - n(e_1); e_2, \ldots, e_t), \frac{c(e_1)}{n(e_1) + 1} \right).$$

Thus, to show (9.2), it suffices to prove

$$Opt(k + 1; e_1, \ldots, e_t) \geq \max \left(Opt(k - n(e_1); e_2, \ldots, e_t), \frac{c(e_1)}{n(e_1) + 1} \right).$$

For contradiction, suppose

$$Opt(k + 1; e_1, \ldots, e_t) < \max \left(Opt(k - n(e_1); e_2, \ldots, e_t), \frac{c(e_1)}{n(e_1) + 1} \right). \tag{9.3}$$

This implies

$$Opt(k + 1; e_1, \ldots, e_t) < Opt(k; e_1, \ldots, e_t).$$

Let $n^*(e_1)$ denote the number of Steiner points on e_1 in an optimal solution for
$Opt(k + 1; e_1, \ldots, e_t)$. By (9.3), we must have $n^*(e_1) > n(e_1) + 1$ and

$$Opt(k + 1 - n^*(e_1); e_2, \ldots, e_t) < Opt(k; e_1, \ldots, e_t).$$

Note that

$$\frac{c(e_1)}{n^*(e_1) - 1} \le \frac{c(e_1)}{n(e_1) + 1} < Opt(k; e_1, \ldots, e_t).$$

Hence,

$$\max\left(Opt(k + 1 - n^*(e_1); e_2, \ldots, e_t), \frac{c(e_1)}{n^*(e_1) - 1}\right) < Opt(k; e_1, \ldots, e_t),$$

a contradiction.   □

Lemma 9.2.10 *Among steinerized spanning trees, the one induced by minimum spanning tree reaches the minimum of longest edge length.*

Proof It follows immediately from Lemma 9.2.3.   □

Theorem 9.2.11 *The steinerized spanning tree induced by minimum spanning tree is a 2-approximation solution for the bottleneck Steiner tree.*

Proof Consider an optimal Steiner tree T^* for the bottleneck Steiner tree problem. We want to modify T^* into a steinerized spanning tree. Note that every Steiner tree can be decomposed into full components in each of which all terminals are leaves. Since this decomposition is on terminals, it suffices to consider each full component.

Let T be a full component with k Steiner points with edge length at most R. We arbitrarily select a Steiner point s as the root. A path from the root to a leaf is called a *root-leaf path*. Its length is the number of edges on the path which is equal to the number of Steiner points on the path. Let h be the length of a shortest root-leaf path. We will show by induction on the depth d of T that there exists a steinerized spanning tree for all terminals with at most $k - h$ degree-two Steiner points and edge length at most $2R$. Here, the depth of T is the length of a longest root-leaf path.

For $d = 0$, T contains only one terminal, so it is trivial. For $d = 1$, T contains only one Steiner point. We can directly connect the terminals without any Steiner points since, by the triangular inequality, the distance between two terminals is at most $2R$.

Next, we consider $d \ge 2$. Suppose s has t sons $s_1, \ldots, s_t$. For each s_i, there is a subtree T_i rooted at s_i with depth $\le d - 1$. Let k_i be the number of Steiner points in T_i and h_i the length of a shortest root-leaf path in T_i, from s_i to a leaf v_i (Fig. 9.6).

By induction hypothesis, there exists a steinerized spanning tree S_i for all terminals in T_i with at most $k_i - h_i$ degree-two Steiner points and edge length at most $2R$. Without loss of generality, assume $h_1 \ge h_2 \ge \cdots \ge h_t$. Connect all S_i for $i = 1, \ldots, t$ into a spanning tree S with edges $(v_1, v_2), \ldots, (v_{t-1}, v_t)$, and put h_i Steiner points on edge (v_i, v_{i+1}). Note that the path between v_i and v_{i+1} in T contains $h_i + h_{i+1} + 2$ edges. By triangular inequality, the distance between v_i

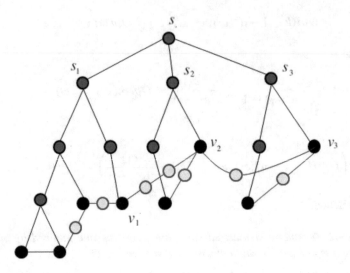

Fig. 9.6 Proof of Theorem 9.2.11

and v_{i+1} is at most $(h_i + h_{i+1} + 2)R \leq 2(h_i + 1)R$. Therefore, h_i Steiner points would break (v_i, v_{i+1}) into $h_i + 1$ pieces each of length $\leq 2R$. Note that S contains $k_1 + \cdots + k_{t-1} + k_t - h_t = k - (h_t + 1)$ Steiner points. Moreover, the path from s to v_t in T contains $h_t + 1$ Steiner points. Hence, $h \leq h_t + 1$.   □

The minimum spanning tree also has many applications in the study the energy efficient problems in wireless ad hoc and sensor networks. Those applications are based on the following property.

Lemma 9.2.12 *Let f be a nonnegative monotone nondecreasing function. Then, the minimum length spanning tree is an optimal solution for the following problem:*

$$\min \sum_{e \in T} f(c(e))$$

subject to T is over all spanning trees.

Proof It follows immediately from Lemma 9.2.3.   □

In wireless ad hoc and sensor networks, each wireless node (sensor or device) u has a communication radius $R_c(u)$. Another wireless node v is able to receive signal from u if and only if the distance from u to v is at most $R_c(u)$. The communication radius $R_c(u)$ is determined by energy consumption at node u, i.e., power at u as follows:

$$p(u) = c R_c(u)^{\alpha}$$

where c and α are positive constants and usually $2 \leq \alpha \leq 6$. Arc (u, v) is said to exist if v is able to receive signals from u. Edge (u, v) is said to exist if v can receive signals from u and u can also receive signals from v.

Suppose a directed graph $G = (V, E)$ is obtained from setting up energy power at every node. Then, denote

$$p(G) = \sum_{u \in V} p(u).$$

In arborescence, if we reverse the direction of every arc, then we obtain an in-arborescence.

Lemma 9.2.13 (In-Arborescence Lemma) *Suppose G contains an in-arborescence T. Then,*

$$p(G) \geq p(T)$$

where

$$p(T) = \sum_{e \in T} c\|e\|^{\alpha}.$$

Proof In every arborescence T, each arc (u, v) is unique arc going from u. Therefore, to make (u, v) exist, we need to set up

$$p = c\|(u, v)\|^{\alpha}.$$

Therefore,

$$p(T) = \sum_{e \in T} c\|e\|^{\alpha}.$$

$\square$

Now, we consider the following two problems.

Problem 9.2.14 (Symmetric Topological Control) *Given a set of wireless nodes in the Euclidean plane, set up power at each node to maintain the existence of a spanning tree interconnecting them and to minimize the total power consumption.*

Problem 9.2.15 (Asymmetric Topological Control) *Given a set of wireless nodes in the Euclidean plane, set up power at each node to maintain the existence of a strongly-connected graph on them and to minimize the total power consumption.*

Theorem 9.2.16 *The minimum length spanning tree is 2-approximation solution for the symmetric topological control problem, and also for the asymmetric topological control problem.*

Proof Suppose G is the optimal (directed) graph maintained in symmetric or asymmetric topological control. Then G must contain an in-arborescence. By In-Arborescence Lemma and Lemma 9.2.3, we have

$$p(G) \geq p(T) \geq \sum_{e \in T^*} c\|e\|^\alpha$$

where T is an in-arborescence and T^* is a minimum length spanning tree. Note that to maintain the existence of an edge (u, v), we need power

$$p(u) + p(v) = 2c\|(u, v)\|^\alpha.$$

Hence, to maintain the existence of T^*, we need power at most

$$2 \cdot \sum_{e \in T^*} c\|e\|^\alpha \leq 2p(G).$$

□

In the next problem, in-arborescence is not required to exist, but out-arborescence exists.

Problem 9.2.17 (Min-Energy Broadcasting) *Given a set of points S in the Euclidean plane and a source node $s \in S$, find a broadcasting routing from s to minimize the total energy consumption of the routing.*

Again, a naive approximation is to turn a minimum (Euclidean length) spanning tree T into a broadcasting routing. Its total energy consumption is

$$c \sum_{e \in T} \|e\|^\alpha$$

where $\|e\|$ is the Euclidean length of edge e. To establish the performance ratio of this approximation induced by the minimum spanning tree, we first prove the following:

Lemma 9.2.18 *Let C be a disk with center x and radius R. Suppose P is a set of points lying in C and contains x. Then, for minimum spanning tree T on P,*

$$\sum_{e \in T} \|e\|^\alpha \leq 8R^\alpha$$

where $\alpha \geq 2$.

Proof Since $x \in P$, every edge of T has length at most R. Let T_r be the subgraph of T, induced by all edges with length at most r. Let $n(T, r)$ denote the number of connected components of T_r. Then,

Fig. 9.7 For each connected component, those disks form a connected region, and those regions for different connected components are disjoint

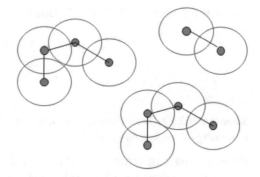

$$\sum_{e\in T} \|e\|^{\alpha} = \alpha \int_{0}^{R} (n(T,r) - 1)r^{\alpha-1}dr.$$

Associate each node $u \in P$ with a disk with center u and radius $r/2$. Then, for each connected component, those disks form a connected region, and those regions for different connected components are disjoint (Fig. 9.7). Moreover, since each such region contains at least one disk with radius $r/2$, its area is at least $\pi(r/2)^2$. Therefore, the boundary of each region has length at least πr. This is because for surrounding a certain amount area, circle gives the shortest length. Let $a(P,r)$ denote the total area covered by those disks with radius $r/2$. Now, we have

$$a(P,R) = \int_{0}^{R} d(a(P,r))$$

$$\geq \int_{0}^{R} n(T,r)\pi r d(r/2)$$

$$= \frac{\pi}{2} \int_{0}^{R} (n(T,r) - 1)r dr + \frac{\pi R^2}{4}$$

$$= \frac{\pi}{4} \sum_{e\in T} \|e\|^2 + \frac{\pi R^2}{4}.$$

Note that $a(P,R)$ is contained by a disk at x with radius $1.5R$. Therefore,

$$\pi(1.5R)^2 \geq \frac{\pi}{4} \sum_{e\in T} \|e\|^2 + \frac{\pi R^2}{4}.$$

Hence,

$$\sum_{e\in T} \|e\|^2 \leq 8R^2.$$

Note that for every $e \in T$, $\|e\| \leq R$. Hence,

$$\sum_{e \in T}(\|e\|/R)^\alpha \leq \sum_{e \in T}(\|e\|/R)^2 \leq 8.$$

$\square$

Theorem 9.2.19 *The minimum spanning tree induces an 8-approximation for the min-energy broadcasting problem.*

Proof Consider an optimal solution T^* for the min-energy broadcasting problem, i.e., T^* is a minimum energy broadcasting routing. For each internal node u of T^*, we draw a smallest disk D_u to cover all out-arc at u. Let $\mathcal{D}$ be the set of such disks and R_u is the radius of disk D_u. Those disks will cover all points in input set S and the total energy consumption of T^* is

$$\sum_{D_u \in \mathcal{D}} cR_u^\alpha.$$

Now, for each disk D_u, construct a minimum spanning tree T_u on u and all endpoints of out-arcs at u (Fig. 9.8). By Lemma 9.2.18,

$$\sum_{e \in T_u}\|e\|^\alpha \leq 8 \cdot R_u^\alpha.$$

Note that $\cup_{D_u \in \mathcal{D}} T_u$ is a spanning tree on all nodes of T^*. Let T be a minimum length spanning tree on all nodes of T^*. However, the energy consumption of broadcasting routing induced by T is at most

$$\sum_{e \in T}c\|e\|^\alpha \leq \sum_{D_u \in \mathcal{D}}\sum_{e \in T_u}c\|e\|^\alpha \leq \sum_{D_u \in \mathcal{D}} c \cdot R_u^\alpha \leq 8p(T^*).$$

$\square$

Fig. 9.8 Construct a minimum spanning tree T_u on u and all endpoints of out-arcs at u

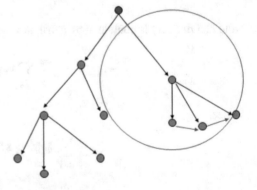

Fig. 9.9 This example shows that constant 6 is tight

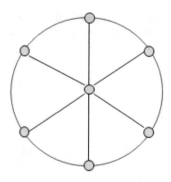

Lemma 9.2.18 can be improved by more careful argument, so that the constant 8 is brought down to 6 [3], which is tight (Fig. 9.9).

9.3 Rectilinear Steiner Minimum Tree

In a coordinate plane, distance $d(p_1, p_2) = |x_1 - x_2| + |y_1 - y_2|$ for two points $p_1 = (x_1, y_1)$ and $p_2 = (x_2, y_2)$ is called the rectilinear distance. The plane with rectilinear distance is called the rectilinear plane. In this section, we study the Steiner tree in the rectilinear plane.

Problem 9.3.1 (Rectilinear Steiner Minimum Tree) *Given a set P of n points, called terminals, in the rectilinear plane, find a minimum length tree interconnecting all terminals in P.*

We will design a PTAS for the rectilinear Steiner minimum tree. First, we draw a minimum square Q including all terminals inside, then divide the square into an $n^2 \times n^2$ grid, and move each terminal to the center of the cell which contains the terminal (Fig. 9.10).

Lemma 9.3.2 *If PATs exists for terminals located at centers of cells, so does for terminals located anywhere.*

Proof Let L be the edge length of square Q. Then, all terminals moving to cell centers spend totally at most $n \times \frac{L}{n^2} = \frac{L}{n}$. Since Q is the minimum square containing all terminals, every rectilinear Steiner tree has length at least L. Suppose P' be the set of cell centers receiving moved terminals. Let $opt(P)$ denote the length of rectilinear Steiner minimum tree for terminal set P. Then, we must have

$$opt(P') \leq opt(P) + \frac{L}{n} \leq \left(1 + \frac{1}{n}\right) \cdot opt(P).$$

Since for P', PTAS exists, we have that for any $\varepsilon > 0$, there exists a polynomial-time approximation solution A for P' with length

Fig. 9.10 Each terminal
moves to the center of a cell

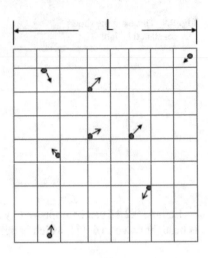

$$length(A) \le (1 + \varepsilon) \cdot opt(P').$$

Connecting each cell center back to terminal in P, we obtain a solution for P with
length at most

$$length(A) + \frac{L}{n} \le \left[(1 + \varepsilon) \left(1 + \frac{1}{n} \right) + \frac{1}{n} \right] \cdot opt(P).$$

When n is sufficiently large and ε is sufficiently small, we can make

$$(1 + \varepsilon) \left(1 + \frac{1}{n} \right) + \frac{1}{n} - 1$$

smaller than any positive number.   □

By this lemma, we consider all terminals lying at cell centers.

Next, we define a type of partition, $(\frac{1}{3}, \frac{2}{3})$-partition. Consider a rectangle R lying
in square Q. As shown in Fig. 9.11, a $(\frac{1}{3}, \frac{2}{3})$-cut is a guillotine cut satisfying the
following conditions:

- Cut line is parallel to shorter edge.
- Cut line is located between $\frac{1}{3}$ and $\frac{2}{3}$ break points on longer edge.
- Cut line must lie on grid line, i.e., pass through lattice points.

The $(\frac{1}{3}, \frac{2}{3})$-cut has an important property.

Lemma 9.3.3 *Suppose that for rectangle R, the length of shorter edge is at least
one third of the length of longer edge. Then, for every rectangle obtained from a
$(\frac{1}{3}, \frac{2}{3})$-cut, the length of shorter edge is at least one third of the length of longer
edge.*

Fig. 9.11 A $\left(\frac{1}{3}, \frac{2}{3}\right)$-cut

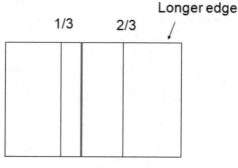

Fig. 9.12 Each $\left(\frac{1}{3}, \frac{2}{3}\right)$-partition results in a binary tree with depth $O(\log n)$

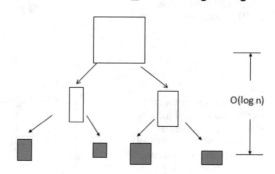

Proof Trivial. □

A $(\frac{1}{3}, \frac{2}{3})$-partition is a sequence of $(\frac{1}{3}, \frac{2}{3})$-cuts. Each cut divides a rectangle into two smaller ones so that all obtained rectangles form a binary tree (Fig. 9.12). Since each cut line is located at a grid line, every obtained rectangle has area at least L^2/n^4. Therefore, this binary tree has a depth at most

$$\lceil \log_2 n^4 \rceil + 1 = O(\log n).$$

For each cut segment, we put on m portals such that m portals divide the cut segment equally (Fig. 9.13). Now, we are ready to describe a restriction as follows:

A rectilinear Steiner tree T is restricted if there exists a $(\frac{1}{3}, \frac{2}{3})$-partition such that if a segment of T passes through a cut line, then it passes at a portal (Fig. 9.13).

Lemma 9.3.4 *Minimum restricted rectilinear Steiner tree can be computed in time $n^{26}2^{O(m)}$ by dynamic programming.*

Proof Each cut has $O(n^2)$ choices. It takes $O(n^2)$ time to select the best one. To show the lemma, it suffices to prove that the number of subproblems is $O(n^{24}2^{O(m)})$. Each subproblem can be determined by the following four facts:

1. Determine a rectangle. (There are $O(n^8)$ possibilities.)
2. Determine position of portals at each edge. (There are $O(n^4)$ possibilities as shown in Fig. 9.14.)

Fig. 9.13 m portals divide a cut segment equally

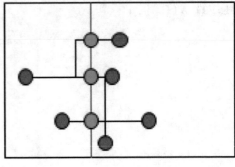

Fig. 9.14 On each rectangle edge, portal position has $O(n^4)$ possibilities

$O(n^2)$ $O(n^2)$

Fig. 9.15 Using portals are partition into several parts in each of which portals is connected and every terminal in the rectangle is connected to some tree containing a portal

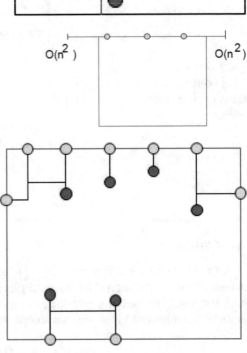

3. Determine the set of using portals. (There are $2^{O(m)}$ possibilities.)
4. There are $2^{O(m)}$ possible partitions for using portals such that in each partition, portals in every part are connected and every terminal in rectangle is connected to a tree containing at least one portal (Fig. 9.15).

For fact 4, we may need a detail proof as follows: Let $N(k)$ denote the number of partitions for k using-portals. Then, $N(0) = 1$ and

$$N(k) = N(k-1) + N(k-2)N(1) + \cdots + N(1)N(k-2) + N(k-1)$$
$$= N(k-1)N(0) + N(k-2)N(1) + \cdots + N(1)N(k-2)$$
$$+ N(0)N(k-1).$$

Define

$$f(x) = N(0) + N(1)x + N(2)x^2 + \cdots + N(k)x^k + \cdots .$$

Then,

$$xf^2(x) = f(x) - 1.$$

Therefore,

$$f(x) = \frac{1 \pm \sqrt{1 - 4x}}{2x}.$$

Since $f(0) = 1$, we have

$$f(x) = \frac{1 - \sqrt{1 - 4x}}{2x} = -\sum_{k=1}^{\infty} \binom{k}{1/2} \cdot \frac{(-4x)^k}{2x}.$$

Thus,

$$\begin{aligned}
N(k) &= -\binom{k+1}{1/2} \cdot \frac{(-4)^{k+1}}{2} \\
&= \frac{0.5(0.5-1)(0.5-2)\cdots(0.5-k)}{(k+1)!} \cdot (-1)^k \cdot 2^{2k+1} \\
&= 2^{O(k)}.
\end{aligned}$$

Finally, note that $k \leq 4m$. $\square$

Let us use the minimum restricted Steiner tree as an approximation solution of the rectilinear Steiner minimum tree. What is its performance ratio? To analyze, consider an optimal solution T^* lying in Hanan grid. The Hanan grid is constructed by drawing vertical and horizontal lines through terminals. Hanan's theorem states that there exists a rectilinear Steiner minimum tree lying in Hanan grid. (We leave the proof as an exercise.) This means that T^* has no segment lying on any possible cut line.

We next need to modify T^* to satisfy the restriction. First, we select a $(\frac{1}{3}, \frac{2}{3})$-partition. The rule is as follows:

- Consider any rectangle R containing a tree-edge segment. If there exists a possible $(\frac{1}{3}, \frac{2}{3})$-cut, then select a cut.
- Selected cut should minimize the number of cross-points among all possible $(\frac{1}{3}, \frac{2}{3})$-cuts.

Let n_R denote the number of cross-points on the cut line selected to cut R. Let $\ell(T_R^*)$ denote the total length of segments of T^* lying in R. Let $a(R)$ be the length of longer edge of R. From the rule of selecting cut, we can see immediately that

Fig. 9.16 Move each
cross-point to closest portal

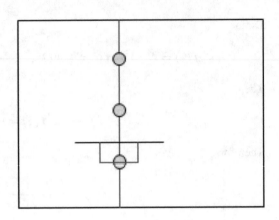

$$n_R \cdot \frac{a(R)}{3} \leq \ell(T_R^*). \tag{9.4}$$

Now, we move each cross-point to closest portal as shown in Fig. 9.16. In each
rectangle R in the binary tree resulting from the selected $(\frac{1}{3}, \frac{2}{3})$-partition, such
moving will increase the total length within

$$n_R \cdot \frac{b(R)}{m+1}$$

where $b(R)$ is the length of selected cut, i.e., the length of the shorter edge of R. By
(9.4), we have that moving cross-points increases the total length inside R, upper
bounded by

$$n_R \cdot \frac{b(R)}{m+1} \leq \frac{3}{m+1} \cdot \ell(T_R^*).$$

Note that as shown in Fig. 9.12, at each level of the binary tree, all rectangles have
disjoint interiors. Therefore, the sum of $\ell(T_R^*)$ for R overall rectangles at each
level is at most $\ell(T^*)$, the length of the optimal solution. Since the binary tree has
$O(\log n)$ levels, the total increased length in moving cross-points to portals is at
most

$$\frac{3}{m+1} \cdot O(\log n) \cdot \ell(T^*). \tag{9.5}$$

Choose $m+1 = \frac{3 \cdot O(\log n)}{\varepsilon}$. Then, the totally increased length will be at most

$$\varepsilon \cdot \ell(T^*).$$

This means that the minimum restricted rectilinear Steiner tree has performance ratio $1 + \varepsilon$ when it is considered as an approximation of the rectilinear Steiner minimum tree. Moreover, for the choice of m in (9.5), we have

$$2^{O(m)} = n^{O(1)}.$$

Therefore, the minimum restricted Steiner tree can be computed in polynomial-time. This completes the proof of the following theorem:

Theorem 9.3.5 *The rectilinear Steiner minimum tree problem has a PTAS.*

9.4   Connected Dominating Set

Consider a graph $G = (V, E)$. A subset of vertices, C, is called a *dominating set* if every vertex is either in C or adjacent to a vertex in C. C is called a *connected dominating set* (CDS) if, moreover, the subgraph $G[C]$ induced by C is connected.

CDS has an interesting relationship with spanning tree. In a spanning tree, deleting all leaves results in a CDS. Conversely, every CDS can induce a spanning tree by adding leaves. Moreover, a CDS with minimum cardinality induces a spanning tree with maximum number of leaves.

Suppose there are n points lying in the Euclidean plane. Add an edge between two points x and y if and only if the distance between them, $d(u, v)$, is at most one. The graph constructed in this way is called a *unit disk graph*. In fact, such a graph can also be induced in the following way: Place a unit disk (i.e., a disk with diameter 1) at each point, and take the point as disk center. Connect two points with an edge if and only if two disks associated with the two points intersect each other.

In this section, we study the following problem:

Problem 9.4.1 (Minimum CDS in Unit Disk Graphs) *Given a connected unit disk graph $G = (V, E)$, find a CDS with minimum cardinality.*

Our goal is to construct a PTAS for this problem. To do so, we first put input unit disk graph G into the interior of a square $Q = \{(x, y) \mid 0 \leq x \leq q, 0 \leq y \leq q\}$. We next construct a grid covering Q. This grid consists of cells with edge length a, and its left-bottom corner has coordinates (x, y) (Fig. 9.17). This grid will partition Q into small areas. Denote this partition by $P(x, y)$.

Next, for each cell e, construct two squares with identical center and edge lengths $a + 2$ and $a + 2h + 2$, respectively, as shown in Fig. 9.18. Let S_e and S'_e denote the interiors of the first square and the second square, respectively. Let us call S_e the central area, $S'_e \setminus e$ the boundary area, and $S'_e \setminus S_e$ the outer boundary area.

Now, let us construct an approximation solution for the minimum CDS in unit disk graph G:

Fig. 9.17 Partition $P(x, y)$

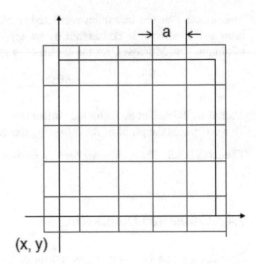

Fig. 9.18 Central area,
boundary area, and outer
boundary area of a cell

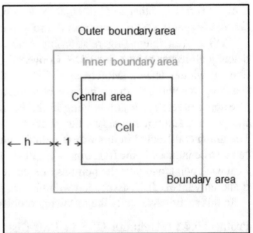

- For each cell e, the part of G lying in central area of cell may be broken
  into several connected components. Let $\mathcal{H}_e$ denote set of those connected
  components.
- For each such connected component H, find a minimum subset C_H of nodes in
  S'_e which dominates nodes in H and induces a connected subgraph (Fig. 9.19).
  C_H will be called a CDS for H in S'_e.
- Let C_e denote the union of C_H for H over $\mathcal{H}_e$.
- Let $C_{x,y}$ denote the union of C_e for e over all cells of partition $P(x, y)$.

Let us estimate the running time for computing $C_{x,y}$.

Fig. 9.19 C_H dominates nodes in H, but is connected in S'_e

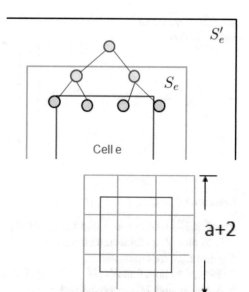

Fig. 9.20 S_e is partitioned into $\lceil (a+2)\sqrt{2} \rceil^2$ small areas

Lemma 9.4.2 $C_{x,y}$ *can be computed in time* $n^{O(a^2)}$.

Proof Let us first estimate the computation time for C_e. Partition S_e into $\lceil (a + 2)\sqrt{2} \rceil^2$ small areas which are squares or rectangles with edge or longer-edge length $\sqrt{2}/2$ (Fig. 9.20). Note that the diameter of each small area is at most one. If a small area contains a node, then choose one of them which can dominate others. Therefore, the minimum dominating set for nodes inside S_e contains at most $\lceil \sqrt{2}(a + 2) \rceil^2$ nodes.

In a connected component H, if D is a dominating set, then we can connect D into a CDS for H by adding at most $2(|D| - 1)$ nodes. (We leave the proof of this fact as an exercise.) From this fact, it follows immediately that

$$|C_e| \le 2(\lceil \sqrt{2}(a + 2) \rceil^2 - 1) = O(a^2).$$

Denote by n_e the number of nodes lying in central area S'_e. Then, by exhausting search, we can find C_e in $n_e^{O(a^2)}$.

For $a > 2(h + 1)$, each node can lie in S'_e for at most four cells e. Therefore, $\sum_{e \in P(x,y)} n_e \le 4n$ where $n = |V|$. Thus, total time for computing $C_{x,y}$ is at most

$$\sum_{e \in P(x,y)} n_e^{O(a^2)} \le (4n)^{O(a^2)} = n^{O(a^2)}.$$

□

Fig. 9.21 For contradiction,
suppose $H \cap H' = \emptyset$

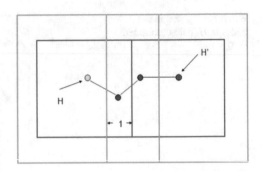

Lemma 9.4.3 $C_{x,y}$ *is a CDS for G.*

Proof It is sufficient to prove the following: Let H be a connected component of $G \cap S_e$ and H' a connected component of $G \cap S_{e'}$. If H and H' are connected, then C_H and $C_{H'}$ are connected.

First, we claim that $H \cap H' \neq \emptyset$ (Fig. 9.21). For contradiction, suppose $H \cap H' = \emptyset$. Since H and H' are connected, there exist nodes $u \in H$ and $v \in H'$ such that $d(u, v) \leq 1$. Then, either u or v lies in the union of inner boundary areas of cells e and e', which must belong to $H \cap H'$, a contradiction.

Now, suppose u lies in $H \cap H'$. Then, there exist $v \in C_H$ and $w \in C_{H'}$ such that v dominates u and w dominates v. Then, either v or w lies in $(S_e \setminus e) \cup (S_{e'} \setminus e')$. Without loss of generality, assume v lies in $(S_e \setminus e) \cup (S_{e'} \setminus e')$. Then, v must be in H' and hence dominated by $C_{H'}$. Hence, C_H and $C_{H'}$ are connected. $\square$

Choose $a = m(h+1)$ and $x = y = 0, -(h+1), -2(h+1), \ldots, -(m-1)(h+1)$. Then, among m partitions $P(x, y)$, choose the one to have the smallest $C_{x,y}$. This $C_{x,y}$ is an approximation solution satisfying the following restriction:

Restriction   There exists a partition $P(x, y)$ with $x = y \in \{0, -(h + 1), \ldots, -(m - 1)(h + 1)\}$ such that for every cell e and every connected component H of $G \cap S_e$, the CDS in S'_e contains connected component dominating H.

It is important to note that $C(x, y)$ may not be the smallest one satisfying the restriction. However, $C(x, y)$ is the one to reach the minimum $\sum_{e \in P(x,y)} |C_e|$ among all CDS satisfying the restriction for partition $P(x, y)$. Let $X = \{0, -(h + 1), -2(h + 1), \ldots, -(m - 1)(h + 1)\}$. Then, $|C(x, y)|$ and $\sum_{e \in P(x,y)} |C_e|$ for $x = y \in X$ may reach the minimum by different (x, y). However, we have

$$\min_{x \in X} |C(x, x)| \leq \min_{x \in X} \sum_{e \in P(x,x)} |C_e|.$$

Therefore, for analysis of approximation performance, we may consider $\sum_{e \in P(x,x)} |C_e|$ instead of $|C(x, x)|$.

Let C^* be a minimum CDS for input graph G. We need to find a partition $P(x, x)$ such that C^* be modified into a CDS S' to satisfy the restriction and

$$\sum_{e \in P(x,x)} |C'_e| \leq (1 + \varepsilon)|C^*|$$

where $C'_e = C' \cap S'_e$.

Lemma 9.4.4 *Suppose $h \geq 2$. For any partition $P(x, x)$, C^* can be modified to a CDS C' such that*

$$\sum_{e \in P(x,x)} |C'_e| \leq |C^*| + 14 \cdot |B_x|$$

where B_x is the set of nodes in C^*, lying in boundary area $\cup_{e \in P(x,x)} (S'_e \setminus e)$.

Proof Consider a cell e and a connected component H of $G \cap S_e$. Suppose that $C^* \cap S'_e$ does not have a connected component dominating H. Let us consider those connected components of $C^* \cap S'_e$, say $C_1, \ldots, C_k$, each of which dominates at least one node in H. Since H is connected, there must exist C_i and C_j such that their distance is at most three, i.e., adding a most two nodes will connect C_i and C_j. Therefore, adding totally at most $2(k - 1)$ nodes will connect all $C_1, \ldots, C_k$ into a connected one. Note that $h \geq 2$ and all $C_1, \ldots, C_k$ are connected outside of S'_e through C^*. Therefore, every C_i must contain a node lying in outer boundary area $S'_e \setminus S_e$. We choose the one adjacent to a node in H. We may charge 2 to each such node in $k - 1$ of them (Fig. 9.22). Moreover, each such node can be charged at most five times since each node can be adjacent to at most five connected components of $G \cap S_e$.

Finally, every node in boundary area can lie in S'_e for at most four cells e. Summarize all above, each node in the boundary area can be repeatedly counted for at most 14 times.   □

Now, we show the result of analysis.

Theorem 9.4.5 *There exists a PTAS for the minimum CDS problem in unit disk graphs.*

Fig. 9.22 Charge 2 for nodes of C^* lying in $S'_e \setminus S_e$ and adjacent to H

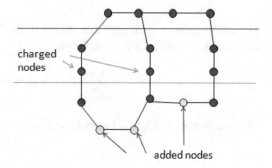

Fig. 9.23 Boundary area can
be covered by horizontal and
vertical strips

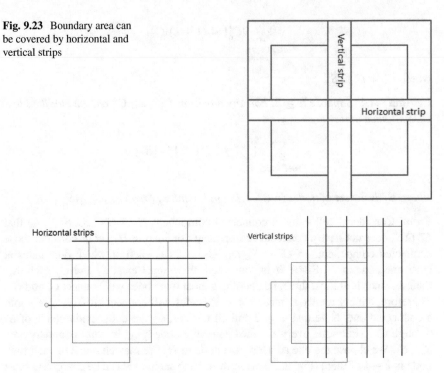

Fig. 9.24 Each node appears in horizontal strip once and in vertical strip once

Proof As shown in Fig. 9.23, for each x, the boundary area $\cup_{e \in P(x,x)}(S'_e \setminus e)$ can be covered by horizontal strips and vertical strips. All horizontal strips are disjoint, and all vertical strips are disjoint for x over X (Fig. 9.24). Therefore, each node appears in at most one horizontal strip and at most one vertical strip. Therefore, each node appears in B_x at most twice for x over X. This implies

$$\sum_{x \in X} |B_x| \leq 2|C^*|.$$

Thus, there exists $x \in X$ such that $|B_x| \leq 2/m$. Choose $m \geq 38/\varepsilon$. For this x, we can modify C^* into a CDS C' such that

$$\sum_{e \in P(x,x)} |C'_e| \leq (1+\varepsilon)|C^*|.$$

This means that the $C_{x,y}$ in Lemma 9.4.3 is $(1+\varepsilon)$-approximation. $\square$

Exercises

1. (L. Kou, G. Markowsky, and L. Berman [265]) Consider an undirected graph
   $G = (V, E)$ with edge weight $d : E \to R_+$. Let $P \subseteq V$ be a set of terminals.
   Prove that the following algorithm gives a Steiner tree with performance ratio
   $2(1 - \frac{1}{\ell})$ for the Steiner minimum tree where ℓ is the number of leaves in an
   optimal tree:

   - Construct the complete undirected distance graph $G_1 = (P, E_1, d_1)$ where
     for any two nodes $u, v \in P$, $d_1(u, v)$ is equal to the shortest distance between
     u and v in G.
   - Compute a minimum spanning tree T_1 of G_1.
   - Construct subgraph G_P of G by replacing each edge in T_1 by its correspond-
     ing shortest path in G.
   - Compute a minimum spanning tree T_P of G_P.
   - Simplify T_P by deleting useless edges, so that all leaves are terminals.

2. Show that for every binary tree, there exist edge-disjoint paths from all internal
   nodes to leaves.

3. A Steiner tree is called a full tree if every terminal is a leaf in it. Every Steiner
   tree can be decomposed at terminal with degree more than one. Obtained full
   trees are called *full components*. A Steiner tree is *k-restricted* if every full
   component contains at most k terminals. Show that for any $\varepsilon > 0$, there exists
   a sufficiently large k such that the length of the k-restricted Steiner minimum
   tree is within a factor of $1 + \varepsilon$ from the network Steiner minimum tree.

4. Consider a set P of terminals on the rectilinear plane. Draw vertical and
   horizontal lines through every terminal. They will form a grid. This is called
   Hanan grid. Show that there exists a rectilinear Steiner minimum tree on P,
   lying in Hanan grid.

5. Consider a dominating set D for a connected graph G. Show that D can be
   connected into a CDS by adding at most $2(|D| - 1)$ nodes.

6. (Minimum Diameter Spanning Tree) Given an edge-weighted undirected graph
   $G = (V, E, w)$, design a polynomial-time algorithm to find a spanning tree
   with the minimum diameter.

7. (Minimum Diameter Steiner Tree) Given an edge-weighted undirected graph
   $G = (V, E, w)$ and a set $P \subseteq V$ of terminals, find a Steiner tree with the
   minimum diameter. Suppose that edge weight w is nonnegative. Show that this
   problem is NP-hard, and design a polynomial-time approximation algorithm.

8. (Group Spanning Tree) Consider an edge-weighted undirected graph
   $G = (V, E, w)$. Suppose that w is nonnegative. Given k disjoint groups
   $V_1, V_2, \ldots, V_k$ of nodes, find a minimum weight tree spanning all groups, i.e.,
   the tree contains at least one node from each group. Please determine whether
   this problem is polynomial-time solvable or not.

9. (Group Spanning Forest) Consider an edge-weighted undirected graph
   $G = (V, E, w)$. Suppose that w is nonnegative. Given k disjoint groups

$V_1, V_2, \ldots, V_k$ of nodes, find a minimum weight forest spanning all groups, i.e., the forest contains at least one node from each group such that if every group is contracted into a node, then the forest becomes a tree. Design a polynomial-time algorithm for this problem.

10. (Group Steiner Tree) Consider an edge-weighted undirected graph $G = (V, E, w)$. Suppose that w is nonnegative. Given k disjoint groups $V_1, V_2, \ldots, V_k$ of nodes, find a minimum weight Steiner forest interconnecting all groups, i.e., the forest contains at least one node from each group such that if every group is contracted into a node, then the forest becomes a tree. Show that the group Steiner tree is NP-hard and the minimum solution for the group spanning forest problem is not c-approximation for the group Steiner tree for any constant $c > 0$.

11. (Interconnecting High Ways) Given n straight segments in the Euclidean plane, find a shortest tree interconnecting them. Design a PTAS for this problem.

12. (1-Guillotine Partition) Consider the minimum length rectangular partition problem. A line cut is called a *1-guillotine cut* if it incompletely partitions a rectangle into two parts such that these two parts contain possibly two windows connected to each other in two sides of the cut (Fig. 9.25). A *1-guillotine* partition is a rectangular partition which results from a sequence of 1-guillotine cuts. A guillotine partition must be a 1-guillotine partition. However, a 1-guillotine partition may not be a guillotine. An example is given in Fig. 9.26. Please prove the following:

   (a) The minimum length 1-guillotine rectangular partition can be computed in $O(n^{15})$ time by a dynamic programming where n is the number of holes in input rectangle.

   (b) The minimum length 1-guillotine rectangular partition can be a 2-approximation for the minimum length rectangular partition.

   (Hint: To show (b), consider an optimal rectangular partition P^*, and we show that P^* can be modified into a 1-guillotine partition by adding some 1-guillotine cuts. To do so, we introduce two concepts, vertical 1-dark point and horizontal

Fig. 9.25 1-guillotine cut

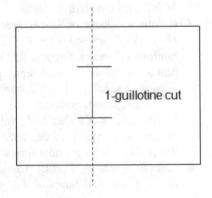

Fig. 9.26 An example for
1-guillotine partition which is
to a guillotine partition

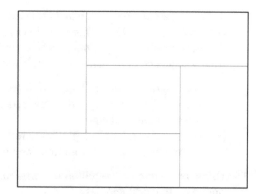

Fig. 9.27 A vertical
1-guillotine cut is found such
that its length can be charged
with 0.5 to those vertical
segments of P^*, which
directly face to horizontal
1-dark point on the cut line

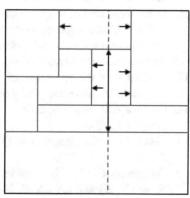

1-dark point. A point in input rectangle x is called a *vertical 1-dark point* if going up and going down from x, we will meet horizontal cut segments in P^*. Similarly, a point is called a *horizontal 1-dark point* if going to the left, and if going to the right from x, we will meet vertical cut segments in P^*. Show that one of the following cases occurs: (*Case 1*). There exists a vertical line on which the length of segments consisting of vertical 1-dark point is not longer than the total length of segments consisting of horizontal 1-dark point. (*Case 2*) There exists a vertical line L in which the length of segments consisting of vertical 1-dark point is not longer than the total length of segments consisting of horizontal 1-dark point. In *Case 1*, a vertical 1-guillotine cut is found such that its length can be charged to those vertical segments of P^*, which directly face to horizontal 1-dark point on line L (Fig. 9.27). Each side can be charged with 0.5. In *Case 2*, a horizontal 1-guillotine cut can be found similarly. Finally, note that every segment of P^* can be charged at most twice.)

13. (Variation of 1-Guillotine Partition). In previous problem about 1-guillotine partition, let us allow two windows lying in the same side or two sides of real cut segment. Show that (a) and (b) are still true.

14. (*m*-Guillotine Partition) Consider the minimum length rectangular partition problem. A line cut is called a *m-guillotine cut* if it incompletely partitions a

rectangle into two parts such that these two parts contain possibly $2m$ windows connected to each other. Those windows are located arbitrarily on two sides of the cut segment. A *m-guillotine* partition is a rectangular partition which results from a sequence of m-guillotine cuts. Please prove the following:

(a) The minimum length 1-guillotine rectangular partition can be computed in $O(n^{10m+5})$ time by a dynamic programming where n is the number of holes in input rectangle.

(b) The minimum length 1-guillotine rectangular partition can be a $(1 + \frac{1}{m})$-approximation for the minimum length rectangular partition.

15. Using technique of m-guillotine partition, show that the rectilinear Steiner minimum tree problem has a PTAS.

16. Using technique of m-guillotine partition, show that the following problem has a PTAS: Given a rectangular polygon with rectangular holes inside (i.e., each hole is a smaller rectangular polygon), partition it into hole-free rectangles with the minimum total length of cuts.

17. Explain why the technique of portals is unable to establish a PTAS for the minimum length rectangular partition problem.

18. Given a unit disk graph $G = (V, E)$, find a minimum vertex cover. Please design a PTAS.

19. Given a unit disk graph $G = (V, E)$, find a minimum connected vertex cover. Please design a PTAS.

20. (Minimum CDS with Routing Cost Constraint) Consider a unit disk graph $G = (V, E)$ and a CDS C. For any two nodes u and v, denote by $d_C(u, v)$ the shortest distance between u and v passing through intermediate nodes in C and by $d(u, v)$ the shortest distance between u and v in G. Please find a constant $\alpha > 0$ and a PTAS for the minimum CDS such that for any two nodes u and v, $d_C(u, v) \le \alpha d(u, v)$.

21. Design a PTAS for the minimum CDS problem in unit ball graphs where a graph is called a unit ball graph if all nodes can be placed in the Euclidean three-dimensional space such that there exists an edge between two nodes u and v if and only if the unit ball with center u intersects the unit ball at center v.

22. Could you design a PTAS for the following problems in unit n-dimensional ball graphs, where a graph is called a unit n-dimensional ball graph if all nodes can be placed in the Euclidean n-dimensional space such that there exists an edge between two nodes u and v if and only if the unit n-dimensional ball with center u intersects the unit n-dimensional ball at center v.

(a) The minimum CDS.
(b) The minimum CDS with routing cost constraint.
(c) The minimum vertex cover.
(d) The minimum connected vertex cover.

Historical Notes

Restriction is an important technique for designing of approximation algorithms [100], which includes two major developments, (nonadaptive) partition and (adaptive) guillotine partition.

Partition is a naive idea. The first theoretical analysis of partition was given by Karp [237] who gave probabilistic analysis for partition with application in Euclidean TSP. Komolos and Shing [262] applied this approach to RSMT. The shafting technique is discovered by Baker [16, 17] and Hochbaum and Maass [213]. This technique enables us to design PTAS for a family of problems in covering and packing. Later, partition has been studied extensively to design PTAS for many problems [25, 51, 220, 229, 315, 386, 396, 448] and also to develop various variations, such as multilayer partition [127] and partition on tree alignment [230, 397, 398].

Guillotine partition was initiated in the study of approximation algorithms for minimum edge length rectangular partition (MELRP). The MELRP was first proposed by Lingas, Pinter, Rivest, and Shamir in 1982 [291]. They showed that the MELRP in general is NP-hard, but for hole-free inputs, it can be solved in time $O(n^4)$ where n is the number of vertices in the input rectilinear polygon. Lingas in 1983 [292] gave the first constant-bounded approximation; its performance ratio is 41. Later, this ratio is improved step-by-step by Du [96], Levcopoulos [279], and Mitchell [317] until PTAS appeared with growth up of guillotine partition [318].

Motivated from a work of Du, Hwang, Shing, and Witbold [102] on the application of dynamic programming to optimal routing trees, Du, Pan, and Shing in 1986 [104] initiated an idea of guillotine partition on a special case of the MELRP problem. This special case has the input consisting of a rectangle and some point holes in its interior, which was first studied by Gonzalez and Zheng [182, 183]. Du, Hsu, and Xu in 1987 [98] extended the idea of guillotine cuts to the convex partition problem. In 1996, Mitchell [317, 320] made a key development. He proposed 1-guillotine partition and later extended to m-guillotine partition [318] which enables guillotine partition to be used for many geometric optimization problems.

Arora in 1996 [6] proposed another technique for adaptive partition, different from m-guillotine partition in the sense that there exist some problems on which m-guillotine partition works, but Arora's technique does not, and vice versa. Later, both Arora [7, 8] and Mitchell [319] noted that combination of two techniques can improve running time. Of course, such combinations can apply only to problems on which both techniques work, such as rectilinear Steiner minimum tree. On rectilinear Steiner minimum tree, Rao and Smith [344] gave $(1 + \varepsilon)$-approximation with running time $O(\varepsilon^\varepsilon n)$ by using techniques on about spanner and banyan.

Restriction together with partition and guillotine partition has been applied to many combinatorial optimization problems in planar graphs [9, 12], in geometric planar [5, 60, 62, 298], in disk graphs [61, 113, 162], in unit ball graphs [447, 448], and in growth-bounded graphs [417].

There are three classic Steiner minimum tree problems, the Euclidean Steiner minimum tree, the rectilinear Steiner minimum tree, and the network Steiner minimum tree. The Euclidean Steiner tree is the first one appearing in the literature. It was initiated by studying Fermat problem [356]: Given three points in the Euclidean plane, find a point connecting them with shortest total distance. Fermat problem has two generalizations for more than three given points. One of them was found by Gauss [356] and unfortunately named Steiner tree by Crourant and Robbins [75]. The detail story can be found in Schreiber's article [356].

All three classic Steiner minimum tree problems are NP-hard [148, 163, 165, 166, 236]. Therefore, one has to put a lot of efforts to study approximation solutions. The minimum spanning tree is the first candidate. Therefore, determination of the performance ratio of the minimum spanning tree becomes an attractive research problem. Hwang [221] determined this ratio in the rectilinear plane. However, for this ratio on the Euclidean Steiner tree, a tortuous story passed a sequence of publications [67–69, 101, 172, 186, 352].

Does there exist a polynomial-time approximation with worst case performance ratio better than that of the minimum spanning tree? For the network Steiner minimum tree, Zelikovsky [437] gave a yes answer. In general, Du, Zhang, and Feng [106, 107] showed that such approximations exist in all metric space as long as Steiner minimum tree for a fixed number of points is polynomial-time computable. Now, much better approximation algorithms have designed. But all designs include restriction technique. For the network Steiner tree, the k-restricted Steiner tree is always involved [40, 350], and hence the k-Steiner ratio [33] plays an important role. For the Euclidean and rectilinear Steiner minimum tree, PTAS can be constructed with guillotine partition.

The Steiner tree has many applications in the real world. Often, various applications also generate variations of Steiner tree, such as terminal Steiner trees [94, 289], Steiner trees with minimum number of Steiner points [288, 310], acyclic directed Steiner trees [438], bottleneck Steiner trees [394], k-generalized Steiner forest [159], Steiner networks [173], and selected internal Steiner trees [217]. The phylogenetic tree alignment can also be considered as a Steiner tree problem with a given topology in a special metric space [346, 355, 395]. For all of them, restriction plays an important role in the study of their approximation.

Is there a polynomial-time constant-approximation for weighted dominating set in unit disk graphs? This open problem was solved by Ambühl, Erlebach, Mihalák, and Nunkesser [4]. Using partition, they constructed a polynomial-time 72-approximation. Gao, Huang, Zhang, and Wu [160] introduced a new technique, called the double partition, and improved the ratio to $(6 + \varepsilon)$. Following this work, through a few efforts [77, 128, 462, 463], this ratio is reduced to $4 + \varepsilon$. Ding et al. [85] note that above techniques can also be used for the weighted sensor cover problem in unit disk graphs, which solves a long-standing open problem. Actually, the unit disk graph is the mathematical formulation of homogeneous wireless sensor networks. Coverage is an important issue in the study of wireless sensor networks [58, 59, 201, 300, 301, 421, 423]. In 2005, Cardei et al. [43] studied a sensor scheduling problem, called the maximum lifetime coverage, and proposed an open problem on the existence of its constant-approximation. Since the work gets a large number of citations, the open problem gets very well-recognized [250, 390, 391, 452]. The success of Ding et al. [85] is based on not only above double partition technique but also previous efforts [26, 167]. However, their success moved one's attention to sensor cover issues with partition techniques [110, 126, 286, 409, 410, 416, 451].

The map labeling is another interesting problem, which is a platform for partition to play. The maximum independent set problem in rectangle intersection graphs is its mathematical formulation [1, 33, 34, 44, 127]. Using partition, PTAS can be obtained under restriction that the ratio of high and the wide is in certain range $[a, b]$ for positive constant a and b, $0 < a < b$. However, for arbitrary rectangles, no constant-approximation has been found. The best known approximation has a performance ratio $O(\log n)$ [1, 45, 243, 328].

Chapter 10
Greedy Approximation and Submodular Optimization

Hell has three gates: lust, anger, and greed.

—Bhagavad Gita

Greedy is an important strategy to design approximation algorithms, especially in the study of submodular optimization problems. In this chapter, we will explore this strategy together with important results in submodular optimization.

10.1 What Is the Submodular Function?

Consider the following problem:

Problem 10.1.1 (Maximum Set Coverage) *Given a collection C of subsets of a finite set X and a positive integer k, find k subsets from C to cover the maximum number of elements in X.*

This problem is NP-hard since its decision version is the same as the decision version of the set cover problem, which is already proved to be NP-complete. This problem also has a greedy approximation algorithm. To describe algorithm, for any subcollection A of C, define

$$\mu(A) = |\cup_{S \in A} S|.$$

This function has a property that for any two subcollections A and B,

$$\mu(A) + \mu(B) \geq \mu(A \cup B) + \mu(A \cap B). \tag{10.1}$$

In fact, comparing $\mu(A) + \mu(B)$ with $\mu(A \cup B)$, the difference is the number of elements appearing in both A and B, i.e.,

$$|(\cup_{S \in A} S) \cap (\cup_{S \in B} S)|.$$

Note that each element appearing in $\mathcal{A} \cap \mathcal{B}$ must appear in both $\mathcal{A}$ and $\mathcal{B}$. Therefore, we obtain the inequality. The equality sign may not hold since there may exist some element appearing in a subset S in $\mathcal{A}$ and another subset S' in $\mathcal{B}$. However, $S \neq S'$.

The function μ with inequality (10.1) is called a submodular function. In general, consider a function f defined over all subsets of a set X, i.e., 2^X. f is called a *submodular function* if for any two subsets A and B of X,

$$f(A) + f(B) \geq f(A \cup B) + f(A \cap B).$$

f is said to be *monotone nondecreasing* if for any two subsets A and B of X,

$$A \subset B \Rightarrow f(A) \leq f(B).$$

The submodular function has a lot of properties. The following two are important ones:

Lemma 10.1.2 *For any subset A and element x, denote $\Delta_x f(A) = f(A \cup \{x\}) - f(A)$. Then, the following holds:*

(a) *A set function $f : 2^X \to R$ is submodular if and only if for any two subsets A and B with $A \subseteq B$ and for any $x \in X \setminus B$, $\Delta_x f(A) \geq \Delta_x f(B)$.*

(b) *A set function $f : 2^X \to R$ is monotone nondecreasing if and only if for any two subsets A and B with $A \subseteq B$ and for any $x \in B \setminus A$, $\Delta_x f(A) \leq \Delta_x f(B)$.*

Proof

(a) Suppose f is submodular. Consider two subsets A and B with $A \subset B$ and an element $x \in X \setminus B$. By modularity, we have

$$f(A \cup \{x\}) + f(B) \geq f(B \cup \{x\}) + f(A)$$

that is,

$$\Delta_x f(A) \geq \Delta_x f(B). \tag{10.2}$$

Conversely, suppose inequality (10.2) holds for any two subsets A and B with $A \subset B$ and for any element $x \in X \setminus B$. Consider any two subsets U and V. Suppose $U \setminus V = U \setminus (U \cap V) = (U \cup V) \setminus V = \{y_1, y_2, \ldots, y_k\}$. Denote $Y_i = \{y_1, \ldots, y_i\}$. Then, we have

$$f(U) - f(U \cap V)$$
$$= \Delta_{y_1} f(U \cap V) + \Delta_{y_2} f((U \cap V) \cup Y_1) + \cdots + \Delta_{y_k} f((U \cap V) \cup Y_{k-1})$$
$$\geq \Delta_{y_1} f(V) + \Delta_{y_2} f(V \cup Y_1) + \cdots + \Delta_{y_k} f(V \cup Y_{k-1})$$
$$= f(U \cup V) - f(V),$$

that is,

$$f(U) + f(V) \geq f(U \cup V) + f(U \cap V).$$

(b) If f is monotone nondecreasing, then for two subsets A and B with $A \subset B$ and $x \in B \setminus A$,

$$\Delta_x f(A) \geq 0 = \Delta_x f(B). \tag{10.3}$$

Conversely, suppose that for two subsets A and B with $A \subset B$ and $x \in B \setminus A$, (10.3) holds. Let $B \setminus A = \{x_1, x_2, \ldots, x_k\}$. Then

$$f(B) - f(A)$$
$$= (f(A \cup \{x_1\}) - f(A)) + (f(A \cup \{x_1, x_2\}) - f(A \cup \{x_1\})) + \cdots$$
$$+ (f(B) - f(A \cup \{x_1, \ldots, x_{k-1}\}))$$
$$\geq 0.$$

□

Lemma 10.1.3 *For any submodular function f,*

$$f(A \cup B) - f(B) \leq \sum_{x \in A} \Delta_x f(B).$$

Proof Since for $x \in B$, $\Delta_x f(B) = 0$, we may assume $A \cap B = \emptyset$ without loss of generality. Assume $A = \{x_1, x_2, \ldots, x_k\}$. Note that

$$f((A \setminus \{x_1\}) \cup B) + f(\{x_1\} \cup B) \geq f(A \cup B) + f(A).$$

Therefore,

$$f(A \cup B) - f(B) \leq f((A \setminus \{x_1\}) \cup B) - f(B) + \Delta_{x_1} f(B).$$

Applying this inequality recursively, we obtain the inequality in the statement of the lemma.   □

The maximum set coverage problem can be formulated as

$$\max \ \mu(\mathcal{A})$$
$$\text{subject to} \ \ |\mathcal{A}| \leq k,$$
$$\mathcal{A} \subseteq \mathcal{C}.$$

This is an instance of the following submodular maximization problem:

Algorithm 29 Greedy Algorithm for Submodular Maximization

Input: A polymatroid function f over 2^X and a positive integer k.
Output: A set A_k of k elements in X.

1: $A_0 \leftarrow \emptyset$;
2: **for** $i = 1$ to k **do**
3:     choose $v \in X \setminus A_{i-1}$ to maximize $\Delta_v f(A_{i-1})$ and
4:     $A_i \leftarrow A_{i-1} \cup \{v\}$;
5: **end for**
6: **return** A_k.

$$\max \quad f(A) \tag{10.4}$$

$$\text{subject to} \quad |A| \le k,$$

$$A \in 2^X,$$

where f is a *polymatroid function*, i.e., monotone nondecreasing submodular function over 2^X with $f(\emptyset) = 0$.

This problem has a greedy approximation as shown in Algorithm 29 with performance indicated in Theorem 10.1.4.

Theorem 10.1.4 *Let A_k be produced by Algorithm 29. Then, $f(A_k) \ge (1 - e^{-1}) \cdot$ opt where opt is the optimal value of objective function.*

Proof Suppose $x_1, x_2, \ldots, x_k$ are obtained by the above greedy algorithm. Denote $A_i = \{x_1, x_2, \ldots, x_i\}$ for $i = 1, \ldots, k$ and $A_0 = \emptyset$. Suppose $A^* = \{u_1, u_2, \ldots, u_k\}$ is an optimal solution. Then, for $i = 0, 1, \ldots, k - 1$, we have

$$f(A^*) \le f(A^* \cup A_i)$$

$$= f(A_i) + \Delta_{u_1} f(A_i) + \cdots + \Delta_{u_k} f(A_i \cup \{u_1, u_2, \ldots, u_{k-1}\})$$

$$\le f(A_i) + \Delta_{u_1} f(A_i) + \cdots + \Delta_{u_k} f(A_i)$$

$$\le f(A_i) + k \cdot \Delta_{x_{i+1}} f(A_i), \tag{10.5}$$

where the first inequality is due to the monotonicity of f, the second inequality is due to the submodularity of f, and the third inequality is due to the greedy rule in the algorithm.

Denote $a_i = f(A^*) - f(A_i)$. Then, it follows from (10.5) that $a_i \le k(a_i - a_{i+1})$. Hence,

$$a_{i+1} \le a_i(1 - 1/k) \le a_i \cdot e^{-1/k}.$$

Iteratively using this inequality, we have $a_k \le a_0 e^{-1}$. Note that $a_0 = f(A^*) - f(\emptyset) = f(A^*) = opt$. Therefore, $f(A_k) \ge (1 - e^{-1})opt$. $\qquad\square$

Theorem 10.1.4 has a lot of applications, especially in the study of social networks. Let us mention a simpler one.

Problem 10.1.5 (Influence Maximization) *Given a directed graph $G = (V, E)$ and a positive integer k, find k nodes to maximize the total number of influenced nodes by selected nodes, where a node u is* influenced *by another node v if there is a directed path from v to u.*

Theorem 10.1.6 *The influence maximization problem is NP-hard.*

Proof We reduce the maximum set coverage problem to the influence maximization problem in the following way: Consider an instance of the maximum set coverage problem, consisting of a collection C of subsets of a finite set X and a positive integer k. Construct a bipartite directed graph $G = (C, X, E)$ by connecting from $S_j \in C$ to $u_i \in X$ if and only if $u_i \in S_j$ (Fig. 10.1). Then, k subsets in C cover the maximum number of elements in X if and only if corresponding k nodes in G influence the maximum number of nodes. This means that the maximum set coverage problem can be solved by solving the influence maximization problem through the above reduction. Since the reduction is constructed in polynomial-time and the maximum set coverage problem is NP-hard, the influence maximization problem is NP-hard.
□

In the above proof, we did not consider decision versions of two optimization problems and construct a polynomial-time many-one reduction between them. Instead, we directly built a reduction between two optimization problems. Such a reduction is called a polynomial-time Turing reduction. Generally speaking, a problem A is said to be polynomial-time Turing reducible to another problem B if A can be solved in polynomial-time by using B as an oracle (i.e., a subroutine).

Lemma 10.1.7 *For any node set A, let $\sigma(A)$ denote the total number of nodes influenced by A. Then, $\sigma(A)$ is a polymatroid function.*

Proof We show that for two node subsets A and B with $A \subset B$ and $v \notin A$,

$$\Delta_v \sigma(A) \geq \Delta_v \sigma(B).$$

Let $I(A)$ denote the set of nodes influenced by node subset A. Then

Fig. 10.1 Construction of directed graph G

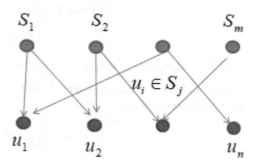

Fig. 10.2 σ is a polymatroid function

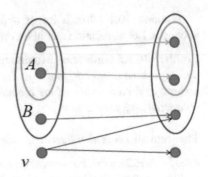

$$\Delta_v \sigma(A) = |I(v) \setminus I(A)| \subseteq |I(v) \setminus I(B)| = \Delta_v \sigma(B),$$

as shown in Fig. 10.2.

□

Theorem 10.1.8 *The influence maximization problem has a greedy $(1 - 1/e)$-approximation.*

Proof It follows immediately from Theorem 10.1.4 and Lemma 10.1.7.   □

10.2   Submodular Set Cover

In submodular optimization, the generalization of the set cover problem has more applications. This problem is called the submodular set cover problem described as follows:

Problem 10.2.1 (Submodular Set Cover (Standard Form)) *Let f be a polymatroid function over 2^X where X is a finite set and c be a nonnegative cost function on X. Consider the minimization problem:*

$$\min \ c(A) = \sum_{x \in A} c(x) \tag{10.6}$$

$$subject \ to \ \ f(A) = f(X),$$

$$A \in 2^X$$

This problem has a greedy approximation as shown in Algorithm 30.

Theorem 10.2.2 *Greedy Algorithm 30 is an $H(\gamma)$-approximation for the minimum submodular cover problem where*

$$\gamma = \max_{x \in X} f(\{x\})$$

Algorithm 30 Greedy Algorithm for Minimum Submodular Set Cover

---

Input: A polymotroid function f over 2^X and a nonnegative cost function c on X.
Output: A subset A of X.

1: $A \leftarrow \emptyset$;
2: **while** $f(A) < f(X)$ **do**
3:      choose $x \in X \setminus A$ to maximize $\Delta_x f(A)/c(x)$ and
4:      $A \leftarrow A \cup \{x\}$
5: **end while**
6: **return** A.

---

and

$$H(\gamma) = \sum_{i=1}^{\gamma} \frac{1}{i}.$$

Proof We claim that the minimum submodular cover problem can be formulated as the following integer LP:

$$\min \sum_{v \in X} c(v)x_v \tag{10.7}$$

$$\text{s.t.} \sum_{v \in X-S} \Delta_v f(S)x_v \geq \Delta_{X-S} f(S) \text{ for all } S \in 2^X,$$

$$x_v \in \{0, 1\} \text{ for } v \in X.$$

To show the claim, we first prove that for any set $A \in 2^X$ satisfying $f(A) = f(X)$, its indicator vector $\mathbf{1}_A$ is a feasible solution of LP (10.7), where $\mathbf{1}_A = (x_v)_{v \in X}$ is defined by

$$x_v = \begin{cases} 1 \text{ if } v \in A, \\ 0 \text{ otherwise.} \end{cases}$$

In fact, for any $S \in 2^X$,

$$\sum_{v \in X-S} \Delta_v f(S)x_v = \sum_{v \in A \setminus S} \Delta_v f(S)$$

$$\geq \Delta_{A \setminus S} f(S)$$

$$= f(A) - f(S)$$

$$= f(X) - f(S)$$

$$= \Delta_{X-S} f(S),$$

where the inequality comes from Lemma 10.1.3.

Next, we show that if x is a feasible solution of LP (10.7), then $A = \{v \mid x_v = 1\}$ satisfies $f(A) = f(X)$. In fact, the inequality constraint for $S = A$ is

$$\sum_{v \in X-A} \Delta_v f(A) x_v \geq \Delta_{X-A} f(A),$$

that is,

$$0 \geq f(X) - f(A).$$

So, $f(A) \geq f(X)$. Since f is monotone nondecreasing, we must have $f(X) = f(A)$.

To sum up, the constraints of (10.7) and (10.6) are the same. We shall use dual-fitting method to analyze the performance of the greedy algorithm.

The dual LP of the relaxed LP of (10.7) is:

$$\max \sum_{S \in 2^X} \Delta_{X-S} f(S) y_S \tag{10.8}$$

$$\text{s.t.} \sum_{S:v \notin S} \Delta_v f(S) y_S \leq c(v) \text{ for } v \in X,$$

$$y_S \geq 0 \text{ for } S \in 2^X.$$

Suppose the above greedy algorithm selects $x_1, \ldots, x_g$ in turn. Denote $A_0 = \emptyset$ and $A_k = \{x_1, \ldots, x_k\}$ for $k = 1, \ldots, g$. Then, A_g is the output. Denote $r_k = \Delta_{x_k} f(A_{k-1})$ and $c_i = c(x_i)$. Define a set of dual variables

$$y_S = \begin{cases} \frac{1}{H(\gamma)} \left(\frac{c_{k+1}}{r_{k+1}} - \frac{c_k}{r_k} \right) & \text{if } S = A_k, 0 \leq k \leq g - 1, \\ 0 & \text{otherwise.} \end{cases}$$

where c_0/r_0 is viewed as 0.

It can be shown that $\{y_S\}$ is a feasible solution to (10.8). By the greedy rule and the submodularity of f, we have

$$\frac{c_k}{r_k} \leq \frac{c_{k+1}}{\Delta_{v_{k+1}} f(A_{k-1})} \leq \frac{c_{k+1}}{\Delta_{v_{k+1}} f(A_k)} = \frac{c_{k+1}}{r_{k+1}}.$$

Hence, $y_S \geq 0$ for any $S \in 2^X$. Next, consider any $v \in X$, we have

$$\sum_{S:v \notin S} \Delta_v f(S) y_S = \frac{1}{H(\gamma)} \sum_{k=0}^{k_v-1} \Delta_v f(A_k) \left(\frac{c_{k+1}}{r_{k+1}} - \frac{c_k}{r_k} \right), \tag{10.9}$$

where $k_v = i$ if $v = x_i$ and $k_v = g$ if $v \notin A_g$. The summation term in the right-hand side can be rewritten as

$$\sum_{k=0}^{k_v-1} \Delta_v f(A_k) \left(\frac{c_{k+1}}{r_{k+1}} - \frac{c_k}{r_k} \right)$$

$$= \sum_{k=1}^{k_v-1} \frac{c_k}{r_k} (\Delta_v f(A_{k-1}) - \Delta_v f(A_k)) + \frac{c_{k_v}}{r_{k_v}} \Delta_v f(A_{k_v-1})$$

$$= \sum_{k=1}^{k_v} \frac{c_k}{r_k} (\Delta_v f(A_{k-1}) - \Delta_v f(A_k)),   \tag{10.10}$$

where $\Delta_v f(A_{k_v}) = 0$ is used in absorbing the second term into the summation. For any $k = 0, 1, \ldots, k_v - 1$, since $v \notin A_k$, we have

$$\frac{c_k}{r_k} \leq \frac{c(v)}{\Delta_v f(A_k)}$$

by the greedy rule. Furthermore, by the submodularity of f, we have $\Delta_v f(A_{k-1}) - \Delta_v f(A_k) \geq 0$. Hence, (10.10) becomes

$$\sum_{k=0}^{k_v-1} \Delta_v f(A_k) \left(\frac{c_{k+1}}{r_{k+1}} - \frac{c_k}{r_k} \right) \leq \sum_{k=1}^{k_v} c(v) \frac{\Delta_v f(A_{k-1}) - \Delta_v f(A_k)}{\Delta_v f(A_k)}$$

$$\leq c(v) H(\Delta_v f(A_0)) \leq c(v) H(\gamma).$$

Combining this with (10.9), the constraint of dual LP (10.8) is satisfied.

Since $f(A_g) = f(X)$, it can be calculated that

$$c(A_g) = \sum_{k=1}^{g} c_k = \sum_{k=1}^{g} \frac{c_k}{r_k} \Delta_{x_k} f(A_{k-1}) = \sum_{k=1}^{g} \frac{c_k}{r_k} (f(A_k) - f(A_{k-1}))$$

$$= \sum_{k=1}^{g-1} f(A_k) \left(\frac{c_k}{r_k} - \frac{c_{k+1}}{r_{k+1}} \right) + \frac{c_g}{r_g} f(A_g)$$

$$= \sum_{k=1}^{g-1} f(A_k) \left(\frac{c_k}{r_k} - \frac{c_{k+1}}{r_{k+1}} \right) + \sum_{k=1}^{g-1} \left(\frac{c_{k+1}}{r_{k+1}} - \frac{c_k}{r_k} \right) f(X)$$

$$= \sum_{k=1}^{g-1} \left(\frac{c_{k+1}}{r_{k+1}} - \frac{c_k}{r_k} \right) (f(X) - f(A_k))$$

$$= H(\gamma) \sum_{k=1}^{g-1} \Delta_{X-A_k} f(A_k) \cdot y_{A_k}$$

$$= H(\gamma) \sum_{S \in 2^X} \Delta_{X-S} f(S) \cdot y_S$$

$$\leq H(\gamma) opt,$$

where *opt* is the optimal value of the integer LP (10.7), which is an upper bound for the objective value of $\{y_S\}$ in the dual LP (10.8).   □

Now, we present an example.

Problem 10.2.3 (Positively Dominating Set) *Given a graph* $G = (V, E)$, *find a minimum positively dominating set where a node* v *is positively dominated by a node set* A *if* $deg_A(v) \geq deg(x)$. *Here,* $deg(v)$ *is the degree of* v *in* G *and* $deg_A(v) = |\{u \mid (u, v) \in E, u \in A \ or \ v \in A\}|$.

For any $A \subseteq V$, define

$$g(A) = \sum_{v \in V} \min(\lceil deg(V)/2 \rceil, deg_A(v)).$$

We will show that g is a polymatroid function. To do so, we first show two properties of the polymatroid function.

Lemma 10.2.4 *Suppose* f *is a polymatroid function. Then, for any constant* $c \geq 0$, $\zeta(A) = \min(c, f(A))$ *is a polymatroid function.*

Proof Note that for $A \subset B$,

$$f(A) \leq f(B) \Rightarrow \min(c, f(A)) \leq \min(c, f(B)),$$

$$f(\emptyset) = 0 \Rightarrow \min(c, f(\emptyset)) = 0.$$

Thus, it suffices to prove the modularity of $\zeta(A)$. Consider two node subsets A and B with $A \subset B$ and $x \notin B$. We divide the proof into three cases:

Case 1. $f(A \cup \{x\}) > c.$

$$\zeta(A \cup \{x\}) - \zeta(A) = c - \zeta(A)$$

$$\geq c - \zeta(B)$$

$$= \zeta(B \cup \{x\}) - \zeta(B).$$

Case 2. $f(A \cup \{x\}) \leq c$ and $f(B) \leq c.$

$$\zeta(A \cup \{x\}) - \zeta(A) = f(A \cup \{x\}) - f(A)$$
$$\geq f(B \cup \{x\}) - f(B)$$
$$\geq \zeta(B \cup \{x\}) - \zeta(B).$$

Case 3. $f(A \cup \{x\}) \leq c$ and $f(B) > c$.

$$\zeta(A \cup \{x\}) - \zeta(A) = f(A \cup \{x\}) - f(A)$$
$$\geq f(B \cup \{x\}) - f(B)$$
$$\geq c - c$$
$$= \zeta(B \cup \{x\}) - \zeta(B).$$

$\square$

Lemma 10.2.5 *If f_i for $i = 1, 2, \ldots, n$ are polymatroid functions, then $f = \sum_{i=1}^{n} f_i$ is a polymatroid function.*

Proof It follows immediately from the following derivations:

$$f_i(\emptyset) = 0 \text{ for } i = 1, 2, \ldots, n \Rightarrow f(\emptyset) = 0$$
$$\Delta_x f_i(A) \geq \Delta_x f_i(B) \text{ for } i = 1, 2, \ldots, n \Rightarrow \Delta_x f(A) \geq \Delta_x f(B).$$

$\square$

Lemma 10.2.6 *g is a polymatroid function.*

Proof By Lemmas 10.2.4 and 10.2.5, it is sufficient to prove that $h_v(A) = deg_A(v)$ with respect to A is a polymatroid function for any fixed v. Clearly, $deg_\emptyset(v) = 0$. Consider two node subsets A and B with $A \subset B$ and node $x \notin A$. We divide the remaining proof into three cases:

Case 1. $v \in A \cup \{x\}$.

$$\Delta_x h_v(A) = 0 = \Delta_x h_v(B).$$

Case 2. $v \in B \setminus A$.

$$\Delta_x h_v(A) \geq 0 = \Delta_x h_v(B).$$

Case 3. $v \notin B \cup \{x\}$. If $(x, v) \in E$, then

$$\Delta_x h_v(A) = 1 = \Delta_x h_v(B).$$

If $(x, v) \notin E$, then

$$\Delta_x h_v(A) = 0 = \Delta_x h_v(B).$$

$\square$

Now, we show that the positively dominating set problem can be formulated in the following form:

$$\min \ |A|$$
$$\text{subject to} \quad g(A) = g(V)$$
$$A \in 2^V.$$

Lemma 10.2.7 *$A \subseteq V$ is a positively dominating set if and only if $g(A) = g(X)$.*

Proof A is a positively dominating set if and only if for any $v \in V$, $deg_A(v) \geq \lceil deg(v)/2 \rceil$ if and only if

$$g(A) = \sum_{v \in V} \lceil deg(v)/2 \rceil = g(V).$$

$\square$

$g(A)$ has an interesting property.

Lemma 10.2.8 *If $g(A) < g(V)$, then for any v not positively dominated by A, $g(A \cup \{v\}) > g(A)$.*

Proof Since v is not positively dominated by A, we have

$$deg_A(v) < \lceil deg(v)/2 \rceil.$$

Therefore,

$$\min(\lceil deg(v)/2 \rceil, deg_{A \cup \{v\}}(v)) = \lceil deg(v)/2 \rceil > \min(\lceil deg(v)/2 \rceil, deg_A(v)).$$

Hence, $g(A \cup \{v\}) > g(A)$. $\square$

Theorem 10.2.9 *Greedy Algorithm 30 gives a $(1 + \ln\lceil \frac{3}{2}\Delta \rceil)$-approximation for the positively domating set problem where Δ is the maximum node degree.*

Proof By Theorem 10.2.2, the performance ratio is determined by

$$\gamma = \max_{v \in V} g(\{v\})$$
$$\leq \max_{v \in V} \left(deg(v) + \lceil \frac{1}{2} deg(v) \rceil \right)$$
$$= \max_{v \in V} \lceil \frac{3}{2} deg(v) \rceil$$

$$\leq \lceil \frac{3}{2} \Delta \rceil.$$

□

Lemma 10.2.4 indicates how to deal with the following general form of the submodular set cover problem.

Problem 10.2.10 (Submodular Set Cover (General Form)) *Let f be a polymatroid function over 2^X where X is a finite set and c be a nonnegative cost function on X. Consider the minimization problem:*

$$\min \ c(A) = \sum_{x \in A} c(x) \tag{10.11}$$

$$subject \ to \ \ f(A) \geq d,$$

$$A \in 2^X,$$

where d is a positive constant.

Define $g(A) = \min(f(A), d)$. Then, condition $f(A) \geq d$ will become $g(A) = g(X)$, so that this becomes the submodular set cover problem in previous form.

10.3 Monotone Submodular Maximization

In Sect. 10.1, we studied a monotone submodular maximization with size constraint. The size constraint is a special case of knapsack constraint and also a special case of matroid constraint. In this section, we study the monotone submodular maximization with a knapsack constraint or several matroid constraints.

First, we study the following problem:

Problem 10.3.1 *Let f be a polymatroid function over 2^X:*

$$\max \ f(S) \tag{10.12}$$

$$subject \ to \ \sum_{x \in S} b(x)x \leq B$$

$$S \in 2^X$$

where X is the universe set, $b(x)$ is the budget cost of item x, and B is the total budget.

This problem can also be seen as a generalization of the knapsack problem. Algorithm 31 is an extension of greedy 1/2-approximation algorithm for the knapsack problem. Denote $b(A) = \sum_{x \in A} b(x)$.

Algorithm 31 $((1 - e^{-1})/2)$-Approximation for Problem 10.3.1

Input: A polymatroid function f on 2^X and knapsack constraint coefficients $0 \leq b(x) \leq B$ for $x \in X$.
Output: A subset A of X and $f(A)$.

1: $S \leftarrow \emptyset$;
2: **if** $b(X) \leq B$ **then**
3:      $A \leftarrow X$
4: **else**
5:      **while** $b(S) \leq B$ **do**
6:          choose $x \in X \setminus S$ to maximize $\frac{\Delta_x f(S)}{b(x)}$
7:          and set $S \leftarrow S \cup \{x\}$;
8:      **end while**
9:      $A \leftarrow \operatorname{argmax}(f(S \setminus \{x\}), f(\{x\}))$;
10: **end if**
11: **return** A and $f(A)$.

Theorem 10.3.2 *Algorithm 31 is $((1 - e^{-1})/2)$-approximation for problem (10.12).*

Proof Let $x_1, x_2, \ldots, x_{k+1}$ be generated by the algorithm. Denote $S_i = \{x_1, x_2, \ldots, x_i\}$. Then

$$x_{i+1} = \operatorname{argmax}_{x \in X \setminus S_i} \frac{\Delta_x f(S_i)}{b(x)},$$

and $b(S_k) \leq B < b(S_{k+1})$. Let $S^* = \{y_1, y_2, \ldots, y_h\}$ be an optimal solution. Denote $S_j^* = \{y_1, y_2, \ldots, y_j\}$. Then

$$
\begin{aligned}
f(S^*) &\leq f(S_i \cup S^*) \\
&= f(S_i) + \Delta_{y_1} f(S_i) + \Delta_{y_2} f\left(S_i \cup S_1^*\right) + \cdots + \Delta_{y_h} f\left(S_i \cup S_{h-1}^*\right) \\
&\leq f(S_i) + \Delta_{y_1} f(S_i) + \Delta_{y_2} f(S_i) + \cdots + \Delta_{y_h} f(S_I) \\
&\leq f(S_i) + \sum_{j=1}^{h} \frac{b(y_j)}{b(x_{i+1})} \cdot \Delta_{x_{i+1}} f(S_i) \\
&= f(S_i) + \frac{b(S^*)}{b(x_{i+1})} \cdot (f(S_{i+1}) - f(S_i)).
\end{aligned}
$$

Denote $\alpha_i = opt - f(S_i)$. Then, we have

$$\alpha_i \leq \frac{b(S^*)}{b(x_{i+1})} \cdot (\alpha_i - \alpha_{i+1}).$$

Hence,

$$\alpha_{i+1} \leq \left(1 - \frac{b(x_{i+1})}{b(S^*)}\right) \alpha_i \leq \alpha_i \cdot e^{-\frac{b(x_{i+1})}{b(S^*)}}.$$

Therefore,

$$opt - f(S_{k+1}) = \alpha_{k+1} \leq \alpha_0 \cdot e^{-\frac{b(S_{k+1})}{b(S^*)}} \leq opt \cdot e^{-1}.$$

Hence,

$$(1 - e^{-1}) \cdot opt \leq f(S_{k+1}) = f(S_{k+1}) + f(\emptyset) \leq f(S_k) + f(\{x_{k+1}\}).$$

Thus,

$$(1 - e^{-1})/2 \cdot opt \leq \max(f(S_k), f(\{x_{k+1}\})).$$

□

Next, we extend the PTAS for the knapsack problem to an algorithm for Problem 10.3.1 which has a better performance ratio.

Theorem 10.3.3 *Algorithm 32 is a* $(1 - 1/e)$-*approximation for Problem 10.3.1.*

Proof Suppose optimal solution $S^* = \{u_1, u_2, \ldots, u_h\}$ in ordering

$$u_i = \mathrm{argmax}_u (f(\{u_1, u_2, \ldots, u_{i-1}, u\}) - f(\{u_1, u_2, \ldots, u_{i-1}\})).$$

Algorithm 32 $(1 - e^{-1})$-Approximation for Problem 10.3.1

Input: A polymatroid function f on 2^X and knapsack constraint coefficients $0 \leq b(x) \leq B$ for $x \in X$.
Output: A subset A of X and $f(A)$.
1: **for** every $I \subseteq V$ with $|I| \leq 3$ **do**
2:     $S \leftarrow I$;
3:     $T \leftarrow V \setminus I$;
4:     **while** $T \neq \emptyset$ **do**
5:         choose $x \in T$ to maximize $\frac{\Delta_x f(S)}{b(x)}$
6:         and set $T \leftarrow T \setminus \{x\}$;
7:         **if** $b(S \cup \{x\}) \leq B$ **then**
8:             $S \leftarrow S \cup \{x\}$
9:         **end if**
10:     **end while**
11:     $S(I) \leftarrow S$;
12: **end for**
13: $A \leftarrow \mathrm{argmax}_I f(S(I))$;
14: **return** A and $f(A)$.

Let $I = \{u_1, u_2, u_3\}$. Define $g(A) = f(A \cup I) - f(I)$. Then computation of $S(I)$ can be seen as a greedy algorithm applying to a polymatroid function g. Suppose this computation produces $S \setminus I = \{v_1, v_2, \ldots, v_i\}$.

If $S \setminus I \supseteq S^* \setminus I$, then $S \setminus I = S^* \setminus I$ and $g(S \setminus I) = g(S^* \setminus I)$. If there exists $v_{i+1} \in (S^* \setminus I) \setminus (S \setminus I)$, then $b(I \cup \{v_1, v_2, \ldots, v_i\}) \leq B$, but $b(I \cup \{v_1, v_2, \ldots, v_{i+1}\}) > B$. By the proof of Theorem 10.3.2, we will have

$$g(\{v_1, v_2, \ldots, v_{i+1}\}) \geq (1 - e^{-1})g(S^* \setminus I).$$

Thus,

$$f(I \cup \{v_1, v_2, \ldots, v_{i+1}\}) \geq (1 - e^{-1})f(S^*) + e^{-1}f(I),$$

and

$$f(S(I)) \geq f(I \cup \{v_1, v_2, \ldots, v_i\})$$
$$\geq (1 - e^{-1})f(S^*) + e^{-1}f(I) - \Delta_{v_{i+1}} f(I \cup \{v_1, v_2, \ldots, v_i\}).$$

Since $v_{i+1} \in S^*$, we have

$$\Delta_{v_{i+1}} f(I \cup \{v_1, v_2, \ldots, v_i\}) \leq \Delta_{v_{i+1}} f(I)$$
$$\leq \Delta_{v_{i+1}} f(\{u_1, u_2\})$$
$$\leq \Delta_{u_3} f(\{u_1, u_2\})$$
$$\leq \Delta_{u_2}(\{u_1\})$$
$$\leq \Delta_{u_1} f(\emptyset) = f(\{u_1\}).$$

Thus,

$$\Delta_{v_{i+1}} f(I \cup \{v_1, v_2, \ldots, v_i\})$$
$$\leq \frac{1}{3}(\Delta_{u_3} f(\{u_1, u_2\}) + \Delta_{u_2} f(\{u_1\}) + \Delta_{u_1} f(\emptyset))$$
$$= \frac{1}{3}f(I).$$

Hence,

$$f(S(I)) \geq f(I \cup \{v_1, v_2, \ldots, v_i\})$$
$$\geq (1 - e^{-1})f(S^*) + e^{-1}f(I) - \frac{1}{3}f(I)$$
$$\geq (1 - e^{-1})f(S^*).$$

$\square$

Clearly, Algorithm 31 runs faster than Algorithm 32 although the performance ratio is smaller.

Next, we study the monotone submodular maximization with matroid constraints.

Problem 10.3.4 *Let* f *be a polymatroid function. Let* $(X, \mathcal{I}_i)$ *be a matroid for every* $1 \leq i \leq k$. *Consider the following problem:*

$$\max \quad f(A)$$

$$\text{subject to} \quad A \in \mathcal{I}_i \text{ for every } 1 \leq i \leq k.$$

Let $\mathcal{I} = \cap_{i=1}^k \mathcal{I}_i$. Then $(X, \mathcal{I})$ is an independent system.

Lemma 10.3.5 *Consider an independent system* $(X, \mathcal{I})$ *which is the intersection of* k *matroids* $(X, \mathcal{I})$. *Suppose* $A, B \in \mathcal{I}$ *and* I *is a maximal independent subset of* $A \cup B$, *containing* B. *Then,* $|A \setminus I| \leq k |B \setminus A|$.

Proof For each i, add $B \setminus A$ to A. We can remove at most $|B \setminus A|$ elements from A to keep the independence in matroid $(X, \mathcal{I}_i)$. Let A_i be the set of those removed elements. Then, $|A_i| \leq |B \setminus A|$ and $(B \cup A) \setminus A_i$ are independent in $(X, \mathcal{I}_i)$. Clearly, $I \supseteq \cap_{i=1}^k ((B \cup A) \setminus A_i)$. Hence,

$$|A \setminus I| \leq |\cup_{i=1}^k A_i| \leq \sum_{i=1}^k |A_k| \leq k |B \setminus A|.$$

$\square$

A greedy approximation for Problem 10.3.4 is presented in Algorithm 33.

Theorem 10.3.6 *Algorithm 33 gives a* $\frac{1}{k+1}$-*approximation for Problem 10.3.4.*

Proof Suppose in iterations of the algorithm, set A is assigned with $A_0, A_1, \ldots, A_g$. Let H_0 be an optimal solution and for $i = 1, 2, \ldots, g$, H_i is a maximal independent set of $H_{i-1} \cup A_i$. By Lemma 10.3.5,

---

Algorithm 33 Greedy Approximation for Problem 10.3.4

Input: A polymatroid function f on 2^X and k matroids $(X, \mathcal{I}_i)$ for $1 \leq i \leq k$. Let $\mathcal{I} = \cap_{i=1}^k \mathcal{I}_i$.
Output: A subset A of X and $f(A)$.

1: $A \leftarrow \emptyset$;
2: **while** A is not a maximal independent set in $(X, \mathcal{I})$ **do**
3:     choose $x \in X \setminus A$ to maximize $\Delta_x f(A)$
4:     subject to $A \cup \{x\} \in \mathcal{I}$;
5:     $A \leftarrow A \cup \{x\}$
6: **end while**
7: **return** A and $f(A)$.

$$|H_{i-1} \setminus H_i| \leq k|A_i \setminus H_{i-1}| \leq |A_i \setminus A_{i-1}| \leq k.$$

Suppose $H_{i-1} \setminus H_i = \{v_1, \ldots, v_r\}$. Denote $H^j_{i-1} = H_{i-1} \setminus \{v_1, \ldots, v_j\}$. Then

$$f(A_i) - f(A_{i-1}) \geq f(\{v_j\} \cup A_{i-1}) - f(A_{i-1})$$
$$= \Delta_{v_j} f(A_{i-1})$$
$$\geq \Delta_{v_j} f(H^j_{i-1} \cup A_{i-1}).$$

Thus,

$$r(f(A_i) - f(A_{i-1})) \geq \sum_{j=1}^{r} \Delta_{v_j} f\left(H^j_{i-1} \cup A_{i-1}\right)$$
$$= f\left(H^0_{i-1} \cup A_{i-1}\right) - f\left(H^r_{i-1} \cup A_{i-1}\right)$$
$$= f(H_{i-1}) - f(H_i).$$

Note that $r \leq k$. Thus,

$$k[f(A_g) - f(A_0)] = r \sum_{i=1}^{g} [f(A_i) - f(A_{i-1})]$$
$$\geq \sum_{i=1}^{g} [f(H_{i-1}) - f(H_i)]$$
$$= f(H_0) - f(H_g).$$

Note that $A_g = H_g$ and $f(A_0) \geq 0$. Therefore,

$$(k+1)f(A_g) \geq f(H_0) = opt.$$

$\square$

If the objective function f is linear, then the performance ratio $\frac{1}{k+1}$ can be improved to $\frac{1}{k}$. To see this, let us first show a lemma.

Lemma 10.3.7 *Consider an independent system $(X, \mathcal{I})$ which is the intersection of k matroids $(X, \mathcal{I})$. For any subset F of X, let $u(F)$ and $v(F)$ denote the maximal size and the minimal size of maximal independent set in $(X, \mathcal{I})$, respectively. Then, $u(F)/v(F) \leq k$.*

Proof Consider two maximal independent subsets I and J of F. Let $I_i \supseteq I$ be a maximal independent subset of $I \cup J$ with respect to $(X, \mathcal{I}_i)$. For each $e \in J \setminus I$, if $e \in \cap_{i=1}^{k}(I_i \setminus I)$, then $I \cup \{e\}$ is independent in $(X, \mathcal{I})$, contradicting the maximality of I. Hence, e appears at most $k - 1$ $(I \setminus I_i)$'s. Thus,

$$\sum_{i=1}^{k} |I_i| - k|I| = \sum_{i=1}^{k} |I_i \setminus I| \le (k - 1)|J \setminus I| \le (k - 1)|J|.$$

Now, let $J_i \supseteq J$ be a maximal independent subset of $I \cup J$ with respect to matroid $(X, \mathcal{I}_i)$. Then, $|J_i| = |I_i|$. Therefore,

$$|J| \le \left(\sum_{i=1}^{k} |J_i| - k|J| \right) + |J|$$

$$\le \left(\sum_{i=1}^{k} |I_i| - k|J| \right) + |J|$$

$$\le k|I|.$$

$\square$

Theorem 10.3.8 *Let f be a linear function with nonnegative coefficients. Then, Algorithm 33 gives a $\frac{1}{k}$-approximation for Problem 10.3.4.*

Proof It follows immediately from Lemma 10.3.7 and Theorem 4.2.2 in Chap. 4.

$\square$

A popular matroid constraint is the partition matroid. For example, consider the influence maximization problem (Problem 10.1.5). Suppose the social network (input directed graph) is partitioned into several communities. Instead of the total size constraint, we may ask certain balance for seed distribution, that is, for each community, set a size constraint. This will form a (partition) matroid constraint.

For the monotone submodular maximization with only one matroid constraint, there exists a greedy approximation algorithm with performance ratio $1 - e^{-1}$, which is better than $1/2$. But this algorithm is different from Algorithm 33. We will introduce it in the next chapter.

10.4   Random Greedy

Consider the following problem:

Problem 10.4.1 *Let f be a nonnegative submodular function on subsets of a finite set X and k a positive integer:*

$$\max \ f(S)$$

$$\text{subject to} \ \ |S| \le k$$

$$S \in 2^X.$$

Algorithm 34 Random Greedy Algorithm for Problem 10.4.1

Input: A nonnegative submodular function f on 2^X and a positive integer k.
Output: A subset A_k of X with $|A_k| \leq k$.

1: $A_0 \leftarrow \emptyset$.
2: **for** $i \leftarrow 1$ to k **do**
3:      Let $M_i \subseteq X \setminus A_{i-1}$ be a subset of size at most k maximizing $\sum_{u \in M_i} \Delta_u f(A_{i-1})$.
4:      Choose each element u_i with probability $1/k$ (mutually exclusive, uniform random) from M_i and not choose any element with probability $1 - |M_i|/k$.
5:      **if** u_i is selected **then**
6:          $A_i \leftarrow A_{i-1} \cup \{u_i\}$
7:      **else**
8:          $A_i \leftarrow A_{i-1}$
9:      **end if**
10: **end for**
11: **return** A_k.

In this section, we will study a random greedy algorithm for this problem, as shown in Algorithm 34. This algorithm can be proved not only to have theoretical guaranteed performance ratio $(1 - e^{-1})$ for monotone nondecreasing function f but also to have performance ratio $1/e$ for general f.

Let us first consider the monotone nondecreasing function f.

Theorem 10.4.2 *If f is monotone nondecreasing, then random greedy (Algorithm 34) has approximation performance ratio $1 - e^{-1}$.*

Proof Let us fix all random process until A_i is obtained for $1 \leq i \leq k$. Let OPT be an optimal solution and denote $opt = f(OPT)$. Therefore, we have

$$
\begin{aligned}
E[\Delta_{u_i} f(A_{i-1})] &= \frac{1}{k} \cdot \sum_{u \in M_i} \Delta_u f(A_{i-1}) \\
&\geq \frac{1}{k} \cdot \sum_{u \in OPT \setminus A_{i-1}} \Delta_u f(A_{i-1}) \\
&\geq \frac{1}{k} \cdot (f(OPT \cup A_{i-1}) - f(A_{i-1})) \\
&\geq \frac{opt - f(A_{i-1})}{k}.
\end{aligned}
$$

The first inequality is due to greedy choice of A_i. The second inequality holds because f is submodular. The third inequality is true because f is monotone nondecreasing.

Now, we release the randomness of A_i and A_{i-1}. Then, we have

$$
E[f(A_i)] - E[f(A_{i-1})] \geq \frac{opt - E[f(A_{i-1})]}{k}.
$$

Therefore,

$$opt - E[f(A_i)] \le \left(1 - \frac{1}{k}\right)(opt - E[f(A_{i-1})]).$$

This implies

$$opt - E[f(A_i)] \le \left(1 - \frac{1}{k}\right)^i \cdot (opt - f(A_0))$$

$$\le \left(1 - \frac{1}{k}\right)^i \cdot opt.$$

Thus,

$$E[f(A_k)] \ge \left(1 - \left(1 - \frac{1}{k}\right)^k\right) \cdot opt \ge (1 - e^{-1}) \cdot opt.$$

$\square$

Next, we consider a submodular function f, not-necessarily monotone nondecreasing. First, show two lemmas.

Lemma 10.4.3 *Consider a submodular function f on subsets of a finite set X. Let $A(p)$ be a random subset of A where each element appears with probability at most p. Then*

$$E[f(A(p))] \ge (1 - p)f(\emptyset).$$

Proof Sort element of A in nonincreasing order of probability to appear in $A(p)$, $u_1, u_2, \ldots, u_h$ where $h = |A|$. Denote $p_i = Pr[u_i \in A(p)]$. Then, $p_1 \ge p_2 \ge \cdots \ge p_h$.

Let x_i be an indicator for event that $u_i \in A(p)$, i.e.,

$$x_i = \begin{cases} 1 & \text{if } u_i \in A(p), \\ 0 & \text{otherwise.} \end{cases}$$

Denote $A_i = \{u_1, u_2, \ldots, u_i\}$. Then, we have

$$E[f(A(p))] = E\left[f(\emptyset) + \sum_{i=1}^{h} x_i \cdot \Delta_{u_i} f(A(p) \cap A_{i-1})\right]$$

$$\ge E\left[f(\emptyset) + \sum_{i=1}^{h} x_i \cdot \Delta_{u_i} f(A_{i-1})\right]$$

$$= f(\emptyset) + \sum_{i=1}^{h} E[x_i] \cdot \Delta_{u_i} f(A_{i-1})$$

$$= f(\emptyset) + \sum_{i=1}^{h} p_i \cdot \Delta_{u_i} f(A_{i-1})$$

$$= (1 - p_1) \cdot f(\emptyset) + \sum_{i=1}^{h-1} (p_i - p_{i+1}) f(A_i) + p_h \cdot f(A)$$

$$\geq (1 - p) \cdot f(\emptyset).$$

□

Lemma 10.4.4 *For* $0 \leq i \leq k$,

$$E[f(OPT \cup A_i)] \geq \left(1 - \frac{1}{k}\right)^i \cdot opt$$

where OPT *is an optimal solution and* $opt = f(OPT)$.

Proof In each iteration i, every element of $X \setminus A_{i-1}$ stays outside of A_i with probability at least $1 - 1/k$. Thus, each element is selected to A_i with probability at most $1 - (1 - 1/k)^i$. Define $g(S) = f(S \cup OPT)$. Since f is submodular, so is g. By Lemma 10.4.3,

$$E[f(OPT \cup A_i)] = E[g(A_i)] \geq (1 - 1/k)^i \cdot g(\emptyset) = (1 - 1/k)^i \cdot opt.$$

□

Now, we are ready to show the following:

Theorem 10.4.5 *In general, random greedy (Algorithm 34) for Problem 10.4.1 has approximation performance ratio* $1/e$.

Proof Let us fix all random process until A_i is obtained for $1 \leq i \leq k$. Let OPT be an optimal solution and denote $opt = f(OPT)$. Therefore, we have

$$E[\Delta_{u_i} f(A_{i-1})] = \frac{1}{k} \cdot \sum_{u \in M_i} \Delta_u f(A_{i-1})$$

$$\geq \frac{1}{k} \cdot \sum_{u \in OPT \setminus A_{i-1}} \Delta_u f(A_{i-1})$$

$$\geq \frac{1}{k} \cdot (f(OPT \cup A_{i-1}) - f(A_{i-1}))$$

The first inequality is due to greedy choice of A_i. The second inequality is true because f is submodular.

Release randomness of A_i and A_{i-1} and take expectation. By Lemma 10.4.4, we obtain

$$E[\Delta_{u_i} f(A_{i-1})] \geq \frac{1}{k} \cdot (E[f(OPT \cup A_{i-1})] - E[f(A_{i-1})])$$

$$\geq \frac{1}{k} \cdot \left(\left(1 - \frac{1}{k}\right)^{i-1} \cdot opt - E[f(A_{i-1})] \right).$$

Thus,

$$E[f(A_i)] \geq \frac{1}{k} \cdot \left(1 - \frac{1}{k}\right)^{i-1} \cdot opt + \left(1 - \frac{1}{k}\right) \cdot E[f(A_{i-1})]$$

$$\geq \frac{1}{k} \cdot \left(1 - \frac{1}{k}\right)^{i-1} \cdot opt.$$

Finally, set $i = k$ in the above inequality.

$$E[f(A_k)] \geq \left(1 - \frac{1}{k}\right)^{k-1} \cdot opt \geq e^{-1} \cdot opt.$$

$\square$

Exercises

1. Consider a graph $G = (V, E)$. For any edge subset $A \subseteq E$, define $f(A)$ to be the maximum number of edges in A which does not contain a cycle. Show that $f(A)$ is submodular.
2. Consider a graph $G = (V, E)$. For any node subset $A \subseteq V$, define $f(A)$ to be the number of connected components in subgraph with node set V and edges each with at least one endpoint in A. Show that $|V| - f(A)$ is submodular.
3. Consider a graph $G = (V, E)$. For any node subset A, $f(A)$ is defined to be the number of connected components in the subgraph induced by A. Show that $|V| - f(A)$ is not submodular.
4. Show that if f and g are submodular and $f - g$ is monotone decreasing, then $\min(f, g)$ is submodular.
5. (Rank Functions of Matroids) Consider a matroid $(X, \mathcal{I})$. For any subset $A \subseteq X$, define rank function $r(A)$ to be the cardinality of maximum independent subset of A. Show that r is a polymatroid function with $f(\{x\}) = 1$ for all

$x \in X$. Conversely, suppose r is a polymatroid function with $f(\{x\}) = 1$ for all $x \in X$. Define $\mathcal{I} = \{I \mid r(I) = |I|\}$. Show that $(X, \mathcal{I})$ is a matroid.

6. (Matroid Duality) For any matroid $\mathcal{M} = (X, \mathcal{I})$, the *dual matroid* $\mathcal{M}^* = (X, \mathcal{I}^*)$ is defined by

$$\mathcal{I}^* = \{X \setminus I \mid I \in \mathcal{I}\}.$$

Prove that $\mathcal{M}^*$ has the rank function

$$r^*(S) = |S| - (r(X) - r(X \setminus S))$$

where r is the rank function of $\mathcal{M}$.

7. Consider a non-singular matrix. Prove that for any row index subset I, there exists a column index subset J such that both submatrices, with indices $I \times J$ and $\bar{I} \times \bar{J}$, respectively, are non-singular.

8. (Cut Functions in Graphs) Consider an undirected graph $G = (V, E)$. For any vertex subset $U \subseteq V$, $\delta(U)$ denotes the set of edges each of which has exactly one endpoint in U. Define $f(U) = |\delta(U)|$. Prove that f is submodular, but may not be monotone nondecreasing.

9. Suppose f is a submodular function on 2^V. Show that for any positive constant $c > 0$, $\min(f(A), c)$ is also a submodular function.

10. Consider a submodular function f on subsets of a finite set X. Show that $f(A \cup B) - f(A) \le \sum_{x \in B} \Delta_x f(A)$.

11. Consider a graph $G = (V, E)$. For any node set $A \subseteq V$, let $f(A)$ be the number of nodes not adjacent to any node in A. Show that $f(A)$ is a monotone nonincreasing supermodular function. (A set function is supermodular if $-f$ is submodular.)

12. Show that the following problem has a greedy $(1 - e^{-1})$-approximation: Given a collection C of subsets of a finite set X and an integer $k > 0$, find a subset A of X with $|A| = k$ to maximize the number of subsets in C hit by A where a subset S is said to be hit by A if $A \cap S \ne \emptyset$.

13. Consider a submodular function f on subsets of a finite set X. Show that $f(A) - f(A \cap B) \ge \sum_{x \in A \setminus B} \Delta_x f(A \setminus \{x\})$.

14. Consider an independent system $(X, \mathcal{I})$. A minimal dependent set is called a *circuit*. Let $A_1, A_2, \ldots, A_k$ be all circuits of this system. Define $\mathcal{I}_i = \{B \mid A_i \not\subseteq B\}$. Show the following:

   (a) For any $1 \le i \le k$, $(X, \mathcal{I}_i)$ is a matroid.
   (b) $\mathcal{I} = \cap_{i=1}^{k} \mathcal{I}_i$.

15. Let f be a nonnegative submodular function on subsets of a finite set X. Suppose that S is subset of X such that $f(S) \ge f(S \cup \{x\})$ for any $x \in X \setminus S$ and $f(S) \ge f(S \setminus \{x\})$ for any $x \in S$. Show the following:

   (a) If $T \subset S$ or $T \supset S$, then $f(T) \le f(S)$.
   (b) $\max(f(S), f(X \setminus S)) \ge \frac{1}{3} \cdot \max_{A \subseteq X} f(A)$.

16. Let f be a nonnegative submodular function on subsets of a finite set X and $(X, \mathcal{I})$ a matroid. Consider the problem of maximizing $f(S)$ subject to $S \in \mathcal{I}$. A solution $S \in \mathcal{I}$ is locally maximal if it cannot be improved through any operation of deletion, addition, and swapping as follows:

Addition   : If $S \cup \{x\} \in \mathcal{I}$, then $f(S) \geq f(S \cup \{x\})$.
Deletion   : For any $x \in S$, $f(S) \geq f(S \setminus \{x\})$.
Swapping : If $d \in X \setminus S$ and $e \in S$ such that $(S \setminus \{e\}) \cup \{d\} \in \mathcal{I}$, then $f(S) \geq f((S \setminus \{e\}) \cup \{d\})$.

Show that the following holds:

(a) For a locally maximal solution S and any subset C of X, $2f(S) \geq f(S \cup C) + f(S \cap C)$.
(b) Let S_1 be a locally maximal solution of f on 2^X and S_2 a locally maximal solution on $2^{X \setminus S_1}$. Then

$$\max(f(S_1), f(S_2)) \geq \frac{1}{4} \cdot f(C)$$

where C is the maximum solution of f on 2^X.

17. Let f be a nonnegative submodular function on subsets of a finite set X and k a positive integer. Let D be a set of $2k$ dummy elements, i.e., for any $u \in D$ and any subset A of X, $\Delta_u f(A) = 0$. Show that Algorithm 35 is equivalent to Algorithm 34.

18. (Linear Threshold Model) Consider a directed graph $G = (V, E)$. Every arc (u, v) has a weight w_{uv} such that for each node v,

$$\sum_{u:(u,v)\in E} w_{uv} \leq 1.$$

Every node has two states, active and inactive. Before starting a process, every node is inactive. Initially, for every node v, a threshold θ_v is selected randomly from $[0, 1]$ with uniform distribution, and activate at most k nodes, called them

---

Algorithm 35 Random Greedy Algorithm for Problem 10.4.1

Input: A nonnegative submodular function f on 2^X and a positive integer k.
Output: A subset A_k of X with $|A_k| \leq k$.
1: $X \leftarrow X \cup D$.
2: $A_0 \leftarrow \emptyset$.
3: **for** $i \leftarrow 1$ to k **do**
4:    Let $M_i \subseteq X \setminus A_{i-1}$ be a subset of size k maximizing $\sum_{u \in M_i} \Delta_u f(A_{i-1})$.
5:    Choose each element u_i uniform randomly from M_i.
6:    $A_i \leftarrow A_{i-1} \cup \{u_i\}$
7: **end for**
8: **return** $A_k \leftarrow A_k \setminus D$.

as seeds. Then, an influence process is carried out step-by-step. In each step, every inactive node v checks whether

$$\sum_{\text{active } u:(u,v)\in E} w_{uv} \geq \theta_v.$$

If yes, then v is activated. Otherwise, v is kept inactive. The process ends when no new active node is produced. The influence spread is the expected number of active nodes at the end of the influence process. Prove that the influence spread is a monotone nondecreasing submodular function with respect to the seed set.

19. (Independent Cascade Model) Consider a directed graph $G = (V, E)$. Every node has two states, active and inactive. Every arc (u, v) has a probability p_{uv} which means that u can influence v successfully with probability p_{uv}. Before starting a process, every node is inactive. Initially, activate at most k nodes, called them as seeds. Then, an influence process is carried out step-by-step. In each step, every freshly active node u will activate inactive neighbor through arc (u, v) with success probability p_{uv} where, by a freshly active node, we mean that a node becomes active at last step. When an inactive node receives influence from k ($k \geq 2$) incoming neighbors, we treat them as k independent events. The process ends when no new active node is produced. The influence spread is the expected number of active nodes at the end of the influence process. Prove that the influence spread is a monotone nondecreasing submodular function with respect to the seed set.

20. (Mutually Exclusive Cascade Model) Consider a directed graph $G = (V, E)$. Every node has two states, active and inactive. Every arc (u, v) has a probability p_{uv} which means that u can influence v successfully with probability p_{uv}. Before starting a process, every node is inactive. Initially, activate at most k nodes, called them as seeds. Then, an influence process is carried out step-by-step. In each step, every freshly active node u will activate inactive neighbor through arc (u, v) with success probability p_{uv} where, by a freshly active node, we mean that a node becomes active at last step. When an inactive node receives influence from k ($k \geq 2$) incoming neighbors, we treat them as k mutually exclusive events. The process ends when no new active node is produced. The influence spread is the expected number of active nodes at the end of the influence process. Prove that the mutually exclusive cascade model is equivalent to the linear threshold model, that is, in both models, the influence spread is the same function with respect to the seed set.

21. Let f be a monotone nondecreasing submodular function on 2^X. Assume $A \subset A'$ and $B \subseteq B'$. Prove that

$$f(A \cup B') - f(A) \geq f(A' \cup B) - f(A').$$

22. (General Threshold Model) Consider a directed graph $G = (V, E)$. Every node v has a monotone nondecreasing threshold function f_v on subsets of incoming

neighbors. Every node has two states, active and inactive. Before stating a process, every node is inactive. Initially, for every node v, a threshold θ_v is selected randomly from $[0, 1]$ with uniform distribution, and activate at most k nodes, called them as seeds. Then, an influence process is carried out step-by-step. In each step, every inactive node v checks whether

$$f_v(\{\text{active } u \mid (u, v) \in E\}) \geq \theta_v.$$

If yes, then v is activated. Otherwise, v is kept inactive. The process ends when no new active node is produced. The influence spread is the expected number of active nodes at the end of the influence process. Prove that if at every node v, threshold function f_v is monotone nondecreasing submodular, then the influence spread is a monotone nondecreasing submodular function with respect to the seed set.

23. Consider a directed graph $G = (V, E)$ with general threshold information diffusion model. Suppose at every node v, the threshold function f_v is monotone nondecreasing and supermodular. Can you prove that the influence spread is monotone nondecreasing supermodular. If not, please give a counterexample.

24. In any in-arborescence with deterministic information diffusion model, show that the influence maximization has a polynomial-time solution.

25. In any in-arborescence with the linear threshold model, show that the influence maximization (i.e., the maximization of influence spread) has a polynomial-time solution.

26. In any in-arborescence with the independent cascade model, show that the influence maximization (i.e., the maximization of influence spread) is NP-hard.

27. Consider an undirected graph $G = (V, E)$. A dominating set D is called a *weakly CDS* if all edges incident to D induce a spanning connected subgraph. Prove that the minimum weakly CDS has a polynomial-time $(1 + \ln \Delta)$-approximation where Δ is the maximum degree of a node.

28. Consider a vertex-weighted undirected graph $G = (V, E, w)$. Suppose w is nonnegative. Design a greedy $(1 + \ln \Delta)$-approximation for the minimum weight connected vertex cover where Δ is the maximum vertex degree.

29. Consider an undirected graph $G = (V, E)$. Design a greedy approximation algorithm for the minimum CDS with performance ratio $2 + \ln \Delta$ where Δ is the maximum vertex degree.

30. Consider an undirected graph $G = (V, E)$. Show that for any $\varepsilon > 0$, there exists a greedy approximation algorithm for the minimum CDS with performance ratio $(1 + \varepsilon)(1 + \ln \Delta)$ where Δ is the maximum vertex degree.

31. Consider an undirected graph $G = (V, E)$. For any CDS C and two nodes u and v, denote by $m_C(u, v)$ the minimum number of intermediate nodes on a path between u and v passing through C. Denote by $m(u, v)$ the number of intermediate nodes on the shortest path between u and v in G. Design a greedy $(2+\ln \Delta)$-approximation C for the minimum CDS with constraint that for every two nodes u and v, $m_C(u, v) \leq 7 \cdot m(u, v)$.

32. Consider an undirected graph $G = (V, E)$. Design a greedy algorithm for maximization of cut function.

33. Consider n subsets $A_1, A_2, \ldots, A_n$ of a finite set X. There are k submodular set functions $w_i : 2^X \to R_+$. Design an algorithm to allocate each set A_j to a function w_i to maximize

$$\sum_{i=1}^{k} w_i (\cup_{j \in S_i} A_j)$$

where S_i is the set of indices of subset allocated to w_i.

Historical Notes

In recent development of computer technology, such as wireless networks [56, 255], cloud computing [124, 341], sentiment analysis [19, 226, 290, 323], and machine learning [414], many nonlinear optimization problems come out with discrete structure. They form a large group of new problems, which belong to a research area, *nonlinear combinatorial optimization*. The nonlinear combinatorial optimization has been studied for a long time but recently becomes very active. One of the important fields in this area is the set function optimization. Its development can be roughly divided into three periods:

The first period is before 2000. The research works came mainly from researchers in operations research. Those works are mainly on submodular function optimization, often with monotone nondecreasing property. In this period, major results include the following:

- Unconstrained submodular minimization can be solved in polynomial-time [187, 330, 359].
- For constrained monotone nondecreasing submodular maximization, it has $(1 - 1/e)$-approximation with size constraint [325, 326, 412] or a knapsack constraint [278, 371].
- For nonlinear-constrained linear optimization, the linear maximization with k matroid constraints has $(1/(k + 1))$-approximation [41, 142], and the linear minimization with submodular cover constraint, called the *submodular cover problem*, has $(1 + \ln \gamma)$-approximation where γ is a number determined by the submodular function defining the constraint [413].

The second period is from 2007 to 2012, the research activity occurs mainly in the theoretical computer science. The major results are about nonmonotone submodular optimization, including submodular maximization with knapsack constraints and matroid constraints [38, 135, 276] and submodular minimization with size constraint [372]. (Especially, the random greedy algorithm was proposed in [38].)

Most of them were published in theoretical computer science conferences, such as STOC, FOCS, and SODA, and journals, such as SIAM Journal on Computing.

The third period is starting from 2014. The research is in application-driven. The main focus is on nonsubmodular optimization.

In the study of submodular optimization, the greedy algorithm plays an important role. Especially, during the first two periods, the greedy algorithm appears very often. Actually, many theoretical results on greedy algorithms are built up with submodularity. Of course, there are some exceptional cases. For example, the analysis on greedy algorithm for independence systems (Sect. 4.1) does not need submodularity, which was made by Jenkyns [228] and Korte and Hausmann [263]. Hausmann, Korte, and Jenkyns [208] gave a nice survey for this part of research works.

The study on set cover and related problems can be seen as a part of research activity on the submodular set cover since the result from the study either is able to be generalized to the submodular set cover or can be covered by existing result on the submodular set cover. For example, Wolsey's theorem [414] was built up based on results of D.S. Johnson [231], L. Lovatz [297], Chvátal [70], and Slavik [370] on set covers. The subset interconnection design was proposed by Du and Miller (1988) [103]. The analysis on greedy algorithm [20, 114, 340] for it can be covered by Wolsey's theorem. About greedy approximation, there is an interesting long-standing open problem in the study of Superstring [29, 374, 383].

Lund and Yannakakis [305] proved that for any $0 < \rho < 1/4$, there is no polynomial-time approximation algorithm with performance ratio $\rho \ln n$ for hitting set problem unless NP $\subset$ DTIME $(n^{poly \log n})$. Feige [134] improved this result by relaxing ρ to $0 < \rho < 1$. Today, the condition that NP $\subset$ DTIME $(n^{poly \log n})$ has been replaced by NP = P [90]. This means that it is unlikely for the set cover problem to have polynomial-time approximation with a performance ratio better than $\ln n$.

While theoretical efforts are still to make significant progress [189, 216, 270, 428], more attentions are moved to new subjects with strong application background.

In the study of wireless ad hoc and sensor networks, a lot of optimization problems are formulated [84, 87, 245, 247–249, 251, 252, 284, 284, 285, 307, 399, 400, 418, 419, 446, 449]. To obtain approximations with theoretical guaranteed performance, one made a lot of efforts to establish submodular property [86, 306]. However, not many are successful. Actually, analysis techniques in the study of submodular optimizations are more useful than established theorem since, very often, existing theorems cannot apply but analysis approach in proof of theorems may fly. In particular, this happened in the study of connected dominating sets (CDS).

The CDS is also called virtual backbone in wireless ad hoc and sensor networks. It is an attractive subject to formulate optimization problems [84, 109, 161, 246, 281–283, 364, 441, 450, 453, 454]. For those problems, greedy algorithm often involves in the construction of approximation solution, and analysis of algorithms becomes an interesting research topic. For example, let us consider minimization of cardinality of CDS. The problem can be formulated in the form similar to submodular set cover problem. However, the function in covering constraint is not submodular. Thus, we cannot apply Wolsey's theorem. However, one found that the function allows a similar analysis to obtain a very nice theoretical bound [111, 112, 287, 351, 461]. Another way is to go around the problem with two-stage greedy [191]. For unit disk graphs, although the minimum (size) CDS problem has a PTAS, the PTAS is designed by using partition, and hence the running time is quite high. Therefore, one still make efforts in design of faster approximation for the minimum CDS [253, 314, 420].

In the study of social networks, monotone submodular maximization received a lot of applications on influence maximization [193, 194, 302–304, 343, 424, 436, 459]. When objective function is changed to the profit, the monotone property is lost, and we have to deal with nonmonotone submodular maximization [52, 195, 293–295, 458].

Chapter 11
Relaxation and Rounding

Relax! Life is beautiful.

—David L. Wolper

The relaxation is a powerful technique to design approximation algorithms. It is similar to restriction, in terms of making a change on feasible domain; however, in an opposite direction, i.e., instead of shrinking the feasible domain, enlarge it by relaxing certain constraint. There are various issues about relaxation. In this chapter, we study some of them.

11.1 The Role of Rounding

As shown in Fig. 11.1, consider problem

$$\min \quad f(x)$$

$$\text{subject to} \quad x \in \Omega$$

where Ω is a feasible domain. By relaxation, we mean to remove some constraint on feasible solutions so that the feasible domain Ω is enlarged to a new domain Γ, on which the minimization of objective function $f(x)$ can be solved or approximated easily. Then, modify the optimal or approximate solution of the relaxed problem to obtain an approximation for the original one.

To analyze the approximation performance, suppose $\min\{f(x) \mid x \in \Gamma\}$ can be solved in polynomial-time and x^* is the optimal solution. We modify x^* to obtain an approximation solution x_A for $\min\{f(x) \mid x \in \Omega\}$. Then the performance ratio is

$$\frac{f(x_A)}{opt} \leq 1 + \frac{f(x_A) - opt}{opt} \leq 1 + \frac{f(x_A) - f(x^*)}{opt},$$

$$\frac{f(x_A)}{opt} = 1 + \frac{f(x_A) - opt}{opt} \le 1 + \frac{f(x_A) - f(x^*)}{opt}$$

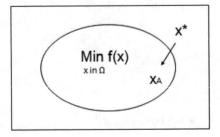

Fig. 11.1 Idea of relaxation

that is, it depends on the estimation of loss during modification from x^* to x_A.

This is similar to approximation performance analysis for using restriction technique. In both cases, the analysis is to estimate the value change of objective function for modifying a solution to satisfy some constraint (or restriction). However, there is an important difference that for relaxation, this modification is a step of construction of approximation solution. Therefore, this modification has to be efficiently computable, which requires more techniques. Hence, we give a special name, *rounding*, for this modification.

The rounding is a very important issue in the study of relaxation methods. There exist various types of rounding in the literature, such as iterated rounding, vector rounding, randomized rounding, etc. Next, we show two examples:

First, consider the following problem:

Problem 11.1.1 (Weighted Vertex Cover) *Given a graph $G(V, E)$ with nonnegative vertex weight, find a minimum total weight vertex cover.*

For unweighted vertex cover problem, we constructed 2-approximation with maximal matching. However, this construction cannot be seen clearly to have an extension to weighed case. Therefore, we introduce a new one as follows:

The weighted vertex cover problem can be formulated into an integer LP as follows:

Suppose $V = \{v_1, v_2, \cdots, v_n\}$. Let x_i be an indicator for vertex v_i belonging to the vertex cover or not, i.e.,

$$x_i = \begin{cases} 1 & \text{if } v_i \text{ is in the vertex cover;} \\ 0 & \text{otherwise.} \end{cases}$$

Let w_i be the weight of vertex v_i. Then, every vertex cover corresponds to a feasible solution in the following 0-1 integer LP, and the minimum weight vertex cover corresponds to the optimal solution of the 0-1 integer LP:

$$\min \quad w_1 x_1 + w_2 x_2 + \cdots + w_n x_n \tag{11.1}$$

$$\text{subject to} \quad x_i + x_j \geq 1 \text{ for } (v_i, v_j) \in E$$

$$x_i = 0 \text{ or } 1 \text{ for } i = 1, 2, \cdots, n.$$

By relaxing integer constraint $x_i = 0$ or 1 to real number $0 \leq x_i \leq 1$, this integer LP is turned to an LP, called an *LP relaxation*:

$$\min \quad w_1 x_1 + w_2 x_2 + \cdots + w_n x_n \tag{11.2}$$

$$\text{subject to} \quad x_i + x_j \geq 1 \text{ for } (v_i, v_j) \in E$$

$$0 \leq x_i \leq 1 \text{ for } i = 1, 2, \cdots, n.$$

The optimal solution x^* of this LP relaxation can be computed efficiently. From x^*, an approximation solution x^A can be obtained through a rounding procedure, called *threshold rounding*.

Threshold Rounding

$$x_i^A = \begin{cases} 1 \text{ if } x_i^* \geq 0.5, \\ 0 \text{ if } x_i^* < 0.5. \end{cases}$$

Theorem 11.1.2 x^A *gives a polynomial-time 2-approximation for the weighted vertex cover problem.*

Proof For each $(v_i, v_j) \in E$, since $x_i^* + x_j^* \geq 1$, we have at least one of x_i^* and x_j^* not smaller than 0.5. Therefore, at least one of x_i^A and x_j^A equals 1. This guarantees that x^A is a feasible solution of (11.1). Moreover,

$$\sum_{i=1}^n w_i x_i^A \leq 2 \sum_{i=1}^n w_i x_i^*$$

and the optimal solution of (11.1) has objective function value not smaller than $\sum_{i=1}^n w_i x_i^*$. $\qquad \square$

Next, we consider the following problem:

Problem 11.1.3 (MAX-SAT) *Given a CNF F, find an assignment to maximize the number of satisfied clauses.*

Suppose F contains m clauses $C_1, \ldots, C_m$ and n variables $x_1, \ldots, x_n$. We also first formulate the MAX-SAT problem into an integer LP:

$$\max \quad z_1 + z_2 + \cdots + z_m$$

$$\text{subject to} \quad \sum_{x_i \in C_j} y_i + \sum_{\bar{x}_i \in C_j} (1 - y_i) \geq z_j \text{ for } j = 1, 2, \ldots, m,$$

$$y_i \in \{0, 1\} \text{ for } i = 1, 2, \ldots, n,$$

$$z_j \in \{0, 1\} \text{ for } j = 1, 2, \ldots, m,$$

where $y_i = 1$ if $x_i = 1$ and $y_i = 0$ if $x_i = 0$. (Note that x_i is a Boolean variable, but y_i is a real number variable.)

Its LP relaxation is as follows:

$$\max \quad z_1 + z_2 + \cdots + z_m$$

$$\text{subject to} \quad \sum_{x_i \in C_j} y_i + \sum_{\bar{x}_i \in C_j} (1 - y_i) \geq z_j \text{ for } j = 1, 2, \ldots, m,$$

$$0 \leq y_i \leq 1 \text{ for } i = 1, 2, \ldots, n,$$

$$0 \leq z_j \leq 1 \text{ for } j = 1, 2, \ldots, m.$$

Let (y^*, z^*) be an optimal solution of above LP. We may do a random rounding as follows:

Independent Rounding

Set $x_i = 1$ with probability y_i^* independently.

Let us first show an inequality since it will be employed not only once.

Lemma 11.1.4 *Let* $f(z) = 1 - \left(1 - \frac{z}{k}\right)^k$. *Then,*

$$f(z) \geq z(1 - e^{-1})$$

for $z \in [0, 1]$.

Proof Note that $f'(z) = (1 - \frac{z}{k})^{k-1} \geq 0$ and $f''(z) = -\frac{k-1}{k} \cdot (1 - \frac{z}{k})^{k-2} \leq 0$ for $0 \leq z \leq 1$. Therefore, $f(z)$ is monotone increasing and concave in interval $[0, 1]$ (Fig. 11.2). Moreover, $f(0) = 0$. Hence, $f(z) \geq z f(1)$ for $z \in [0, 1]$. Note that

$$f(1) = 1 - \left(1 - \frac{1}{k}\right)^k \geq 1 - e^{-1}.$$

Fig. 11.2 Function $f(z)$

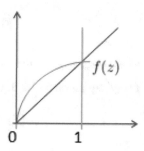

Thus,

$$f(z) \geq z(1 - e^{-1})$$

for $z \in [0, 1]$.

□

Let Z_j be a random variable which indicates whether clause C_j is satisfied.

Lemma 11.1.5 *For any clause C_j, its expectation*

$$E[Z_j] \geq z_j^* \left(1 - \frac{1}{e}\right).$$

Proof

$$E[Z_j] = 1 - \prod_{x_i \in Z_j} (1 - y_i^*) \prod_{\bar{x}_i \in Z_j} y_i^*$$

$$\geq 1 - \left(\frac{\sum_{x_i \in Z_j}(1 - y_i^*) + \sum_{\bar{x}_i \in Z_j} y_i^*}{k}\right)^k \quad (k = |Z_j|)$$

$$\geq 1 - \left(1 - \frac{\sum_{x_i \in Z_j} y_i^* + \sum_{\bar{x}_i \in Z_j}(1 - y_i^*)}{k}\right)^k.$$

By Lemma 11.1.4,

$$E[Z_j] \geq (1 - e^{-1}) \left(\sum_{x_i \in Z_j} y_i^* + \sum_{\bar{x}_i \in Z_j}(1 - y_i^*)\right) = (1 - e^{-1}) \cdot z_j^*.$$

□

Theorem 11.1.6 *Denote $Z_F = Z_1 + Z_2 + \cdots + Z_m$. Then,*

$$E[Z_F] \geq opt_{max\text{-}sat} \cdot \left(1 - \frac{1}{e}\right),$$

where $opt_{max\text{-}sat}$ is the optimal objective function value of the MAX-SAT problem.

Proof By Lemma 11.1.5, we have

$$E[Z_F] = E[Z_1] + E[Z_2] + \cdots + E[Z_m]$$

$$\geq \left(1 - \frac{1}{e}\right) \left(z_1^* + z_2^* + \cdots + z_m^*\right)$$

$$\geq opt_{lp} \cdot \left(1 - \frac{1}{e}\right)$$

$$\geq opt_{max\text{-}sat} \cdot \left(1 - \frac{1}{e}\right)$$

where opt_{lp} is the optimal objective function value of LP relaxation.   □

Note that

$$E[Z_F] = E[Z_F \mid x_1 = 1]y_1^* + E[Z_F \mid x_1 = 0](1 - y_1^*).$$

Therefore, we have

$$E[Z_{F|_{x_i=1}}] = E[Z_F \mid x_1 = 1] \geq opt_{lp} \cdot \left(1 - \frac{1}{e}\right)$$

or

$$E[Z_{F|_{x_i=0}}] = E[Z_F \mid x_1 = 0] \geq opt_{lp} \cdot \left(1 - \frac{1}{e}\right).$$

In the former case, it means that among assignments with $x_1 = 1$, the expectation of the number of satisfied clauses not less than $opt_{lp} \cdot (1 - \frac{1}{e})$. In the latter case, it means that among assignments with $x_1 = 0$, the expectation of the number of satisfied clauses is not less than $opt_{lp} \cdot (1 - \frac{1}{e})$. Motivated from this observation, we can find the following way to derandomization procedure:

Derandomization
    for $i = 1$ **to** n **do**
        if $E[Z_F \mid x_i = 1] \geq opt_{lp} \cdot (1 - \frac{1}{e})$
        then $x_i \leftarrow 1$ and
            $F \leftarrow F|_{x_i=1}$
        else $x_i \leftarrow 0$ and
            $F \leftarrow F|_{x_i=0}$
    end-for.

Theorem 11.1.7 *The MAX-SAT problem has a polynomial-time* $(1 - e^{-1})$-*approximation.*

Proof Note that assign $x_i = 1$ ($x_i = 0$) if and only if $E[Z_F \mid x_i = 1] \geq (1 - 1/e)opt_{lp}$ ($E[Z_F \mid x_i = 0] \geq (1 - 1/e)opt_{lp}$). This means that there exists an assignment for remaining variables such that the number of satisfied clauses is at least $(1 - 1/e)opt_{lp}$. Since this is true also at step $i = n$, the assignment obtained from independent rounding would make at least $(1 - 1/e)opt_{lp}$ ($\geq (1 - 1/e)opt_{\text{max-sat}}$) clauses being true. $\qquad \square$

11.2 Group Set Coverage

A lot of combinatorial optimization problems can be formulated into a 0-1 mathematical programming. A popular technique is to relax 0-1 variables to real variables, and after obtaining the solution for relaxed mathematical programming, find an approximation solution for the original 0-1 one with rounding techniques. This relaxation method seems more powerful than greedy strategy.

In this section, we present an example for which the greedy algorithm can give an approximation solution with performance ratio 1/2. However, an LP relaxation algorithm can produce $(1 - e^{-1})$-approximation.

Problem 11.2.1 (Group Set Coverage) *Given m groups* $\mathcal{G}_1, \mathcal{G}_2, \ldots, \mathcal{G}_m$ *of subsets of a finite set X, select at most one subset from each group to maximize the total number of elements covered by selected subsets.*

Theorem 11.2.2 *The group set coverage problem is* NP-*hard.*

Proof Consider the maximum set coverage problem: Given a collection $\mathcal{C}$ of subsets of X and an integer $k > 0$, find k subsets from $\mathcal{C}$ to maximize the total number of elements covered by selected subsets. If we set $\mathcal{G}_1 = \cdots = \mathcal{G}_m = \mathcal{C}$ and $m = k$. Then, the maximum set coverage problem is reduced to the maximum group set coverage problem. Since the maximum set coverage problem is NP-hard, so is the maximum group set coverage problem. $\qquad \square$

Let us first study a greedy algorithm as shown in Algorithm 36.

Theorem 11.2.3 *Algorithm 36 gives a greedy 1/2-approximation for the group set coverage problem.*

Proof Let subsets $S_1, S_2, \ldots, S_g$ be selected in turn by Algorithm 36. Relabel all groups such that $S_1 \in \mathcal{G}_1, S_2 \in \mathcal{G}_2, \ldots, S_g \in \mathcal{G}_g$.

Let $S_{1'}^*, S_{2'}^*, \ldots, S_{k'}^*$ be an optimal solution. Assume $1' < 2' < \cdots < k'$ and $S_{i'}^* \in \mathcal{G}_{i'}$ for $1 \leq i \leq k$. Let us consider two cases:

Case 1. $g \geq k$. Note that $1 \leq 1', 2 \leq 2', \ldots, k \leq k'$. Thus, we have

---

Algorithm 36 Greedy Approximation for Group Set Coverage

Input: m groups $\mathcal{G}_1, \mathcal{G}_2, \ldots, \mathcal{G}_m$ of subsets of a finite set X.
Output: A collection $\mathcal{C}$ of subsets of X.
1: $\mathcal{C} \leftarrow \emptyset$;
2: $\Gamma \leftarrow \{\mathcal{G}_1, \mathcal{G}_2, \ldots, \mathcal{G}_m\}$;
3: **repeat**
4:     pick up a group $\mathcal{G}$ from which contains the subset S that covers the maximum number of
       uncovered elements, mark elements in S as covered, put S into $\mathcal{C}$ and delete $\mathcal{G}$ from Γ
5: **until** there is no uncovered element which can be covered by any subset in a group in Γ;
6: **return** $\mathcal{C}$.

$$|S_i \setminus (S_1 \cup \cdots \cup S_{i-1})| \geq |S_{i'}^* \setminus (S_1 \cup \cdots \cup S_{i-1})|$$

for $1 \leq i \leq k$. Therefore,

$$
\begin{aligned}
|S_1 \cup \cdots \cup S_k| &= |S_1| + |S_2 \setminus S_1| + \cdots + |S_k \setminus (S_1 \cup \cdots S_{k-1})| \\
&\geq |S_{1'}^*| + |S_{2'}^* \setminus S_1| + \cdots + |S_{k'}^* \setminus (S_1 \cup \cdots \cup S_{k-1})| \\
&\geq |S_{1'}^* \setminus (S_1 \cup \cdots \cup S_k)| + |S_{2'}^* \setminus (S_1 \cup \cdots \cup S_k)| + \cdots \\
&\quad + |S_{k'}^* \setminus (S_1 \cup \cdots \cup S_k)| \\
&\geq |(S_{1'}^* \cup \cdots \cup S_{k'}^* \setminus (S_1 \cup \cdots \cup S_k)| \\
&\geq |S_{1'}^* \cup \cdots \cup S_{k'}^*| - |S_1 \cup \cdots \cup S_k|.
\end{aligned}
$$

Hence,

$$|S_1 \cup \cdots \cup S_g| \geq |S_1 \cup \cdots \cup S_k| \geq |S_{1'}^* \cup \cdots \cup S_{k'}^*|/2.$$

Case 2. $g < k$. Select S_i arbitrarily from $\mathcal{G}_i$ for $g + 1 \leq i \leq k$. Note that

$$S_1 \cup \cdots \cup S_i = S_1 \cup \cdots \cup S_g$$

for $g \leq i \leq k$. Since $i' > g' \geq g$ for $g + 1 \leq i \leq k$, we have

$$|S_i \setminus (S_1 \cup \cdots \cup S_{i-1})| = 0 = |S_i^* \setminus (S_1 \cup \cdots \cup S_{i-1})|.$$

Therefore, we can use the same argument as that in Case 1 to show that

$$|S_1 \cup \cdots \cup S_g| = |S_1 \cup \cdots \cup S_k| \geq |S_{1'}^* \cup \cdots \cup S_{k'}^*|/2.$$

□

Next, we present the second approximation algorithm with LP relaxation. First, we formulate the group set coverage problem into an integer LP as follows: Let x_{iS} be 0-1 variable which indicates if subset S is selected from group $\mathcal{G}_i$. Let $|X| = n$

and $\mathcal{G} = \mathcal{G}_1 \cup \cdots \cup \mathcal{G}_m$. Use y_j to indicate whether element j appears in a selected subset or not. The following is the integer LP for the group set coverage problem:

$$\max \sum_{i=j}^{n} y_j$$

$$\text{s.t. } y_j \leq \sum_{i=1}^{m} \sum_{S:j\in S\in\mathcal{G}_i} x_{iS} \quad \forall j = 1, \ldots, n,$$

$$\sum_{S:S\in\mathcal{G}_i} x_{iS} \leq 1 \quad \forall i = 1, \ldots, m,$$

$$y_j \in \{0, 1\} \quad \forall j = 1, \ldots, n,$$

$$x_{iS} \in \{0, 1\} \quad \forall S \in \mathcal{G} \text{ and } i = 1, 2, \ldots, m.$$

Its relaxation is as follows:

$$\max \sum_{i=j}^{n} y_j$$

$$\text{s.t. } y_j \leq \sum_{i=1}^{m} \sum_{S:j\in S\in\mathcal{G}_i} x_{iS} \quad \forall j = 1, \ldots, n,$$

$$\sum_{S:S\in\mathcal{G}_i} x_{iS} \leq 1 \quad \forall i = 1, \ldots, m,$$

$$0 \leq y_j \leq 1 \quad \forall j = 1, \ldots, n,$$

$$0 \leq x_{iS} \leq 1 \quad \forall S \in \mathcal{G} \text{ and } i = 1, 2, \ldots, m.$$

Let (y_j^*, x_{iS}^*) be an optimal solution of this LP. We do a randomized rounding as follows:

Mutually Exclusive Rounding: For each group $\mathcal{G}_i$, make a mutually exclusive selection to choose one subset S with probability x_{iS}^* and not select any subset with probability $1 - \sum_{S\in\mathcal{G}_i} x_{iS}^*$. Set $y_j = 1$ if element j appears in a selected subset, and $y_j = 0$, otherwise.

Let (x_{iS}, y_j) be a solution obtained from the randomized rounding. We show properties of this solution.

Lemma 11.2.4 $E[y_j] \geq (1 - e^{-1})y_j^*$.

Proof For each $i = 1, \ldots, n$,

$$\text{Prob}[y_j = 0] = \prod_{i=1}^{m} \prod_{S:j\in S\in\mathcal{G}_i} \left(1 - x_{iS}^*\right)$$

$$\leq \left(\frac{\sum_{i=1}^{m} \sum_{S:j\in S\in \mathcal{G}_i}(1 - x_{iS}^*)}{K_j} \right)^{K_j}$$

$$(\text{where } K_j = |\{(i, S) \mid j \in S \in \mathcal{G}_i\}|)$$

$$= \left(1 - \frac{\sum_{i=1}^{m} \sum_{S:j\in S\in \mathcal{G}_i} x_{iS}^*}{K_j} \right)^{K_j}$$

$$\leq \left(1 - \frac{y_j^*}{K_j} \right)^{K_j}.$$

Hence,

$$\text{Prob}[y_i = 1] \geq 1 - \left(1 - \frac{y_i^*}{K_j} \right)^{K_j}.$$

By Lemma 11.1.4, we obtain

$$\text{Prob}[y_j = 1] \geq (1 - e^{-1})y_j^*.$$

Hence, $E[y_j] = \text{Prob}[y_j = 1] \geq (1 - e^{-1})y_j^*.$ □

Theorem 11.2.5 *Let (y_j, x_{iS}) be approximation solution obtained by randomized rounding. Then,*

$$E\left[\sum_{j=1}^{n} y_j \right] \geq (1 - e^{-1})opt$$

where opt is the objective function value of optimal solution for the group set coverage problem.

Proof By Lemma 11.2.4,

$$E\left[\sum_{i=1}^{n} y_j \right] = \sum_{i=1}^{n} E[y_j]$$

$$\geq (1 - e^{-1}) \cdot \sum_{j=1}^{n} y_j^*$$

$$\geq (1 - e^{-1}) \cdot opt.$$

□

Algorithm 37 $(1 - 1/e)$-approximation for group set coverage

---

input m groups $\mathcal{G}_1, \mathcal{G}_2, \ldots, \mathcal{G}_m$ of subsets of a finite set X.
output a collection $\mathcal{C}$ of subsets of X.
1: $\mathcal{C} \leftarrow \emptyset$;
2: Solve LP relaxation to obtain optimal solution (x_{iS}^*, y_j^*);
3: **for** $i \leftarrow 1$ to m **do**
4:     make mutually exclusive selection to choose at most one subset S from $\mathcal{G}_i$ with probability x_{iS}^*;
5:     $\mathcal{C} \leftarrow \mathcal{C} \cup \{S\}$
6: **end for**
7: **return** $\mathcal{C}$.

Now, let us summarize the designed algorithm into Algorithm 37.

11.3 Pipage Rounding

In this section, we introduce a rounding technique, called the *pipage rounding* since it can be applied to submodular optimization.

Consider the following problem:

Problem 11.3.1 (Maximum Weight Hitting) *Given a collection $\mathcal{C}$ of subsets of a finite set X with nonnegative weight function w on $\mathcal{C}$ and a positive integer p, find a subcollection A of X with $|A| = p$ to maximize the total weight of subsets hit by A.*

Assume $X = \{1, 2, \ldots, n\}$ and $\mathcal{C} = \{S_1, S_2, \ldots, S_m\}$. Denote $w_i = w(S_i)$. Let x_i be a 0-1 variable to indicate whether element i is in subset A. Then, this problem can be formulated into the following integer LP:

$$\max \sum_{j=1}^{m} w_j z_j \tag{11.3}$$

$$\text{s.t.} \sum_{i \in S_j} x_i \geq z_j, j = 1, \ldots, m,$$

$$\sum_{i=1}^{n} x_i = p$$

$$x_i \in \{0, 1\}, i = 1, 2, \ldots, n$$

$$z_j \in \{0, 1\}, j = 1, 2, \ldots, m.$$

There are two equivalent formulations as follows:

$$\max L(x) = \sum_{j=1}^{m} w_j \min\{1, \sum_{i \in S_j} x_i\} \tag{11.4}$$

$$\text{s.t.} \qquad \sum_{i=1}^{n} x_i = p$$

$$x_i \in \{0, 1\}, i = 1, 2, \ldots, n$$

$$\max F(x) = \sum_{j=1}^{m} w_j (1 - \prod_{i \in S_j} (1 - x_i)) \tag{11.5}$$

$$\text{s.t.} \qquad \sum_{i=1}^{n} x_i = p$$

$$x_i \in \{0, 1\}, i = 1, 2, \ldots, n$$

$L(x)$ and $F(x)$ have the same value when each x_i takes value 0 or 1. But when x_i is relaxed to $0 \leq x_i \leq 1$, they may have different values. The following gives a relationship between them:

Lemma 11.3.2 $F(x) \geq (1 - 1/e) L(x)$ for $0 \leq x \leq 1$.

Proof Note that

$$1 - \prod_{i \in S_j} (1 - x_i) \geq 1 - \left(\frac{\sum_{i \in S_j} (1 - x_i)}{k} \right)^k \qquad (k = |S_j|)$$

$$\geq 1 - \left(1 - \frac{\sum_{i \in S_j} x_i}{k} \right)^k.$$

By Lemma 11.1.4, we have

$$1 - \prod_{i \in S_j} (1 - x_i) \geq (1 - e^{-1}) \sum_{i \in S_j} x_i \geq \min \left(1, \sum_{i \in S_j} x_i \right).$$

$\square$

Now, we consider LP relaxation of (11.4):

$$\max L(x) = \sum_{j=1}^{m} w_j \min \left\{ 1, \sum_{i \in S_j} x_i \right\} \tag{11.6}$$

$$\text{s.t.} \quad \sum_{i=1}^{n} x_i = p$$

$$0 \leq x_i \leq 1, i = 1, 2, \ldots, n$$

It is equivalent to the following LP:

$$\max \sum_{j=1}^{m} w_j z_j \tag{11.7}$$

$$\text{s.t.} \quad \sum_{i \in S_j} x_i \geq z_j, j = 1, \ldots, m,$$

$$\sum_{i=1}^{n} x_i = p$$

$$0 \leq x_i \leq 1, i = 1, 2, \ldots, n$$

$$0 \leq z_j \leq 1, j = 1, 2, \ldots, m.$$

Therefore, it is polynomial-time solvable. Let x^* be an optimal solution of (11.6). We will use the following rounding to find an integer solution $\bar{x}$ from x^*:

Pipage Rounding
$x \leftarrow x^*$;
while x has an noninteger component **do begin**
    choose $0 < x_k < 1$ and $0 < x_j < 1$ $(k \neq j)$;
    define $x(\varepsilon)$ by setting

$$x_i(\varepsilon) = \begin{cases} x_i & \text{if } i \neq k, j, \\ x_j + \varepsilon & \text{if } i = j, \\ x_k - \varepsilon & \text{if } i = k; \end{cases}$$

    define $\varepsilon_1 = \min(x_j, 1 - x_k)$ and $\varepsilon_2 = \min(1 - x_j, x_k)$;
    if $F(x(-\varepsilon_1)) \geq F(x(\varepsilon_2))$
        then $x \leftarrow x(-\varepsilon_1)$
        else $x \leftarrow x(\varepsilon_2)$;
end-while;
return $\bar{x} = x$.

The existence of x_k and x_j is due to the fact that when x has a noninteger component, x has at least two noninteger components since $\sum_{i=1}^{n} x_i = p$.

The following is an important property of $F(x(\varepsilon))$:

Lemma 11.3.3 $F(x(\varepsilon))$ *is convex with respect to ε.*

Proof If S_j contains only one of k and j, then the jth term of $F(x(\varepsilon))$, corresponding to $1 - \prod_{i \in S_j}(1 - x_i)$, is linear and hence convex with respect to ε. If S_j contains both k and j, then the jth term of $F(x(\varepsilon))$, corresponding to $1 - \prod_{i \in S_j}(1 - x_i)$, is in the form

$$g(\varepsilon) = 1 - a(b + \varepsilon)(c - \varepsilon)$$

where a, b, and c are nonnegative constants with respect to ε. If $a = 0$, then this term is a constant 1 and hence convex. For $a > 0$, since $g''(\varepsilon) = a > 0$, $g(\varepsilon)$ is convex. Finally, we note that the sum of several convex functions is convex.   □

By Lemma 11.3.3, the value of $F(x)$ is nondecreasing during the pipage rounding process. Therefore, $F(\bar{x}) \geq F(x^*)$.

Theorem 11.3.4 *The maximum weight hitting problem has polynomial-time an $(1 - e^{-1})$-approximation.*

Proof $L(\bar{x}) = F(\bar{x}) \geq F(x^*) \geq (1 - 1/e)L(x^*) \geq (1 - e^{-1}) \cdot opt.$ □

From above example, we may get a little impression on pipage rounding. Next, we give a general description.

Consider a bipartite graph $G = (U, V, E)$ and an integer programming with 0-1 variables x_e each associated with an edge e and each constraint is in the form

$$\sum_{e \in \delta(v)} x_e \leq p_e$$

or

$$\sum_{e \in \delta(v)} x_e = p_e$$

or

$$\sum_{e \in \delta(v)} x_e \geq p_e$$

where $\delta(v)$ is the set of all edges incident to $v \in U \cup V$ and p_e is a nonnegative integer. For example, we may consider the following integer programming:

$$\max \quad L(x) \tag{11.8}$$

$$\text{s.t.} \quad \sum_{e \in \delta(v)} x_e \leq p_v \text{ for } v \in U \cup V$$

$$x_e \in \{0, 1\} \text{ for } e \in E.$$

As shown in Fig. 11.3, suppose $L(x)$ has a company $F(x)$ such that

(A1) $L(x) = F(x)$ for $x_e \in \{0, 1\}$.
(A2) $L(x) \leq cF(x)$ for $0 \leq x_e \leq 1$.

We also assume the following:

(A3)   The relaxation of integer programming (11.8) is equivalent to an LP or is polynomial-time solvable.

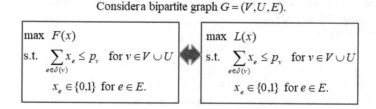

Consider a bipartite graph $G = (V, U, E)$.

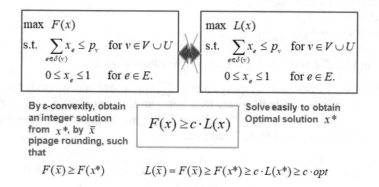

Fig. 11.3 Framework of pipage rounding

Suppose x^* is an optimal solution of the relaxation of (11.8), i.e.,

$$\max \quad L(x) \tag{11.9}$$

$$\text{s.t.} \quad \sum_{e \in \delta(v)} x_e \leq p_v \text{ for } v \in U \cup V$$

$$0 \leq x_e \leq 1 \text{ for } e \in E.$$

We next explain a pipage rounding procedure to obtain an integer solution $\bar{x}$ from x^*.

Initially, set $x \leftarrow x^*$. While x is not an integer solution, we will do the following:

1. Consider the subgraph H_x of G induced by all edges e with $0 < x_e < 1$. Let R be a cycle or a maximal path of H_x. Then, R can be decomposed into two matchings M_1 and M_2.
2. Define $x(\varepsilon)$ by

$$x_e(\varepsilon) = \begin{cases} x_e & \text{if } e \notin R, \\ x_e + \varepsilon & \text{if } e \in M_1, \\ x_e - \varepsilon & \text{if } e \in M_2. \end{cases}$$

Define

$$\varepsilon_1 = \min\left(\min_{e \in M_1} x_e, \ \min_{e \in M_2} (1 - x_e) \right)$$

$$\varepsilon_2 = \min\left(\min_{e \in M_1} (1 - x_e), \ \min_{e \in M_2} x_e \right).$$

3. If $F(x(-\varepsilon_1)) \geq F(x(\varepsilon_2))$
   then $x \leftarrow x(-\varepsilon_1)$
   else $x \leftarrow x(-\varepsilon_2)$.

Lemma 11.3.5 *For $\varepsilon \in [-\varepsilon_1, \varepsilon_2]$, $x(\varepsilon)$ is feasible for (11.9).*

Proof If R is a cycle, then $\sum_{e \in \delta(v)} x_e(\varepsilon) = \sum_{e \in \delta(v)} x_e$ and hence $x(\varepsilon)$ is feasible.
If R is a maximal path, then only for v being an endpoint of R, $\sum_{e \in \delta(v)} x_e(\varepsilon) \neq \sum_{e \in \delta(v)} x_e$. Suppose $e' \in \delta(v) \cap R$. Since R is a maximal path, we have that for $e \in \delta(v) \setminus \{e'\}$, x_e is an integer. Therefore,

$$p_v - \sum_{e \in \delta(v)} x_e(\varepsilon) = p_v - \sum_{e \in \delta(v) \setminus \{e'\}} x_e - x_{e'}(\varepsilon) \geq 1 - x_{e'}(\varepsilon) \geq 0.$$

$\square$

Assume $F(x)$ is ε-convex, i.e.,

(A4) For any R, $F(x(\varepsilon))$ is convex with respect to ε.

Then by pipage rounding, we would obtain an integer solution $\bar{x}$ such that $F(\bar{x}) \geq F(x^*)$. Therefore,

$$F(\bar{x}) \geq F(x^*) \geq cL(x^*) \geq c \cdot opt.$$

For the maximum weight hitting problem, we faced a star (which is a bipartite graph) $G = (U, V, E)$ where $U = \{u\}$, $V = \{v_1, v_2, \ldots, v_n\}$ and $E = \{(u, v_1), (u, v_2), \ldots, (u, v_2)\}$. Each variable x_i corresponds to an edge (u, v_i). Therefore, in each iteration of pipage rounding, we deal with a maximal path consisting of two edges.

11.4 Continuous Greedy

Why the pipage rounding can be applied to set function optimization? We may get some idea from relaxation by expectation for set functions.

Consider a set function $f(S)$ on subsets of a finite set X. Let $X = \{1, 2, \ldots, n\}$. For each element i, let x_i be an indicator which indicates whether i belongs to subset S or not, i.e.,

$$x_i = \begin{cases} 1 \text{ if } i \in S, \\ 0 \text{ otherwise.} \end{cases}$$

Then $f(S)$ can also be seen as a function $F(x) = f(S)$ defined on $x = (x_1, \ldots, x_n) \in \{0, 1\}^n$. Now, we extend $F(x)$ to $x \in [0, 1]^n$ through expectation as follows:

For each $x \in [0, 1]^n$, let S be a random set that element i belongs to S with probability x_i. Define $f(x)$ to be the expectation of function value of f, that is,

$$F(x) = E_{S \sim x}[f(S)] = \sum_{S \in 2^X} f(S) \left(\prod_{i \in S} x_i \right) \left(\prod_{i \notin S} (1 - x_i) \right).$$

This is called the *multilinear extension* of f, which has the following property:

Theorem 11.4.1 *Suppose F is the multilinear extension of f. Then,*

1. *If f is monotone nondecreasing, then F is monotone nondecreasing along any direction $d \geq 0$.*
2. *If f is submodular, then F is concave along any line $d \geq 0$.*
3. *If f is submodular, then F is convex along line $e_i - e_j$ ($i \neq j$) where e_i has its ith component equal 1 and others equal 0.*

Proof

1. Note that

$$\frac{\partial F(x)}{\partial x_i} = F(x_1, .., x_{i-1}, 1, x_{i+1}, \ldots, x_n) - F(x_1, .., x_{i-1}, 0, x_{i+1}, \ldots, x_n)$$

$$= E[f(R \cup \{i\})] - E[f(R)]$$

where R is the random subset of $X \setminus \{i\}$ such that R contains an element $j \in X \setminus \{i\}$ with probability x_j. If f is monotone nondecreasing, we have

$$\frac{\partial F(x)}{\partial x_i} \geq 0.$$

Now, let $x \in [0, 1]^n$ and $d \in [0, +\infty)^n$. Then,

$$\frac{dF(x + \alpha d)}{d\alpha} = \langle d, \nabla F(x + \alpha d) \rangle \geq 0.$$

Therefore, along direction d, $F(x)$ is monotone nondecreasing.

2. Next, assume f is submodular. Let R be a random subset of $X \setminus \{i, j\}$ such that R contains an element $k \in X \setminus \{i, j\}$ with probability x_k. Then, for $i \neq j$,

$$\frac{\partial^2 F(x)}{\partial x_i \partial x_j} = E[f(R \cup \{i, j\})] - E[f(R \cup \{i\})] - E[f(R \cup \{j\})] + E[f(R)]$$

$$= E[f(R \cup \{i, j\}) - f(R \cup \{i\}) - f(R \cup \{j\}) + f(R)]$$

$$\leq 0.$$

Moreover,

$$\frac{\partial^2 F(x)}{\partial x_i^2} = 0.$$

Therefore, for $x \in [0, 1]^n$ and $d \in [0, +\infty)^n$,

$$\frac{d^2 F(x + \alpha d)}{d\alpha^2} = d^T H_f(x + \alpha d)d \leq 0$$

where $H_f(x + \alpha d)$ is Hessian matrix of f at point $x + \alpha d$. It implies that $F(x)$ is concave along line d.

3. Note that for $d = e_i - e_j$, we have

$$\frac{d^2 F(x + \alpha d)}{d\alpha^2} = d^T H_f(x + \alpha d)d = -2\frac{\partial^2 F(x)}{\partial x_i \partial x_j} \geq 0.$$

Therefore, F is convex along $e_i - e_j$.

□

Next, we present an application of multilinear extension.
Consider the following problem:

Problem 11.4.2 (Monotone Submodular Maximization with a Matroid Constraint)

$$\max \; f(S)$$

$$subject\ to\ S \in \mathcal{I}$$

where f is a monotone nondecreasing, submodular function on 2^X for a finite set X
and $\mathcal{I}$ is the family of independent sets of a matroid $\mathcal{M} = (X, \mathcal{I})$.

Let $F(x)$ be the multilinear extension of $f(S)$ and $P(\mathcal{M})$ the polytope of matroid
$\mathcal{M}$, i.e.,

$$P(\mathcal{M}) = \{x \geq 0 \mid \forall S \in 2^X : \sum_{j \in S} \leq r_{\mathcal{M}}(S)\}$$

and

$$r_{\mathcal{M}}(S) = \max\{|I| \mid I \subseteq S, I \in \mathcal{I}\}.$$

Problem 11.4.2 can be relaxed into the following:

$$\max \quad F(x) \tag{11.10}$$

$$subject\ to\ x \in P(\mathcal{M})$$

This relaxation can be solved by a continuous greedy algorithm as follows:

Algorithm 38 Continuous Greedy Algorithm

---

input $F(x)$ and $P(\mathcal{M})$.
output $x(1)$.
1: Define $v_{\max}(x) = \text{argmax}_{v \in P(\mathcal{M})} \langle v, \nabla F(x) \rangle$.
2: $x(0) \leftarrow 0 \in R^n$.
3: **for** $t \in [0, 1]$ **do**
4:    $x'(t) = v_{\max}(x(t))$
5: **end for**
6: **return** $x(1)$.

This algorithm contains a step to solve a differential equation. Therefore, it needs an explanation about implementation. Before doing so, we first analyze it.

Lemma 11.4.3 *For any $x \in R^n$, there exists $v \in P(\mathcal{M})$ such that $\langle v, \nabla F(x) \rangle \geq opt - F(x)$, where opt is the objective function value of optimal solution for the relaxation (11.10).*

Proof Let v be an optimal solution of the relaxation (11.10), i.e., $F(v) = opt$. We show that $\langle v, \nabla F(x) \rangle \geq F(v) - F(x)$.

To do so, consider $d = (v - x) \vee 0$, where $x \vee y$ denotes the coordinate-wise max of x and y, i.e., $(x \vee y)_i = \max(x_i, y_i)$. Since $d \geq 0$, F is concave along direction d. Thus,

$$\langle d, \nabla F(x) \rangle \geq F(x + d) - F(x).$$

Since F is nondecreasing along d and $x + d = v \vee x \geq v$, we have

$$F(x + d) \geq F(x).$$

Moreover, $d \leq v$ and $\nabla F(x) \geq 0$ imply

$$\langle d, \nabla F(x) \rangle \leq \langle v, \nabla F(x) \rangle.$$

Therefore,

$$\langle v, \nabla F(x) \rangle \geq \langle d, \nabla F(x) \rangle \geq F(x + d) - F(x) \geq F(v) - F(x).$$

$\square$

Theorem 11.4.4 *Let F be the multilinear extension of a nondecreasing submodular function f. Then, $x(1)$ obtained by Algorithm 38 satisfies the following:*

1. *$x(1) \in P(\mathcal{M})$.*
2. *$F(x(1)) \geq (1 - e^{-1}) \cdot opt$, where opt is the value of F at an optimal solution of the relaxation (11.10).*

Proof

1. Note that

$$x(1) = \int_0^1 x'(t)dt = \int_0^1 v_{\max}(x(t))dt$$

$$= \lim_{n \to \infty} \frac{1}{n} \sum_{i=1}^n v_{\max}\left(x\left(\frac{i}{n}\right)\right).$$

Since $v_{\max}(x(\frac{i}{n})) \in P(\mathcal{M})$ and $P(\mathcal{M})$ are convex and closed, we have $x(1) \in P(\mathcal{M})$.

2. Note that

$$\frac{d}{dt}F(x(t)) = \langle x'(t), \nabla F(x(t)) \rangle = \langle v_{\max}(x(t)), \nabla F(x(t)) \rangle.$$

By Lemma 11.4.3, there exists $v \in P(\mathcal{M})$ such that

$$\langle v, \nabla F(x(t)) \rangle \geq opt - F(x(t)).$$

Therefore,

$$\langle v_{\max}(x(t)), \nabla F(x(t)) \rangle \geq opt - F(x(t)).$$

Thus,

$$\frac{d}{dt}F(x(t)) \geq pt - F(x(t)).$$

Denote $g(t) = F(x(t))$. Then, we have

$$g'(t) + g(t) \geq opt \text{ and } g(0) = 0.$$

Define $h(t) = g'(t) + g(t)$. Solve this differential equation. Then, we obtain

$$g(t) = \int_0^t e^{x-1} h(x)dx.$$

Hence,

$$F(x(1)) = g(1) \geq opt \cdot \int_0^1 x^{x-1}dx = opt \cdot (1 - e^{-1}).$$

$\square$

Now, let us discuss issues about implementation:

(1) How to compute $F(x)$ and $\nabla F(x)$? Note that in the definition of $F(x)$, the sum is over all subsets of X, i.e., it has $2^{|X|}$ terms. Therefore, computing $F(x)$ for a single x may take exponential time.

To overcome this trouble, we may note its representation through expectation $F(x) = E_{S \sim x}[f(S)]$. This representation suggests a computation with random sampling. By Chernoff bound, we can obtain

$$\left| \frac{1}{t} \sum_{i=1}^{t} f(S_i) - F(x) \right| \le \varepsilon \cdot f(X)$$

with probability at least $1 - e^{t\varepsilon^2/4}$, where $S_1, \ldots, S_t$ are random subsets based on element selection probability x. Therefore, if $t = O(\frac{1}{\varepsilon^2})$, then with a constant probability, we can compute an approximation of $F(x)$ within $\varepsilon F(X)$ error.

Similarly, the sampling method can be employed for computing $\nabla F(x)$ since

$$\frac{\partial F(x)}{\partial x_i} = E[f(R \cup \{i\})] - E[f(R)].$$

(2) How to compute $v_{\max}(x) = \mathrm{argmax}_{v \in P(\mathcal{M})} \langle v, \nabla F(x) \rangle$? It looks like a trouble, but not a real trouble. This is a linear programming. Why it looks like a trouble? This is because $P(\mathcal{M})$ is not described by a constant number of constraints. From matroid theory, we see several ways to represent $P(\mathcal{M})$. However, in general, everyone involves a large number of constraints. For example, there are exponential number of inequalities in the following representation:

$$P(\mathcal{M}) = \{x \ge 0 \mid \forall S \in 2^X : \sum_{j \in S} \le r_{\mathcal{M}}(S)\}.$$

Why it is not a real trouble? The reason is that $P(\mathcal{M})$ is the convex hull of 1_I for $I \in \mathcal{I}$ where

$$(1_I)_i = \begin{cases} 1 \text{ if } i \in I, \\ 0 \text{ otherwise.} \end{cases}$$

Since optimal solution of linear programming can be found in vertices, $v_{\max}(x)$ can be a solution of $\max\{\langle 1_I, \nabla F(x) \rangle \mid I \in \mathcal{I}\}$. This can be solved by a greedy algorithm since $\nabla F(x) \ge 0$.

(3) How to solve differential equation $x'(t) = v_{\max}(x(t))$ numerically? Algorithm 39 shows a simple numerical computational solution.

Using Algorithm 39 to solve differential equation, we can obtain the following:

Lemma 11.4.5 $x(1) \in P(\mathcal{M})$ and

$$F(x(1)) \ge \left(1 - \left(1 - \frac{1}{n}\right)^n\right) \cdot opt - \frac{c}{n}$$

for some constant c.

Proof Note that

Algorithm 39 Solving the differential equation

input $F(x)$ and $v_{\max}(x)$.
output $x(1)$.

1: $\alpha \leftarrow \frac{1}{n}$.
2: $x(0) \leftarrow 0 \in R^n$.
3: **for** $t \leftarrow 1$ to n **do**
4:    $v \leftarrow v_{\max}(x((t-1)\alpha))$
5:    $x(t\alpha) \leftarrow x((t-1)\alpha) + \alpha v$
6: **end for**
7: **return** $x(1)$.

$$x(1) = \frac{1}{n} \sum_{t=0}^{n-1} v_{\max}\left(x\left(\frac{t}{n}\right)\right),$$

that is, $x(1)$ is a convex combination of points in $P(\mathcal{M})$. Therefore, $x(1) \in P(\mathcal{M})$.

Next, we employ Taylor expansion on $F(x((t+1)\alpha)) = F(x(t\alpha)) + \alpha \cdot v_{\max}(x(t\alpha))$. For some constant c, we have

$$F(x((t+1)\alpha)) \geq F(x(t\alpha)) + \alpha \cdot \langle v_{\max}(x(t\alpha)), \nabla F(x(t\alpha)) \rangle - c\alpha^2$$
$$\geq F(x(t\alpha)) + \alpha[opt - F(x(t\alpha))] - c\alpha^2$$
$$= (1-\alpha)F(x(t\alpha)) + \alpha \cdot opt - c\alpha^2.$$

The second inequality is due to Lemma 11.4.3. Exchanging two sides and adding opt, we obtain

$$opt - F(x((t+1)\alpha)) \leq (1-\alpha)(opt - F(x(t\alpha))) + c\alpha^2$$
$$\leq (1-\alpha)^{t+1}(opt - F(x(0))) + c\alpha^2(t+1)$$
$$= (1-\alpha)^{t+1} \cdot opt + c\alpha^2(t+1)$$

since $F(x(0)) = 0$. Setting $t+1 = n$, we have

$$F(x(1)) \geq (1 - (1-\alpha)^n) \cdot opt - \frac{c}{n}.$$

□

Theorem 11.4.6 *Set* $n = O(1/\varepsilon)$ *in the continuous greedy algorithm. Then, we can obtain*

$$F(x(1)) \geq (1 - e^{-1}) \cdot opt - \varepsilon.$$

Proof It follows immediately from Lemma 11.4.5. □

(4) How to round $x(1)$ into an integer solution? By Theorem 11.4.1, F is convex along line $e_i - e_j$. Therefore, we can apply pipage rounding to obtain an integer solution $\hat{x}$ such that $F(\hat{x}) \geq F(x(1))$.

$\square$

Exercises

1. A collection of subset groups $\mathcal{G}_1, \mathcal{G}_2, \ldots, \mathcal{G}_m$ is called a *group set cover* if there exists a selection, which selects one subset S_i from each group $\mathcal{G}_i$, such that every element is covered by $\cup_{i=1}^m S_i$. Clearly, the group set cover is an extension of the set cover since the set cover can be seen as a special case that each group contains only one subset. Show the following problem is NP-complete: Given a collection of subset groups $\mathcal{G}_1, \mathcal{G}_2, \ldots, \mathcal{G}_m$, determine whether this collection is a group set cover or not.

2. Consider n jobs and m machines. The ith machine time t_{ij} to process the jth job. Using LP relaxation, please construct a 2-approximation for scheduling all jobs on machines to minimize the makespan, i.e., the maximum processing time over all machines.

3. (Minimum Two-Satisfiability) Given a 2CNF F, determine whether F is satisfiable. If the answer is yes, then please design a 2-approximation for the minimum number of true variable in a satisfying assignment.

4. (Maximum 1-in-3 SAT) Given a 3CNF F, find an assignment to maximize the number of 1-in-3 clauses, i.e., there exists exactly one true literal in each such clause. Can you find a polynomial-time constant approximation?

5. (Maximum Not-All SAT) Given a 3CNF F, find an assignment to maximize the number of "not-all" clauses, i.e., in each such clause, there exist a true literal and a false literal. Can you find a polynomial-time constant approximation?

6. Using LP relaxation, design a $(1 - e^{-1})$-approximation for a polymatroid function maximization with a size constraint and a partition matroid constraint.

7. (Matroid Polytope) Show that for a matroid $\mathcal{M} = (X, \mathcal{I})$, the *matroid polytope* $P(\mathcal{M}) = conv(\{x_I \mid I \in \mathcal{I}\})$ can be described by

$$P(\mathcal{M}) = \{x \in R_+^X \mid \text{ for every } S \subseteq X, x(S) \leq r_{\mathcal{M}}(S)\}$$

where $r_{\mathcal{M}}$ is the rank function of $\mathcal{M}$.

8. (Matroid Base Polytope) Consider a matroid $\mathcal{M} = (X, \mathcal{I})$. Show that the matroid base polytope

$$P_{base}(\mathcal{M}) = conv(\{x_B \mid B \text{ is a base of } \mathcal{M}\})$$

can be described by

$$P_{base}(\mathcal{M}) = \left\{ x \in R_+^X \mid \text{ for all } S, x(S) \le r(S), x(X) = r(X) \right\}.$$

9. (Forest Polytope) Show that the forest polytope of a graph $G = (V, E)$ can be given by

$$P_{forest}(G) = \left\{ x \in R_+^E \mid \forall W \subseteq V, x(E[W]) \le |W| - 1 \right\}$$

   where $E[W]$ is the set of edges in subgraph induced by W.

10. (Spanning Tree Polytope) Show that the spanning tree polytope of a graph $G = (V, E)$ can be described by

$$P_{spanning-tree}(G) = \left\{ x \in R_+^E \mid \forall W \subset V, x(E[W]) \le |W| - 1, x(E) = |V| - 1 \right\}$$

   where $E[W]$ is the set of edges in subgraph induced by W.

11. Consider a three-regular undirected graph $G = (V, E)$ and

$$P = conv(\{\chi_F \mid F \subseteq E \text{ is a collection of vertex-disjoint cycles in } V\}).$$

   Please give the description of P in linear inequalities.

12. Consider a matroid base polytope P. Show that $[\chi_B, \chi_C]$ is an edge of P if and only if B and C are bases such that $|B \oplus C| = 2$.

13. (Lovász Extension) For a set function $f : 2^X \to R$, its Lovász extension is defined by

$$\forall x \in [0, 1]^X, f^L(x) = E[f(\{i \mid x_i \ge \theta\})]$$

   where θ is uniformly random in $[0, 1]$. Prove that f is submodular if and only if f^L is convex.

14. Consider an undirected graph $G = (V, E)$. We intend to orient every edge such that each node has at most k incoming edges. Prove that this orientation is possible if and only if $|E[W]| \le k|W|$ for every subset W of V, where $E[W]$ is the edge set of subgraph induced by W.

15. Design a continuous greedy approximation algorithm for submodular maximization with a knapsack constraint.

16. Design a continuous greedy approximation algorithm for submodular maximization with k matroid constraints.

Historical Notes

Analysis of approximation algorithms with linear programming can be found as early as the 1970s. Lovasz [297], Chvatal [70], and Wolsey [411] are pioneer works. Meanwhile, design of approximation algorithms with LP relaxation was also started for the vertex cover problem. A detail survey about it can be found in Hochbaum [212].

Bellare, Goldreich, and Sudan [23] showed that the vertex cover problem has no polynomial-time ρ-approximation for $\rho < 16/15$ unless $NP = P$. However, so far, no polynomial-time ρ-approximation for a constant $\rho < 2$ has been obtained for the vertex cover problem.

The maximum satisfiability problem has no PTAS unless NP=P. Its approximation has been studied extensively in [174, 231, 432].

The pipage rounding was proposed by Ageev and Sviridenko [2]. Gandhi et al. [155] applied this technique to dependent rounding. With pipage rounding, Calinescu et al. [42] studied the maximization of monotone submodular function subject to matroid constraint. In this work, continuous greedy also plays an important role. The continuous greedy is a quite interesting algorithm proposed by Vondrák [388]. The technique is extended to deal with nonmonotone submodular maximization by Feldman et al. [139]. A local search algorithm is discovered to deal with submodular maximization with a matroid constraint in [141], and they obtained the same performance ratio.

There are various rounding techniques in the literature. For examples, the dependent rounding was initiated in [31], and the vector rounding was proposed by Betsman et al. in an earlier version of [31]. Its generalization, the geometric rounding, can be found in [169, 170]. The iterated rounding was proposed by Jain [227] and improved by Gabow et al. [152, 153]. It has received a lot of applications [51, 64, 143, 312].

The 2-approximation for the minimum two-satisfiability problem in exercises was designed by D. Gusfield and L. Pitt [202], and the 2-approximation for the scheduling in UPM was given by Lenstra, Shmoys and Tados [277].

Chapter 12
Nonsubmodular Optimization

True optimization is the revolutionary contribution of modern research to decision processes.

—George Dantzig

We must develop knowledge optimization initiatives to leverage our key learnings.

—Scott Adams

In the real world, there are many set function optimization problems with objective function and/or constraint which is neither submodular nor supermodular. Usually, it is hard to study their approximation solutions. In this chapter, we summarize existing efforts in the literature.

12.1 An Example

The rumor is an important research subject in the study of social networks since its spread can make a lot of negative effects. For example, a rumor on earthquake will cause people's panic, and a rumor on a political leader's health will cause a shaking of stock market. Therefore, there exist many publications in the literature, which proposed many methods to block the spread of rumor. In this section, we introduce one of them, blocking the rumor by cutting at nodes.

Consider a social network represented by a directed graph $G = (V, E)$ with the independent cascade (IC) model for information diffusion. In this model, every node has two states, active and inactive. When active, it means a node is getting influenced. In the current case that we are studying, the spread of the rumor, an active node means a node gets a negative influence, i.e., influenced by the rumor.

In the IC model, the information diffusion process consists of discrete steps. In the initial step, a subset of nodes, called *seeds*, are activated. In the spread of the rumor, seeds are rumor sources. In each subsequent step, every newly active node

© The Author(s), under exclusive license to Springer Nature Switzerland AG 2022
D.-Z. Du et al., *Introduction to Combinatorial Optimization*, Springer Optimization and Its Applications 196, https://doi.org/10.1007/978-3-031-10596-8_12

tries to influence its inactive out-neighbors where a node is called a *newly active node* if it just becomes active in the last step. Suppose u is a newly active node and v is an inactive out-neighbor of u. Then, v gets influenced from u, i.e., v becomes active with the probability p_{uv} which is given with the model. When an inactive node v receives more than one newly active nodes' influence, we assume that all newly active in-neighbors influence v independently. The process ends if no node becomes active in the current step.

Now, consider a situation that a rumor is spreading in a social network G by following the rule of the IC model. There may exist one or more rumor sources. Due to budget limit, there are only k monitors available for screening out the rumor and blocking the rumor passing through the monitor. We meet the following problem:

Problem 12.1.1 (Blocking Rumor by Node Cuts) *Given a set of rumor sources, how to allocate k monitors to maximize the expectation of the number of blocked nodes?*

Let S denote the set of rumor sources spreading the same rumor and $I_G(S)$ the set of active nodes in G, influenced by the information spread from S. Define

$$\sigma^G(S) = E[I_G(S)],$$

the expected number of nodes in $I_G(S)$. Then, the expectation of the number of blocked nodes is

$$\tau(C) = \sigma^G(S) - \sigma^{G \setminus C}(G) \tag{12.1}$$

where C is the set of monitors. The problem of blocking rumor by node cuts can be expressed as

$$\max_{C:|C| \leq k} \tau(C).$$

First, we show the NP-hardness of this problem.

Theorem 12.1.2 *The problem of blocking rumor by node cuts is NP-hard.*

Proof Let us construct a polynomial-time Turing reduction from the well-known NP-hard knapsack problem to the problem of blocking rumor by node cuts.

The knapsack problem can be expressed as follows:

$$\max \quad c_1 x_1 + c_2 x_2 + \cdots + c_n x_n$$

$$\text{subject to} \quad b_1 x_1 + b_2 x_2 + \cdots + b_n x_n \leq B$$

where $c_1, c_2, \ldots, c_n, b_1, b_2, \ldots, b_n, B$ are positive integers and $b_i \leq B$ for $i = 1, 2, \ldots, n$. For this instance of the knapsack problem, we construct a social network G as follows: First, create a rumor source node r. Then, for each i, construct a clique C_i consisting of $c_i + b_i$ nodes u_{ij} for $1 \leq j \leq c_i + b_i$ and also add b_i arcs (r, u_{ij}) for

Fig. 12.1 Construction of
graph G

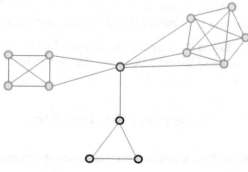

Fig. 12.2 Two
counterexamples

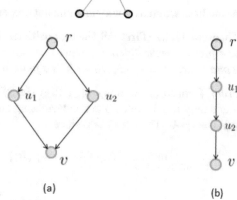

(a)   (b)

$j = 1, \ldots, b_i$ where all C_i are disjoint (Fig. 12.1). In every arc (x, y), set $p_{xy} = 1$. Thus, if we intend to block the rumor to influence C_i, then we must allocate at least b_i monitors at nodes u_{ij} for $1 \leq j \leq b_i$. Clearly, the knapsack problem has a feasible solution with objective function value at least c if and only if B monitors can be allocated into the constructed social network to protect at least $B + c$ nodes from the rumor influence. $\qquad \square$

Recall that a set function $f : 2^X \rightarrow R$ is *submodular* if for any two sets $A \subset B$ and any element $x \notin B$, $\Delta_x f(A) \geq \Delta_x f(B)$ where $\Delta_z f(A) = f(A \cup \{x\}) - f(A)$.

Proposition 12.1.3 *The function $\tau(\cdot)$ defined in (12.1) is not submodular.*

Proof Consider a social network as shown in Fig. 12.2(a). It has four nodes r, u_1, u_2, and v and two paths (r, u_1, v) and (r, u_2, v). For every arc (u, v), assign $p_{uv} = 1$. r is the unique rumor source. Then, $\Delta_{u_1} \tau(\emptyset) = 1$ and $\Delta_{u_1} \tau(\{u_2\}) = 2$. Therefore, $\Delta_{u_1} \tau(\emptyset) < \Delta_{p_1} \tau(\{p_2\})$, contradicting the definition of submodularity. $\qquad \square$

A set function $f : 2^X \rightarrow R$ is *supermodular* if $-f$ is submodular, that is, for any two sets $A \subset B$ and any element $x \notin B$, $\Delta_x f(A) \leq \Delta_x f(B)$.

Proposition 12.1.4 *The function $\tau(\cdot)$ defined in (12.1) is not supermodular.*

Proof Consider a social network as shown in Fig. 12.2(b). It has four nodes r, u_1, u_2, and v and a path (r, u_1, u_2, v). For every arc (u, v), assign $p_{uv} = 1$. r is the unique rumor source. Then, $\Delta_{u_1} \tau(\emptyset) = 3$ and $\Delta_{u_1} \tau(\{u_2\}) = 1$. Hence, $\Delta_{u_1}(\emptyset) > \Delta_{u_1} \tau(\{u_2\})$, contradicting the definition of supermodularity. $\square$

How to study a maximization problem for nonsubmodular and nonsupermodular functions? We introduce some approaches in this chapter.

12.2 Properties of Set Functions

At the first, we study some fundamental properties of set functions.

Theorem 12.2.1 (First DS Decomposition) *Every set function $f : 2^X \to R$ can be expressed as the difference of two monotone nondecreasing submodular functions g and h, i.e., $f = g - h$, where X is a finite set.*

Proof Define $\zeta(f) = \min_{A \subset B \subseteq X, x \in X \setminus A} \{\Delta_x f(A) - \Delta_x f(B)\}$. Note that $\zeta(f) \geq 0$ if and only if f is monotone nondecreasing and submodular. Consider set function $p(A) = \sqrt{|A|}$. Then $\zeta(p) > 0$ since

$$\min_{A \subset B \subseteq X, x \in B \setminus A} \{\Delta_x p(A) - \Delta_x p(B)\} = \min_{A \subset B \subseteq X, x \in B \setminus A} \{\Delta_x p(A)\}$$

$$= \min_{A \subset X} \left\{ \sqrt{|A| + 1} - \sqrt{|A|} \right\}$$

$$\geq \sqrt{|X| + 1} - \sqrt{|X|}$$

$$> 0,$$

and

$$\min_{A \subset B \subset X, x \in X \setminus B} \{\Delta_x p(A) - \Delta_x p(B)\}$$

$$= \min_{A \subset B \subset X} \left\{ \left(\sqrt{|A| + 1} - \sqrt{|A|} \right) - \left(\sqrt{|B| + 1} - \sqrt{|B|} \right) \right\}$$

$$= \min_{A \subset X} \left\{ \left(\sqrt{|A| + 1} - \sqrt{|A|} \right) - \left(\sqrt{|A| + 2} - \sqrt{|A| + 1} \right) \right\}$$

$$\geq 2\sqrt{|X| + 1} - \sqrt{|X|} - \sqrt{|X| + 2}$$

$$> 0,$$

since $\sqrt{n}$ is a strictly concave function. If $\zeta(f) \geq 0$, then f is monotone nondecreasing and submodular, and hence $f = f - 0$ is a trivial decomposition meeting the requirement.

Now, assume $\zeta(f) < 0$. Define set functions $h = 2 \cdot \frac{-\zeta(f)}{\zeta(p)} \cdot p$ and $g = f + h$. Then, $\zeta(h) = 2 \cdot \frac{-\zeta(f)}{\zeta(p)} \cdot \zeta(p) = -2\zeta(f) > 0$ and $\zeta(g) = \zeta(f) + \zeta(h) = -\zeta(f) > 0$. $\square$

Theorem 12.2.2 (Second DS Decomposition) *Every set function $f : 2^X \to R$ can be decomposed into the difference of two monotone nondecreasing supermodular functions g and h, i.e., $f = g - h$.*

Proof Define $\eta(f) = \min_{A \subset B \subset X, x \in X \setminus B} \{\Delta_x f(B) - \Delta_x f(A)\}$. Note that $\eta(f) \geq 0$ if and only if f is supermodular. Define $\tau(f) = \min_{A \subset B \subseteq X} \{f(B) - f(A)\}$. Then $\tau(f) \geq 0$ if and only if f is monotone nondecreasing. Therefore, $\min(\eta(f), \tau(f)) \geq 0$ if and only if f is a monotone nondecreasing supermodular function.

Consider set function $q(A) = |A|^2$. Then,

$$\eta(q) = \min_{A \subset B \subset X, x \in X \setminus B} \{\Delta_x q(B) - \Delta_x q(A)\}$$

$$= \min_{A \subset B \subset X} \{((|B| + 1)^2 - |B|^2) - ((|A| + 1)^2 - |A|^2)\}$$

$$= \min_{A \subset B \subset X} \{2(|B| - |A|)\}$$

$$\geq 2 > 0$$

and

$$\tau(q) = \min_{A \subset B \subseteq X} \{q(B) - q(A)\}$$

$$= \min_{A \subset B \subseteq X} (|B|^2 - |A|^2)$$

$$\geq \min_{A \subset X} \{(|A| + 1)^2 - |A|^2\}$$

$$\geq 1 > 0.$$

If $\min\{\eta(f), \tau(f)\} \geq 0$, then $f = f - 0$ is a trivial required decomposition. Now, assume $\min\{\eta(f), \tau(f)\} < 0$. Define set functions $h = 2 \cdot \frac{-\min\{\eta(f), \tau(f)\}}{\min\{\eta(q), \tau(q)\}} \cdot q$ and $g = f + h$. Then,

$$\eta(h) = 2 \cdot \frac{-\min\{\eta(f), \tau(f)\}}{\min\{\eta(q), \tau(q)\}} \cdot \eta(q) \geq -2 \cdot \min\{\eta(f), \tau(f)\} > 0$$

and

$$\tau(h) = 2 \cdot \frac{-\min\{\eta(f), \tau(f)\}}{\min\{\eta(q), \tau(q)\}} \cdot \tau(q) \geq -2 \cdot \min\{\eta(f), \tau(f)\} > 0.$$

Therefore,

$$\eta(g) = \eta(f) + \eta(h) \geq \eta(f) - 2 \cdot \min\{\eta(f), \tau(f)\} \geq -\min\{\eta(f), \tau(f)\} > 0$$

and

$$\tau(g) = \tau(f) + \tau(h) \geq \tau(f) - 2 \cdot \min\{\eta(f), \tau(f)\} \geq -\min\{\eta(f), \tau(f)\} > 0.$$

$\square$

In the real world, the DS decomposition sometime exists quite naturally. For example, consider viral marketing in social networks with the independent cascade model for information diffusion. To advertise a product, initially, the company has to distribute some free samples or discount coupons to potential buyers. They form seed set S for the marketing. Let $I(S)$ be the set of active nodes at the end of information diffusion process. Then, $E[|I(S)|]$ is the expectation of the number of active nodes, i.e., expected number of customers who will adopt the product. Suppose the price of the product is c. The profit received by the company is

$$c \cdot E[|I(S)|] - d \cdot |S|$$

where d is the cost of each free sample or discount lost on each seed. It can be proved that both terms are monotone nondecreasing and submodular.

For each specific set function, one is always able to find a DS decomposition in some way. However, no efficient approach has been found to do so. Therefore, there exists an important open problem here.

Open Problem 1 *Is there an efficient algorithm to produce a DS decomposition for any given set function?*

A set function m over 2^X is a modular function if for any two sets A and B, $m(A) + m(B) = m(A \cup B) + m(A \cap B)$. The following lemma indicates that the modular function is similar to a linear set function.

Lemma 12.2.3 *For any modular function $m : 2^X \to R$,*

$$m(A) = m(\emptyset) + \sum_{x \in A}(m(x) - m(\emptyset))$$

for any set $A \subseteq X$.

Proof This lemma can be proved by induction on $|A|$. For $|A| = 1$, it is trivial. For $|A| \geq 2$, suppose $y \in A$. Then,

$$m(y) + m(A \setminus y) = m(A) + m(\emptyset).$$

Therefore,

$$m(A) = m(A \setminus y) + (m(y) - m(\emptyset))$$
$$= m(\emptyset) + \sum_{x \in A \setminus y} (m(x) - m(\emptyset)) + (m(y) - m(\emptyset))$$
$$= m(\emptyset) + \sum_{x \in A} (m(x) - m(\emptyset)).$$

□

Theorem 12.2.4 (Decomposition of Submodular Function [76]) *Every submodular function* f *can be expressed as* $f = p + m$ *where* p *is a polymatroid function (i.e., a monotone nondecreasing submodular function with* $p(\emptyset) = 0$) *and* m *is a modular function.*

Proof Define $m(A) = f(\emptyset) - \sum_{x \in A} \Delta_x f(X \setminus x)$ and $p = f - m$. Then, m is a modular function. Hence, p is a submodular function. Moreover, $p(\emptyset) = f(\emptyset) - m(\emptyset) = 0$ and for any set A and $x \in X \setminus A$, $\Delta_x p(A) = \Delta_x f(A) - \Delta_x f(X \setminus x) \geq 0$, i.e., p is monotone nondecreasing. Therefore, p is a polymatroid function. □

Next, we show an interesting result, sandwich theorem.

Theorem 12.2.5 (Sandwich Theorem) *For any set function* $f : 2^X \to R$ *and any set* $Y \subseteq X$, *there are two modular functions* $m_u : 2^X \to R$ *and* $m_l : 2^X \to R$ *such that* $m_u \geq f \geq m_l$ *and* $m_u(Y) = f(Y) = m_l(Y)$.

This theorem states a surprising property of the set function. Why? Note that the modular function is similar to linear function. Theorem 12.2.5 contains two different modular functions passing through the same set and one is always smaller than or equal to the other. This phenomenon cannot occur for continuous linear functions. A continuous linear function with n variables can be expressed as an n-dimensional plane in the $(n+1)$-dimensional space. A pair of different n-dimensional planes with a point in common cannot have a coordinate along which one is always smaller than or equal to the other. Therefore, this theorem states a special property of the set function.

To prove the sandwich theorem, we first show two lemmas:

Lemma 12.2.6 *For any submodular function* $f : 2^X \to R$ *and any set* $Y \subseteq X$, *there exists a modular function* $m_u : 2^X \to R$ *such that* $m_u \geq f$ *and* $m_u(Y) = f(Y)$.

Proof Define

$$m_u(A) = f(Y) + \sum_{j \in A \setminus Y} \Delta_j f(\emptyset) - \sum_{j \in Y \setminus A} \Delta_j f(Y \setminus j).$$

Clearly, m_l is modular and $m_u(Y) = f(Y)$. Next, we show that $m_l \geq f$.

Assume $A \setminus Y = \{j_1, \ldots, j_k\}$. Then,

$$f(A) - f(A \cap Y) = \Delta_{j_1} f(A \cap Y) + \Delta_{j_2} f((A \cap Y) \cup \{j_1\})$$
$$+ \cdots + \Delta_{j_k} f((A \cap Y) \cup \{j_1, \ldots, j_{k-1}\})$$
$$\leq \sum_{j \in A \setminus Y} \Delta_j f(A \cap Y).$$

Assume $Y \setminus A = \{i_1, .., i_k\}$. Then,

$$f(Y) - f(A \cap Y) = \Delta_{i_1} f(A \cap Y) + \Delta_{i_2} f((A \cap Y) \cup \{i_1\})$$
$$+ \cdots + \Delta_{i_k} f((A \cap Y) \cup \{i_1, \ldots, i_{k-1}\})$$
$$\geq \sum_{j \in A \setminus Y} \Delta_j f(Y \setminus j).$$

Therefore,

$$f(A) \leq f(Y) + \sum_{j \in A \setminus Y} \Delta_j f(A \cap Y) - \sum_{j \in A \setminus Y} \Delta_j f(Y \setminus j)$$
$$\leq f(Y) + \sum_{j \in A \setminus Y} \Delta_j f(\emptyset) - \sum_{j \in A \setminus Y} \Delta_j f(Y \setminus j)$$
$$= m_u(A).$$

$\square$

Lemma 12.2.7 *For any submodular function* $f : 2^X \to R$ *and any set* $Y \subseteq X$, *there exists a modular function* $m_l : 2^X \to R$ *such that* $f \geq m_l$ *and* $f(Y) = m_l(Y)$.

Proof Put all elements of X into an ordering $X = \{x_1, x_2, \ldots, x_n\}$ such that $Y = \{x_1, x_2, \ldots, x_{|Y|}\}$. Denote $S_i = \{x_1, x_2, \ldots, x_i\}$. Define $m_l(\emptyset) = f(\emptyset)$ and for $\emptyset \neq A \subseteq X$, define

$$m_l(A) = f(\emptyset) + \sum_{x_i \in A} (f(S_i) - f(S_{i-1})).$$

Clearly, m_l is modular and

$$m_l(Y) = f(\emptyset) + \sum_{x_i \in Y} (f(S_i) - f(S_{i-1})) = f(Y).$$

Moreover, for any set $A \subseteq X$ with $A \neq \emptyset$, suppose $A = \{x_{i_1}, x_{i_2}, \ldots, x_{i_k}\}$, and then we have

$$m_l(A) = f(\emptyset) + (f(S_{i_1}) - f(S_{i_1-1})) + (f(S_{i_2}) - f(S_{i_2-1}))$$
$$+ \cdots + (f(S_{i_k}) - f(S_{i_k-1}))$$
$$\leq f(\emptyset) + (f(\{x_{i_1}\}) - f(\emptyset)) + (f(\{x_{i_1}, x_{i_2}\}) - f(\{x_1\}))$$
$$+ \cdots + (f(A) - f(\{x_{i_1}, \ldots x_{i_{k-1}}\}))$$
$$= f(A).$$

$\square$

Now, we are ready to prove Theorem 12.2.5.

Proof (Theorem 12.2.5). By Theorem 12.2.1, there exist submodular functions g and h such that $f = g - h$. By Lemmas 12.2.6 and 12.2.7, there exist modular functions $m_{gu}, m_{gl}, m_{hu}, m_{hl}$ such that

$$m_{gu} \geq g \geq m_{gl}, \quad m_{gu}(Y) = g(Y) = m_{gl}(Y),$$

and

$$m_{hu} \geq g \geq m_{hl}, \quad m_{hu}(Y) = h(Y) = m_{hl}(Y).$$

Set $m_u = m_{gu} - m_{hl}$ and $m_l = m_{gl} - m_{hu}$. Then,

$$m_u \geq f \geq m_l$$

and

$$m_u(Y) = g(Y) - h(Y) = f(Y) = g(Y) - h(Y) = m_l(Y).$$

$\square$

12.3 Parameterized Methods

To deal with nonsubmodular optimization, one intends to measure how far the function differs from the submodularity. Motivated from this intension, several parameters are introduced, and theoretical results for submodular optimization are extended to nonsubmodular optimization, usually with parameter involving in performance analysis. Let us give two examples in the following:

Consider a set function $f : 2^X \to R$. The *supermodular degree of an element* $u \in X$ by a function f is defined to be $|\mathcal{D}^+(u)|$ where

$$\mathcal{D}_f^+(u) = \{v \in X \mid \exists A \subseteq X : \Delta_u f(A \cup \{v\}) > \Delta_u f(A)\}.$$

The *supermodular degree of function* f is defined by

$$\mathcal{D}_f^+ = \max_{u \in X} |\mathcal{D}^+(u)|.$$

When only function f is studied on submodular degree, we may simply write $\mathcal{D}^+ = \mathcal{D}_f^+$.

With the supermodular degree, a nice theoretical result can be established.

Consider the monotone nonsubmodular maximization with matroid constraints as follows:

Problem 12.3.1 (Monotone Nonsubmodular Maximization) *Let f be a nonnegative and monotone nondecreasing function over 2^X where X is a finite set. Let $(X, \mathcal{C}_i)$ be a matroid for $i = 1, 2, \ldots, k$. Consider the following problem:*

$$\max \ f(A)$$

$$subject \ to \ \ A \in \mathcal{C}_i \ for \ i = 1, 2, \ldots, k,$$

Algorithm 40 is an extension of a greedy algorithm for submodular maximization with matroid constraints.

Theorem 12.3.2 *Greedy Algorithm 40 produces a $\frac{1}{k(\mathcal{D}^+ + 1) + 1}$-approximation solution for the maximization of monotone nondecreasing nonnegative set function with k matroid constraints (Problem 12.3.1).*

Proof Let $S_0, S_1, \ldots, S_g$ be produced by Greedy Algorithm 40. Let H_0 be an optimal solution. For $1 \leq i \leq g$, let H_i be a maximal independent set of $H_{i-1} \cup S_i$, containing S_i. By Lemma 9.3.5,

$$|H_{i-1} \setminus H_i| \leq k|S_i \setminus H_{i-1}| \leq k|S_i \setminus S_{i-1}| \leq k(\mathcal{D}^+ + 1).$$

Suppose $H_{i-1} \setminus H_i = \{v_1, \ldots, v_r\}$. Denote $H_{i-1}^j = H_{i-1} \setminus \{v_1, \ldots, v_j\}$. Then,

---

Algorithm 40 Greedy Approximation for Problem 12.3.1

Input: a nonnegative monotone set function f on 2^X and k matroids $(X, \mathcal{C}_i)$ for $1 \leq i \leq k$. Let $\mathcal{C} = \cap_{i=1}^k \mathcal{C}_i$.
Output: A subset S_i of X.
  1: $S_0 \leftarrow \emptyset$;
  2: $i \leftarrow 0$;
  3: **while** S_i is not a maximal independent set **do**
  4:     $i \leftarrow i + 1$;
  5:     choose $u_i \in X \setminus S_{i-1}$ and $D_{i-1} \in \mathcal{D}^+(u_i)$ to maximize
  6:     $f(D_i \cup \{u_i\} \cup S_{i-1}) - f(S_{i-1})$ subject to $D_i \cup \{u_i\} \cup S_{i-1} \in \mathcal{C}$;
  7:     set $S_i \leftarrow D_i \cup \{u_i\} \cup S_{i-1}$;
  8: **end while**
  9: **return** S_i.

$$f((\mathcal{D}^+(v_j) \cap H_{i-1}^j) \cup \{v_j\} \cup S_{i-1}) - f(S_{i-1})$$

$$= \Delta_{v_j} f((\mathcal{D}^+(v_j) \cap H_{i-1}^j) \cup S_{i-1}) + f((\mathcal{D}^+(v_j) \cap H_{i-1}^j) \cup S_{i-1}) - f(S_{i-1})$$

$$\geq \Delta_{v_j} f((\mathcal{D}^+(v_j) \cap H_{i-1}^j) \cup S_{i-1})$$

$$\geq \Delta_{v_j} f(H_{i-1}^j \cup S_{i-1}).$$

Since

$$\left(\left(\mathcal{D}^+(v_j) \cap H_{i-1}^j\right) \setminus S_{i-1}\right) \cup \{v_j\} \cup S_{i-1} = \left(\mathcal{D}^+(v_j) \cap H_{i-1}^j\right) \cup \{v_j\} \cup S_{i-1}$$

and $(v_j, (\mathcal{D}^+(v_j) \cap H_{i-1}^j) \setminus S_{i-1})$ is a candidate pair for the ith step of the algorithm, we have

$$f(S_i) - f(S_{i-1})$$

$$\geq f\left(\left(\left(\mathcal{D}^+(v_j) \cap H_{i-1}^j\right) \setminus S_{i-1}\right) \cup \{v_j\} \cup S_{i-1}\right) - f(S_{i-1})$$

$$= f\left(\left(\mathcal{D}^+(v_j) \cap H_{i-1}^j\right) \cup \{v_j\} \cup S_{i-1}\right) - f(S_{i-1})$$

$$\geq \Delta_{v_j} f\left(\left(\mathcal{D}^+(v_j) \cap H_{i-1}^j\right) \cup S_{i-1}\right)$$

$$\geq \Delta_{v_j} f\left(H_{i-1}^j \cup S_{i-1}\right).$$

Thus,

$$r(f(S_i) - f(S_{i-1})) \geq \sum_{j=1}^r \Delta_{v_j} f\left(H_{i-1}^j \cup S_{i-1}\right)$$

$$= f\left(H_{i-1}^0 \cup S_{i-1}\right) - f\left(H_{i-1}^r \cup S_{i-1}\right)$$

$$= f(H_{i-1}) - f(H_i).$$

Note that $r \leq k(\mathcal{D}^+ + 1)$. Thus,

$$k(\mathcal{D}^+ + 1)[f(S_i) - f(S_0)] \geq r \sum_{i=1}^g [f(S_i) - f(S_{i-1})]$$

$$\geq \sum_{i=1}^g [f(H_{i-1}) - f(H_i)]$$

$$= f(H_0) - f(H_i).$$

Since $S_g = H_g$, $f(H_0) = opt$, and $f(S_0) \geq 0$, we finally have

$$(k(\mathcal{D}^+ + 1) + 1)f(S_g) \geq opt.$$

$\square$

The supermodular degree has been successfully applied to a few optimization problems, such as committee member selection.

We next introduce another parameter, curvature. There are several definitions about curvature in the literature. What we will study is one of them.

By the first DS decomposition theorem, every set function is the difference of two monotone nondecreasing submodular functions g and h, i.e., $g - h$. Therefore, the minimization of such a set function can be worked out through solving a sequence of problems $\min\{g - c \mid h \geq c\}$ for a sequence of discrete constants c. Note that each monotone nondecreasing submodular function can be represented by the sum of a polymatroid function and a constant. Thus, $\min\{g - c \mid h \geq c\}$ can be transformed into the following form:

Problem 12.3.3 (Submodular Set Cover with Submodular Objective) *Let g and f be polymatroid functions over 2^X for a finite set X:*

$$\min \ g(A)$$

$$subject \ to \ \ f(A) = f(X)$$

This problem has Algorithm 41 which can be analyzed as follows:
Define the *curvature* of g by

$$\chi(g) = \min_{A \subseteq X} \frac{\sum_{x \in A} g(\{x\})}{g(A)}.$$

Then, the following holds:

Theorem 12.3.4 *Greedy Algorithm 41 produces an approximation solution with performance ratio at most $\chi(g)H(\gamma)$ where*

Algorithm 41 Greedy Approximation for Problem 12.3.3

Input: two polymatroid functions g and f.
Output: A subset S of X.
1: $S \leftarrow \emptyset$;
2: **while** $f(S) < f(X)$ **do**
3:     choose x to maximize $\frac{\Delta_x f(S)}{\Delta_x g(S)}$
4:     and set $S \leftarrow S \cup \{x\}$;
5: **end while**
6: **return** S.

$$\gamma = \max_{x \in X} f(\{x\})$$

and $H(\gamma) = \sum_{i=1}^{\gamma} 1/i$.

Proof Define

$$c(A) = \sum_{x \in A} g(x).$$

Then, Algorithm 41 is exactly the greedy algorithm for the submodular set cover (standard form) problem. Let S be the set obtained by the algorithm. Then,

$$c(S) \le H(\gamma) \cdot c(OPT_c)$$

where OPT_c is an optimal solution with respect to objective function c. Let OPT_g be an optimal solution with respect to objective function g. Then,

$$g(S) \le \sum_{x \in S} g(x) = c(S)$$

and

$$c(OPT_c) \le c(OPT_g) \le \sum_{x \in OPT_g} g(x) \le \chi(g) \cdot g(OPT_g).$$

Therefore,

$$g(S) \le \gamma H(\gamma) \cdot g(OPT_g).$$

□

Problem 12.3.3 is also closely related to the generalized hitting set problem as follows:

Problem 12.3.5 (Generalized Hitting Set) *Given m nonempty collections $C_1, C_2, \ldots, C_m$ of subsets of a finite set X, find the minimum subset A of X such that every C_i has a member $S \subseteq A$.*

Let $C = \cup_{i=1}^{m} C_i$. For every subcollection $A \subseteq C$, define

$$g(A) = |\cup_{A \in A} A|$$

and

$$f(A) = |\{C_i \mid A \cap C_i \ne \emptyset, 1 \le i \le m\}|.$$

It is not hard to prove the following:

Lemma 12.3.6 *Both g and f are polymatroid functions.*

Moreover, we have

Theorem 12.3.7 *The generalized hitting set problem is equivalent to the following:*

$$\min \ g(\mathcal{A}) \tag{12.2}$$
$$subject\ to\ \ f(\mathcal{A}) = f(\mathcal{C}),$$
$$\mathcal{A} \subseteq \mathcal{C}.$$

This equivalence means that $\mathcal{A}$ is a minimum solution of problem (12.2) if and only if $\cup_{A\in\mathcal{A}} A$ is the minimum solution of the generalized hitting set problem.

Proof Suppose $\mathcal{A}$ is the minimum solution of problem (12.2). For contradiction, suppose $\cup_{A\in\mathcal{A}} A$ is not a minimum generalized hitting set. Consider a minimum generalized hitting set D. Then, $|D| < |\cup_{A\in\mathcal{A}} A|$. For each $\mathcal{C}_{uv}$, let C_{uv} be a subset of D, contained in $\mathcal{C}_{uv}$. Denote

$$\mathcal{C}_D = \{C_{uv} \mid u, v \in V \text{ with } d(u, v) = 2\}.$$

Then, $h(\mathcal{C}_D) = h(\mathcal{C})$ and $g(\mathcal{C}_D) \leq |D| < g(\mathcal{A})$, a contradiction.

Conversely, suppose $\cup_{A\in\mathcal{A}} A$ is a minimum generalized hitting set. For contradiction, suppose $\mathcal{A}$ is not a minimum solution for above problem of submodular minimization with submodular cover constraint. Consider a minimum solution $\mathcal{B}$ for it. Then, $g(\mathcal{B}) < g(\mathcal{A})$. By above argument, $\cup_{B\in\mathcal{B}} B$ is a generalized hitting set such that

$$|\cup_{B\in\mathcal{B}} B| = g(\mathcal{B}) < g(\mathcal{A}) = |\cup_{A\in\mathcal{A}} A|,$$

a contradiction.   □

Now, we present an application. Consider a wireless sensor network. Each sensor has a communication disk and a sensing disk with itself as common center. If sensor s_1 lies in the communication disk of s_2, then s_1 can receive message from s_2. When all sensors have the same size of communication disks and the same size of sensing disks, they are said to be homogeneous. In a homogeneous wireless sensor system, the communication network is an undirected graph, in which a virtual backbone is a connected dominating set, that is, it is a node subset such that every node is either in the subset or adjacent to the subset. Construction of the virtual backbone is an important issue in the study of wireless sensor networks.

Fig. 12.3 Proof of
Lemma 12.3.9

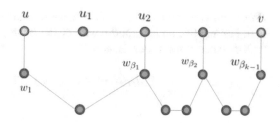

Let $G = (V, E)$ be a homogeneous wireless sensor network and D a subset
of nodes. For each pair of nodes u and v, let $d(u, v)$ denote the shortest distance
between u and v in G, i.e., the minimum number of edges on a path between u and
v and $d_D(u, v)$ the shortest distance between u and v in the subgraph induced by
$D \cup \{u, v\}$. Motivated from reducing routing cost and improving load balancing, the
following problem has been studied in the literature:

Problem 12.3.8 (Routing-cost Constrained CDS) *Given a homogeneous wireless
sensor network $G = (V, E)$, find a minimum connected dominating set D such that*

$$d_D(u, v) \leq \alpha d(u, v), \forall u, v \in V, \tag{12.3}$$

where α is a constant.

Condition (12.3) can be simplified as follows:

Lemma 12.3.9 *To satisfy condition (12.3), it is sufficient to satisfy*

$$d_D(u, v) \leq \alpha + 1, \forall u, v \in V \text{ with } d(u, v) = 2. \tag{12.4}$$

Proof Actually, $(d_D(u, v) - 1) \leq \alpha(d(u, v) - 1)$ for $d(u, v) \geq 2$ if and only
if condition (12.4) holds. We need only to show the backward direction. Suppose
$d(u, v) = k$. Then, there exists a shortest path $(u, u_1, u_2, \ldots, u_{k-1}, v)$ in G.
By condition (12.4), there exist node paths $(u, w_1, \ldots, w_{\beta_1}, u_2)$, $(w_{\beta_1}, w_{\beta_1+1},$
$\ldots, w_{\beta_2}, u_3)$, $\ldots$, $(w_{\beta_{k-2}}, w_{\beta_{k-2}+1}, \ldots, w_{\beta_{k-1}}, v)$ (Fig. 12.3) such that $\beta_1 \leq \alpha$,
$\beta_2 \leq 2\alpha$, $\ldots$, $\beta_{k-1} \leq (k-1)\alpha$, and all w_i for $1 \leq i \leq \beta_{k-1}$ lie in D. Thus,

$$d_D(u, v) \leq \beta_{k-1} + 1 \leq \alpha(k - 1) + 1 \leq \alpha k.$$

□

Lemma 12.3.10 *Suppose D contains a maximal independent set I. If for any two
nodes x and y in I with distance at most four, $d_D(x, y) \leq \alpha$, then for any two nodes
u and v with distance two, $d_D(u, v) \leq \alpha + 2$.*

Proof For any node u, if $U \in I$, then set $u' = u$; otherwise, set $u' \in I$ is adjacent to u. For any two nodes u and v with distance two, we then have that u' and v' in I with distance at most four. Hence,

$$d_D(u, v) \le d_D(u'v') + 2 \le \alpha + 2.$$

$\square$

Motivated from this lemma, we divide the construction into two stages.

In the first stage, we construct a maximal independent set I. It is worth mentioning a well-known conjecture about the maximal independent set.

Conjecture 12.3.11 *Let $\alpha(G)$ be the size of the maximum independent set in graph G and $\gamma_c(G)$ the size of the minimum connected dominating set in graph G. Then, for any unit disk graph G (i.e., the graph structure of any homogeneous wireless sensor network),*

$$\alpha(G) \le 3 \cdot \gamma_c(G) + 3.$$

This conjecture is still open. The best proved result (see [115]) is

$$\alpha(G) \le 3.399 \cdot \gamma_c(G) + 4.874.$$

Since $|I| \le \alpha(G)$ and $\gamma_c(G)$ are lower bound for opt_{acds}, the minimum size of routing-cost constrained connected dominating set with parameter α, we have

$$|I| \le 3.399 \cdot opt_{acds} + 4.874.$$

In the second stage, for every pair of nodes u and v in I with distance at most four, let C_{uv} denote the collections of node subsets each of which is the set of intermediate nodes on a path between u and v with distance at most α ($\alpha \ge 4$). Let D be a node set hitting every C_{uv}. Then, D would satisfy constraint (12.4) and hence (12.3) holds. This would imply that D is a connected dominating set. Thus, the routing-cost constrained CDS problem is equivalent to the generalized hitting set problem with input collections C_{uv}.

Now, we proved that in this example, $\chi(g)$ and γ are bounded by constants:

Lemma 12.3.12

$$\gamma \le 25$$

and

$$\chi(g) \le 420.$$

Proof Note that each node is adjacent to at most five nodes in an independent set. Thus, there exist at most 25 paths with length 2, sharing the same intermediate nodes with endpoints in I. Therefore, $\gamma = \max_{S \in C} f(\{S\}) \leq 25$.

To estimate $\chi(g)$, we first note that there are at most $\frac{\pi(d+0.5)^2}{\pi \cdot 0.5^2} = (2d+1)^2$ independent nodes within distance d from any node.

Suppose u is an intermediate node of a path between x and y in I with at most distance 4. There are two cases:

Case 1. Both x and y are within a distance 2 from u. The number of possible pairs $\{x, y\}$ is at most $25(25-1)/2 = 300$.

Case 2. One of x and y has distance one from u and the other has distance three from u. The number of possible pairs $\{x, y\}$ in this case is at most $5 \times (7^2 - 5^2) = 5 \times 24 = 120$.

Putting two cases together, we obtain $\chi(g) \leq 420$.

□

Theorem 12.3.13 *For any connected unit disk graph G, a connected dominating set D can be constructed, in polynomial-time, to satisfy*

$$|D| \leq (420 \cdot H(25) + 3.399)opt_{4cds} + 4.874$$

and

$$d_D(u, v) \leq 5 \cdot d(u, v).$$

Proof Let A be obtained by the greedy algorithm hitting all C_{uv} for all pairs of nodes u, v in a maximal independent set I with $d(u, v) \leq 4$. Then, by Lemma 12.3.12,

$$|A| \leq 420 \cdot H(25) \cdot opt_{4cds}.$$

Set $D = I \cup A$. Then, we have

$$|D| \leq (420 \cdot H(25) + 3.399)opt_{4cds} + 4.874$$

Moreover, by Lemma 12.3.10, we have that for any two nodes u and v with distance two

$$d_D(u, v) \leq 6.$$

By Lemma 12.3.9, we obtain that for any two nodes u and v,

$$d_D(u, v) \leq 5 \cdot d(u, v).$$

□

In most of nonsubmodular optimization problems, the parameterized method cannot be used successfully due to the lack of significant bound for parameters. For example, the supermodular degree or curvature as above goes to infinity. Therefore, one has to look for more approaches.

12.4 Sandwich Method

The sandwich method has been used quite often for solving nonsubmodular optimization problems in the literature. It runs as follows:

Suppose we face a problem $\max_{A \in \Omega} f(A)$ where Ω is a collection of subsets of 2^X and X is a finite set.

Sandwich Method :

- **Input** a set function $f : 2^X \to R$.
- Initially, find two submodular functions u and l such that $u(A) \geq f(A) \geq l(A)$ for $A \in \Omega$. Then, carry out the following operations:

  - Compute an α-approximation solution S_u for $\max_{A \in \Omega} u(A)$ and a β-approximation solution S_l for $\max_{A \in \Omega} l(A)$.
  - Compute a feasible solution S_o for $\max_{A \in \Omega} f(A)$.
  - Set $S = \text{argmax}(f(S_u), f(S_o), f(S_l))$.

- **Output** S.

The performance ratio of this algorithm is data-dependent as follows: Hence, this algorithm is also called a *data-dependent* approximation algorithm.

Theorem 12.4.1 *The solution S produced by the sandwich method satisfies the following:*

$$f(S) \geq \max \left\{ \frac{f(S_u)}{u(S_u)} \cdot \alpha, \frac{opt_l}{opt_f} \cdot \beta \right\} \cdot opt_f,$$

where opt_f (opt_l) is the objective function value of the minimum solution for $\max_{A \in \Omega} f(A)$ ($\max_{A \in \Omega} l(A)$).

Proof Since S_u is a α-approximation solution for $\max_{A \in \Omega} u(A)$, we have

$$f(S_u) = \frac{f(S_u)}{u(S_u)} \cdot u(S_u)$$

$$\geq \frac{f(S_u)}{u(S_u)} \cdot \alpha \cdot opt_u$$

$$\geq \frac{f(S_u)}{u(S_u)} \cdot \alpha \cdot u(OPT_f)$$

$$\geq \frac{f(S_u)}{u(S_u)} \cdot \alpha \cdot opt_f,$$

where OPT_f is an optimal solution for $\min_{A\in\Omega} f(A)$. Since S_l is an β-approximation solution for $\max_{A\in\Omega} l(A)$, we have

$$f(S_l) \geq l(S_l) \geq \beta \cdot opt_l = \beta \cdot \frac{opt_l}{opt_f} \cdot opt_f.$$

Therefore, the theorem holds.   □

Next, we follow this approach to give a data-dependent approximation for the problem of blocking rumor by cut.

Define

$$\alpha_1(C) = \sum_{c\in C} \tau(\{c\})$$

and

$$\alpha_2(C) = \sum_{c,c'\in C:c\neq c'} \tau(\{c,c'\}).$$

Clearly, α_1 is modular, that is, for any two subsets $A \subset B$ and any element $x \notin B$,

$$\Delta_x\alpha_1(A) = \Delta_x\alpha_1(B).$$

However, we have

Lemma 12.4.2 α_2 *is supermodular.*

Proof By definition of α_2, we have that for any two subsets $A \subset B$ and any element $x \notin B$,

$$\Delta_x\alpha_2(A) = \sum_{y\in A} \tau(\{x,y\}) \leq \Delta_x\alpha_2(B) = \sum_{y\in B} \tau(\{x,y\}).$$

Therefore, α_2 is supermodular.   □

By the inclusive-exclusive formula, we have

Lemma 12.4.3 *For any set C,*

$$\alpha_1(C) \geq \tau(C) \geq \alpha_1(C) - \alpha_2(C).$$

Since α_1 is modular, $\max\{\alpha_1(C) \mid |C| \leq k\}$ can be solved in polynomial-time. Define $\beta = \alpha_1 - \alpha_2$. Then, by Lemma 12.4.2, β is a nonnegative submodular function. There exists a $1/e$-approximation algorithm in Section 10.4. Now, we can describe a data-dependent approximation algorithm as follows:

Data-Dependent Approximation
  Compute an optimal solution C_{α_1} for $\max\{\alpha_1(C) \mid |C| \le k\}$.
  Compute $1/e$-approximation C_β for $\max\{\beta(C) \mid |C| \le k\}$.
  Compute a feasible solution C_τ for $\max\{\tau(C) \mid |C| \le k\}$.
  Choose $C_{data} = \mathrm{argmax}(\tau(C_{\alpha_1}), \tau(C_\beta), \tau(C_\tau))$.

By Theorem 12.4.1, this solution C_{data} has the following performance:

Theorem 12.4.4

$$\tau(C_{data}) \ge \max\left(\frac{\tau(C_{\alpha_1})}{opt_{\alpha_1}}, \frac{1}{e} \cdot \frac{opt_\beta}{opt_\tau}\right) \cdot opt_\tau$$

where opt_τ (opt_{α_1}, and opt_β) is the objective function value of an optimal solution for problem $\max\{\tau(C) \mid |C| \le k\}$ (problem $\max\{\alpha_1(C) \mid |C| \le k\}$, and problem $\max\{\beta(C) \mid |C| \le k\}$ respectively).

Note that τ is monotone nondecreasing, i.e., for $A \subset B$, $\tau(A) \le \tau(B)$. Therefore, C_τ can be obtained by the following greedy algorithm:

Greedy Algorithm
  $C_0 \leftarrow \emptyset$;
  for $i = 1$ **to** k **do**
    $x = \mathrm{argmax}_{x \in V \setminus C_{i-1}}(\tau(C_{i-1} \cup \{x\}) - \tau(C_{i-1}))$ and
    $C_i \leftarrow C_{i-1} \cup \{x\}$;
  end-for
  return $C_\tau = C_k$.

From theoretical point of view, the sandwich method is always applicable since we have the following:

Theorem 12.4.5 *For any set function f on 2^X, there exist two monotone nondecreasing submodular functions u and l such that $u(A) \ge f(A) \ge l(A)$ for every $A \in 2^X$.*

Proof By the first DS decomposition theorem, there exist two monotone nondecreasing submodular functions g and h such that $f = g - h$. Note that for every $A \in 2^X$, $h(\emptyset) \le h(A) \le h(X)$. Set $u(A) = g(A) - h(\emptyset)$ and $l(A) = g(A) - h(X)$ for any $A \in 2^X$. Then, u and l meet our requirement.   □

However, in practice, it is often quite hard to find such an upper-bound u and a lower-bound l which are easily computable since the DS decomposition exists but is unknown to be efficiently computable. Therefore, more efforts are required to construct them for specific real-world problems.

12.5   Algorithm Ending at Local Optimal Solution

For nonsubmodular optimization, there also exists a class of algorithms which end at local optimal solutions. What is the local optimal solution? For set function optimization, there exist several definitions in the literature. However, they have a property in common, that is, all of them are necessary conditions for optimality. In this section, we introduce two of them together with two algorithms which end at these two types of local optimal solutions, respectively.

Here are two necessary conditions for minimality:

1. Let f be a set function on 2^X. Suppose A is a minimum solution of f in 2^X. Then, $f(A) \leq f(A \setminus \{x\})$ and $f(A) \leq f(A \cup \{x\})$ for any $x \in X$.
2. Let $f = g - h$ be a set function and g and h submodular functions on subsets of X. If set A is a minimum solution for $\min_{Y \subseteq X} f(Y)$, then $\partial h(A) \subseteq \partial g(A)$.

Condition 1 is obvious. Condition 2 needs a little explanation. First, let us explain what is the notation $\partial h(A)$. $\partial h(A)$ is the subgradient of function h at set A, defined as

$$\partial h(A) = \{c \in R^X \mid h(Y) \geq h(A) + \langle c, Y - A \rangle\}.$$

Actually, for a submodular set function $h : 2^X \to R$, the subgradient at set A consists of all linear functions $c : X \to R$ satisfying $h(Y) \geq h(A) + c(Y) - c(A)$ where $c(Y) = \sum_{y \in Y} c(y)$. Each linear function c can also be seen as a vector in R^X, i.e., a vector c with components labeled by elements in X. The characteristic vector of each subset Y of X is a vector in $\{0, 1\}^X$ such that the component with label $x \in X$ is equal to 1 if and only if $x \in Y$. Here, for simplicity of notation, we use the same notation Y to represent the set Y and its characteristic vector.

To see Condition 2, note that since A is a minimum solution for

$$\min_{Y \subseteq X} f(Y),$$

we have $f(A) \leq f(Y)$ and hence $g(Y) - g(A) \geq h(Y) - h(A)$ for any $Y \subseteq X$. Therefore, for any $c \in \partial h(A)$, $g(Y) - g(A) \geq h(Y) - h(A) \geq c(Y) - c(A)$. This means that $\partial h(A) \subseteq \partial g(A)$.

Condition 2 implies Condition 1. To see this, we first introduce two lemmas:

Lemma 12.5.1 *Suppose A satisfies condition 2. Then, for any $Y \in \mathcal{U}$, $f(A) \leq f(Y)$ where*

$$\mathcal{U} = \{Y \mid \partial h(Y) \cap \partial g(A) \neq \emptyset\}.$$

Proof Choose $c \in \partial h(Y) \cap \partial g(A)$. Then,

$$h(A) \geq h(Y) + (c(A) - c(Y)) \text{ and } g(Y) \geq g(A) + (c(Y) - c(A)).$$

Hence, $h(Y) - h(A) \leq c(Y) - c(A) \leq g(Y) - g(A)$. Therefore, $f(Y) \geq f(A)$. □

Lemma 12.5.2 (Fujishige [150]) *A point* $c \in R^X$ *is an extreme point of* $\partial f(A)$ *if and only if there is a permutation* σ *for elements in* X, *i.e.,* $X = \{\sigma(1), \sigma(2), \ldots, \sigma(|X|)\}$, *such that* $A = \{\sigma(1), \sigma(2), \ldots, \sigma(|A|)\}$ *and* $c(\{\sigma(i)\}) = f(S_i) - f(S_{i-1})$ *for* $1 \leq i \leq |X|$ *where* $S_0 = \emptyset$ *and* $S_i = \{\sigma(1), \sigma(2), \ldots, \sigma(i)\}$.

Theorem 12.5.3 *Condition 2 implies Condition 1.*

Proof For any $x \in A$, consider permutation $X = \{\sigma(1), \sigma(2), \ldots, \sigma(|X|)\}$ such that $A = \{\sigma(1), \sigma(2), \ldots, \sigma(|A|)\}$ and $\sigma(|A|) = x$. Define linear function c by $c(\{\sigma(i)\}) = h(S_i) - h(S_{i-1})$ for $1 \leq i \leq |X|$ where $S_0 = \emptyset$ and $S_i = \{\sigma(1), \sigma(2), \ldots, \sigma(i)\}$. Then, $c \in \partial h(A \setminus \{x\}) \cap \partial h$. Since $\partial h(A) \subseteq \partial g(A)$, we have $c \in \partial h(A \setminus \{x\}) \cap \partial g(A)$. By Lemma 12.5.1, $f(A) \leq f(A \setminus \{x\})$.

Similarly, we can show that for any $x \in X \setminus A$, $f(A) \leq f(A \cup \{x\})$. □

Now, let us study the submodular-supermodular algorithm.

As shown in Algorithm 42, the algorithm mainly uses the first DS decomposition. Given a set function $f : 2^X \rightarrow R$, initially assume that $f = g - h$ is already known for two submodular functions g and h. Start from an arbitrary set A. At each iteration, replace h by a lower-bound modular function m_{hl} such that $h(A) = m_{hl}(A)$. Then, compute a minimum solution A^+ for $\min_{Y \in 2^X}[g(Y) - m_{hl}(Y)]$. This is possible since the unconstrained submodular minimization can be solved in polynomial-time. Note that $f(A) = g(A) - m_{hl}(A)$. Thus, we must have $f(A^+) \leq f(A)$. If $f(A^+) < f(A)$, then set $A \leftarrow A^+$, and a new iteration will start; otherwise, algorithm stops.

In above iteration, an important remark should be made on the lower-bound modular function m_{hl}. Actually, for each permutation σ of X such that $A = \{\sigma(1), \ldots, \sigma(|A|)\}$, we can construct a modular function m_{hl}^{σ} such that $h(A) =$

Algorithm 42 Submodular-Supermodular Algorithm for Minimization

Input: a set function $f : 2^X \rightarrow R$ and its DS decomposition $f = g - h$ where g and h are submodular functions.
Output: A subset A of X.
 1: choose a set $A \subseteq X$;
 2: **while** $f(A^+) < f(A)$ **do**
 3:      $A \leftarrow A^+$;
 4:      **for** $\sigma \in \Sigma_A$ **do**
 5:          compute a lower-bound modular function m_{hl}^{σ} for h;
 6:          compute a minimum solution A_{σ}^+ for $\min_{Y \in 2^X}[g(Y) - m_{hl}^{\sigma}(Y)]$;
 7:      **end for**
 8:      $\sigma \leftarrow \operatorname{argmin}_{\sigma \in \Sigma_A} f(A_{\sigma}^+)$;
 9:      $A^+ \leftarrow A_{\sigma}^+$
10: **end while**
11: **return** A.

$m^\sigma_{hl}(A)$ and, moreover, $h(S_i) = m^\sigma_{hl}(S_i)$ for any $S_i = \{\sigma(1), \ldots, \sigma(i)\}$. Let $\theta = \max(|A|, |X| - |A|)$. Let Σ_A be a collection of θ permutations σ of X such that $\sigma(|A|)$ goes over all elements of A and $\sigma(|A| + 1)$ goes over all elements of $X \setminus A$, that is, for any element $x \in A$, there exists $\sigma \in \Sigma_A$ such that $A \setminus \{x\} = \{\sigma(1), \ldots, \sigma|A| - 1\}$ and for any $x \in X \setminus A$, there exists $\sigma \in \Sigma_A$ such that $A \cup \{x\} = \{\sigma(1), \ldots, \sigma(|A| + 1)\}$. Now, let A^+_σ denote the minimum solution for $\min_{Y \in 2^X}[g(Y) - m^\sigma_{hl}(Y)]$. Set

$$\sigma^+ = \operatorname{argmin}_{\sigma \in \Sigma_A} f(A^+_\sigma)$$

and

$$A^+ = A^+_{\sigma^+}.$$

Then, we will have

$$f(A^+) \le f(A \setminus \{x\}) \text{ for any } x \in A$$

and

$$f(A^+) \le f(A \cup \{x\}) \text{ for any } x \in X \setminus A.$$

Therefore, for Algorithm 42, we have

Theorem 12.5.4 *Algorithm 42 always ends at a local minimum solution satisfying Condition 1.*

12.6   Global Approximation of Local Optimality

Sometimes, an algorithm may not be able to stop at a local optimal solution, and instead, it stops at a local approximately optimal solution. For example, consider the following problem:

$$\max \ f(A)$$
$$\text{subject to } |A| \le k$$
$$A \in 2^X$$

where f is a set function over 2^X for a finite set X. Algorithm 43 is the submodular-supermodular algorithm for this problem.

Algorithm 43 Submodular-Supermodular Algorithm for Maximization

---

Input: a set function $f : 2^X \to R$ and its DS decomposition $f = g - h$ where g and h are submodular functions.
Output: A subset A of X.

1: choose a set $A \subseteq X$;
2: **while** $f(A^+) > f(A)$ **do**
3:    $A \leftarrow A^+$;
4:    **for** $\sigma \in \Sigma_A$ **do**
5:       compute a lower-bound modular function m_{hl}^σ for h;
6:       compute a $(1 - e^{-1})$-approximation solution A_σ^+ for $\max_{Y:|Y|\le k}[g(Y) - m_{hl}^\sigma(Y)]$;
7:    **end for**
8:    $\sigma \leftarrow \mathrm{argmax}_{\sigma \in \Sigma_A} f(A_\sigma^+)$;
9:    $A^+ \leftarrow A_\sigma^+$
10: **end while**
11: **return** A.

Theorem 12.6.1 *Algorithm 43 stops at a solution A satisfying condition that for any $x \in A$, $f(A) \ge (1 - e^{-1}) \cdot f(A \setminus \{x\})$ and for any $x \in X \setminus A$, $f(A) \ge (1 - e^{-1}) \cdot f(A \cup \{x\})$.*

Proof In each iteration, A^+ satisfies that for any $x \in A$, $f(A^+) \ge (1-e^{-1}) \cdot f(A \setminus \{x\})$ and for any $x \in X \setminus A$, $f(A^+) \ge (1 - e^{-1}) \cdot f(A \cup \{x\})$. When the algorithm stops, we have $f(A) \ge f(A^+)$. Therefore, A has the property. $\square$

Since the algorithm always stops at a local approximately optimal solution, we may call it as a *global optimal approximation* or G-L approximation for a brief name. There is a more important reason for us to pay attention on G-L approximation, that is, the submodular-supermodular is unlikely to run in polynomial-time. Actually, computing a local minimum solution satisfying Condition 1 is PLS (Polynomial Local Search)-complete, which is unlikely to run in polynomial-time. In order to obtain a polynomial-time algorithm, we have to modify the submodular-supermodular algorithm while accepting the G-L approximation solution.

Let us call A as a G-L ρ-approximation for minimization of set function $f : 2^X \to R$ if for any $x \in A$, $f(A) \le (1+\varepsilon) \cdot f(A \setminus \{x\})$ and for any $x \in X \setminus A$, $f(A) \le (1 + \varepsilon) \cdot f(A \cup \{x\})$. Now, we modify the submodular-supermodular algorithm for minimization as shown in Algorithm 44, and obtain the following:

Theorem 12.6.2 *Let f be a nonnegative function. Then, Algorithm 44 ends at G-L $(1 + \varepsilon)$-approximation within $O(\frac{1}{\varepsilon} \ln \zeta)$ iterations where ζ is the ratio of the maximum value and the minimum value of f.*

Proof In each iteration, we have that for any $x \in A$, $f(A^+) \le f(A \setminus \{x\})$ and for any $x \in X \setminus A$, $f(A^+) \le f(A \cup \{x\})$. When the algorithm stops, $f(A) \le (1 + \varepsilon) \cdot f(A^+)$. Hence, for any $x \in A$, $f(A) \le (1 + \varepsilon) \cdot f(A \setminus \{x\})$ and for any $x \in X \setminus A$, $f(A) \le (1 + \varepsilon) \cdot f(A \cup \{x\})$.

Before algorithm stops, in each iteration, we have $f(A^+) \cdot (1 + \varepsilon) < f(A)$. Let k satisfy $opt \cdot (1 + \varepsilon)^k = f(A)$ for initial set A and optimal value opt. Then, the

---

Algorithm 44 Submodular-Supermodular Algorithm for Minimization

---

Input: a set function $f : 2^X \to R$ and its DS decomposition $f = g - h$ where g and h are submodular functions.
Output: A subset A of X.

1: choose a set $A \subseteq X$;
2: **while** $f(A^+) \cdot (1 + \varepsilon) < f(A)$ **do**
3:   $A \leftarrow A^+$;
4:   **for** $\sigma \in \Sigma_A$ **do**
5:     compute a lower-bound modular function m_{hl}^σ for h;
6:     compute a minimum solution A_σ^+ for $\min_{Y \in 2^X}[g(Y) - m_{hl}^\sigma(Y)]$;
7:   **end for**
8:   $\sigma \leftarrow \operatorname{argmin}_{\sigma \in \Sigma_A} f(A_\sigma^+)$;
9:   $A^+ \leftarrow A_\sigma^+$
10: **end while**
11: **return** A.

---

algorithm must stop within $\lfloor k \rfloor$ iterations. Note that $e < (1 + \varepsilon)^{1 + 1/\varepsilon}$ and $1 + 1/\varepsilon < 2/\varepsilon$ for $\varepsilon < 1$. Thus, $e^{k\varepsilon/2} < f(A)/opt$. Hence, $k\varepsilon < 2\ln(f(A)/opt) \le 2\ln \zeta$.   □

12.7 Large-Scale System

Current technology developments and economic activities produce a huge amount of data, which raise a lot of large-scale combinatorial optimization problems. In order to give efficient solution for them, many techniques are generated, which form an important research direction. In this section, we give a brief introduction to let the reader get a taste on its importance in practice and theory.

LP is one of the most frequently applied tools for solving real-world optimization problems. It was initiated by its large amount of applications in economics and industries. At its initial stage, one has developed techniques to deal with large-scale LP. In fact, one of indicators for research progress on LP is the capability for solving large-scale LP. Currently, commercial LP codes can solve LP with about 6000 constraints.

In the large-scale LP, an important property is the sparsity of coefficient matrix. There are certain structures that reappear frequently. For example, in the following applications, zero and nonzero coefficients in constraints are appeared in a pattern as shown in Fig. 12.4. This structure is called the *primal block angular*.

Multicommodity Flow Consider a flow network $G = (V, E)$, where each arc $(i, j) \in E$ has capacity $c(i, j)$. There are K types of commodities. Let a_{ij}^k denote the per-unit cost for moving a type k commodity from node i to node j. Let b_i^k denote the required net commodity of type k at node i. $b_i^k > 0$ means that node i is a source for type k commodity and $b_i^k < 0$ means that node i is a receiver for type

Fig. 12.4 Primal block angular

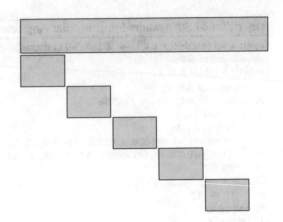

k commodity. The problem is to determine x_{ij}^k, the amount of type k commodity moving from node i to node j, to meet required net commodities at every node, and to minimize the total cost, which can be formulated as an LP problem as follows:

$$\min \sum_{(i,j)\in E} a_{ij}^1 x_j^1 + \sum_{(i,j)\in E} a_{ij}^2 x_{ij}^2 + \cdots + \sum_{(i,j)\in E} a_{ij}^K x_{ij}^K$$

$$\text{subject to } x_{ij}^1 + x_{ij}^2 + \cdots + x_{ij}^K \le c(i,j) \text{ for all } (i,j) \in E$$

$$\left(\sum_j x_{ij}^k - \sum_h x_{hi}^k \right) = b_i^k \text{ for all } i \in V \text{ and } 1 \le k \le K$$

$$x_{ij}^k \ge 0 \text{ for all } i, j, k.$$

Job Assignment There are K groups, J jobs, and I types of resources. For each resource i, availability is b_i. For each job j, if group k completes job j, then consumption of resource i is a_{ik}^i. The cost for group k working on job j is c_{jk}. The problem is to find an assignment for distributing J jobs to K groups, under availability of every resource, to minimize the total cost. Let x_{jk} be an indicator for assigning job j to group k. Then, this problem can be formulated as a 0-1 integer LP as follows:

$$\min \sum_{k=1}^K \sum_{j=1}^J c_{jk} x_{jk}$$

$$\text{subject to } \sum_{k=1}^K \sum_{j=1}^J a_{jk}^i \le b_i \text{ for } 1 \le i \le I$$

$$\sum_{k=1}^{K} x_{jk} = 1 \text{ for } 1 \leq j \leq J$$

$$x_{jk} \in \{0, 1\} \text{ for all } 1 \leq j \leq J, 1 \leq k \leq K.$$

LP-relaxation of this formulation will be an LP with primal block angular structure.

There is a well-known approach to deal with large-scale LP with primal block angular structure, which is Dantzig-Wolfe decomposition [81].

An important recent technology development is the big data, including graph data consisting of a class of social data generated from the online social network, such as Facebook, LinkedIn, and ResearchGate. The widespread use of them leads to an increasing interest in discovering important, useful, and efficient techniques for optimizations about social data, with applications across many domains, including public safety, environment management, election, and viral marketing. The large-scale graph data also has an important sparse property, and the edge sparsity also has in some special certain pattern. Let us mention one of them.

Airline is an important tool for traveling. A lot of readers may have experience to search for a cheaper air ticket from one airport to another airport. This task actually is a shortest path problem on a large-scale social network, the airline network.

Let each air-flight be represented by a directed edge connecting two airports (Fig. 12.5). The edge weight is the price to take this flight. Each airport is represented by two sets, a set of flight start points and a set of flight endpoints. If there is a possibility to transfer from a flight to another flight at an airport, then at this airport, put an edge from the endpoint of the first flight to the start point of the second flight, and also put edge weight to be the transfer cost (Fig. 12.5). At each airport, create one virtual start point with virtual edges connecting it to all start points of flights, and also create one virtual endpoint with virtual edges connecting all endpoints of flights to it. All virtual edges have zero weight (Fig. 12.5). Now, the cheapest ticket from an airport to another airport is equivalent to the shortest path

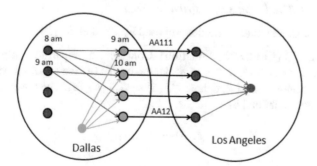

Fig. 12.5 Each flight is represented by a directed edge. Each airport contains a set of ending points and a set of starting points. Each edge between them represents a possible transfer from one air-flight to another one. Add a virtual starting point and a virtual ending point in each airport

from the virtual start point of the former airport to the virtual endpoint of the latter airport.

For each airline, its flight map is a power law graph, that is, the number of nodes with degree k is $\lceil \alpha \cdot k^{-\beta} \rceil$ where α and β are positive constants. Therefore, the fast algorithm should be developed on power law graphs for the shortest path problem. A successful solution can make a big social benefit.

For more complicated problems, such as nonsubmodular optimizations, in more complicated large-scale background, one of the successful algorithms developed recently is *optimization from samples*.

There are different models for optimization from samples. In different models, the same problems may have different computational complexity. In the following, let us show an example.

First, we consider a model proposed by Balkanski et al. [18].

Definition 12.7.1 (Optimization from Samples) Consider a family $\mathcal{F}$ of set functions over 2^L where L is the ground set. $\mathcal{M} \subseteq 2^L$ is a constraint over distribution $\mathcal{D}$ on 2^L. $\mathcal{F}$ is said to be α-optimizable from samples in $\mathcal{M}$ if there exists an algorithm satisfying that for any parameter $\delta > 0$ and sufficiently large L, there exists an integer $t_0 \in poly(|L|, 1/\delta)$ such that for all $t \geq t_0$ and for any set of samples $\{S_i, f(S_i)\}_{i=1}^t$ with $f \in \mathcal{F}$ and S_i selected i.i.d. from $\mathcal{D}$, the algorithm takes samples $\{S_i, f(S_i)\}_{i=1}^t$ as the input and returns $S \in \mathcal{M}$ satisfying that

$$\Pr_{S_1,\ldots,S_t \sim \mathcal{D}}[E[f(S)] \geq \alpha \cdot \max_{T \in \mathcal{M}} f(T)] \geq 1 - \delta,$$

where the expectation is taken over the randomness of the algorithm. (Note that the algorithm runs not necessarily in polynomial-time.)

With this model, a negative result is obtained as follows:

Theorem 12.7.2 *The maximum set coverage problem (Problem 10.1.1) cannot be approximated within a ratio better than $2^{\Omega(\sqrt{\log |N|})}$ using polynomially many samples selected i.i.d. from any distribution $\mathcal{D}$.*

Next, we consider another model proposed by Chen et al. [54].

Definition 12.7.3 (Coverage Function) Consider a bipartite graph $G = (L, R, E)$. For every node $u \in L \cup R$, denote by $N_G(u)$ the set of all neighbors of u. For any subset $S \subseteq L \cup R$, denote $N_G(S) = \cup_{u \in S} N_G(u)$. The coverage function $f_G : 2^L \to R_+$ is defined by

$$f_g(S) = |N_G(S)|$$

for $S \subseteq L$.

Definition 12.7.4 (Optimization from Structured Samples) Consider a family $\mathcal{F}$ of coverage functions on bipartite graphs $G = (L, R, E)$. $\mathcal{F}$ is said to be

α-optimizable in constraint $\mathcal{M} \subseteq 2^L$ over distribution $\mathcal{D}$ on 2^L, if there exists an algorithm satisfying that for any $\delta > 0$ and sufficiently large L, there exists $t_0 \in poly(|L|, |R|, 1/\delta)$ such that for all $t \geq t_0$ and for any sample set $\{S_i, N_G(S_i)\}_{t=1}^t$ with $f_G \in \mathcal{F}$ and S_i selected i.i.d. from $\mathcal{D}$, the algorithm takes samples $\{S_i, N_G(S_i)\}_{t=1}^t$ as the input and return $S \in \mathcal{M}$ satisfying that

$$\Pr_{S_1,\ldots,S_t \sim \mathcal{D}}[E[f_G(S)] \geq \alpha \cdot \max_{T \in \mathcal{M}} f_G(T)] \geq 1 - \delta,$$

where the expectation is taken over the randomness of the algorithm. (Note that the algorithm runs not necessarily in polynomial-time.)

With this model, a positive result is obtained as follows:

Theorem 12.7.5 *Suppose that the distribution $\mathcal{D}$ on 2^L satisfies the following three assumptions:*

(a1) Feasibility *For any sample $S \sim \mathcal{D}$, $|S| \leq k$.*
(a2) Polynomial bounded *For any $u \in L$,*

$$p_u = \Pr_{S \sim \mathcal{D}}[u \in S] \geq 1/|L|^c$$

for some constant c.
(a3) Negative correlation *Over distribution $\mathcal{D}$, the random variables $X_u = \mathbf{1}_{u \in S}$ are "negatively correlated."*

Let A be an α-approximation algorithm for the maximum set coverage problem. Then, under the model of optimization from structured samples, coverage functions are $\frac{\alpha}{2}$-optimizable in the cardinality constraint $\mathcal{M} = \{S \subseteq L \mid |S| \leq k\}$ over $\mathcal{D}$ for $k \leq |L|$. Moreover, the algorithm in the model will use a polynomial number of arithmetic operations and one call of algorithm A.

Furthermore, Chen et al. [55] extended their work to network inference and social influence maximization. In this research direction, there are a lot of unexplored topics which need one's efforts. Hence, it seems quite attractive.

Exercises

1. Consider a set function $f : 2^X \to R$. f is *strictly increasing* if for any two subsets $A \subset B$, $f(A) < f(B)$. f is *strictly submodular* if for any two subsets $A \subset B$ and any element $x \in X \setminus B$, $\Delta_x f(A) > \Delta_x f(B)$. Show that every set function can be decomposed into a difference of two strictly increasing and strictly submodular functions.
2. A set function f on 2^X is *strictly supermodular* if $-f$ is strictly submodular. Show that every set function can be decomposed into a difference of two strictly increasing and strictly supermodular functions.

3. Give a counterexample to show that not every set function can be decomposed into the sum of a monotone nondecreasing submodular function and a monotone nondecreasing supermodular function.

4. Show that for any monotone nondecreasing submodular function $f : 2^X \to R$ and any $Y \subseteq X$, there exist a pair of monotone nondecreasing modular functions $u, l : 2^X \to R$ such that $u(Y) = f(Y) = l(Y)$ and $u(S) \geq f(S) \geq l(S)$ for any $S \subseteq X$.

5. Let $C = \{A \mid |A| \leq k\}$. Then, (X, C) is a matroid. This means that the size constraint is a specific matroid constraint. With this constraint, the monotone nonsubmodular maximization has a better approximation solution. Consider the following maximization problem and a greedy algorithm as shown in Algorithm 45:

$$\max \ f(A)$$

$$\text{subject to } |A| \leq k,$$

where $f : 2^X \to R$ is nonnegative and monotone nondecreasing.

Prove that this algorithm produces a better approximation, that is, Greedy Algorithm 45 produces a $(1 - e^{-1/(\mathcal{D}^+ + 1)})$-approximation solution for maximization of monotone nondecreasing nonnegative set function with a size constraint.

6. (Activity Profit Maximization [401]) Consider a social network $G = (V, E)$ with independent cascade information diffusion model (see the definition in Exercise 19 in Chapter 10). For a seed set S, the influence process will end at a set $I(S)$ of active nodes. Suppose that two active nodes u and v will join an activity, which produces a profit $A(u, v)$. Thus, the activity profit of $I(S)$ is

---

Algorithm 45 Greedy Approximation

Input: a nonnegative monotone set function f on 2^X and a positive integer k.
Output: A subset S of X.

1: $S \leftarrow \emptyset$;
2: **for** $d = 1$ to $\mathcal{D}^+$, $v \in X$ and C with $|C| = k \bmod (d + 1)$ **do**
3:    $i \leftarrow 0$;
4:    **while** $|S_i| < k$ **do**
5:      $i \leftarrow i + 1$;
6:      choose $u_i \in X \setminus S_{i-1}$ and $D_i \subseteq \mathcal{D}^+(u_i)$ to maximize
7:      $f(D_i \cup \{u_i\} \cup S_{i-1}) - f(S_{i-1})$ subject to
8:      $|D_I \cup \{u_i\} \cup S_{i-1}| \leq k$ and $|D_i| \leq d$
9:      set $S_i \leftarrow D_i \cup \{u_i\} \cup S_{i-1}$;
10:    **end while**
11:    $S \leftarrow \text{argmax}(f(S), f(S_i))$;
12: **end for**
13: **return** S.

$$\sum_{u,v \in I(S)} A(u, v).$$

We are interested in the following problem:

$$\max \ \text{profit}(S) = E[\sum_{u,v \in I(S)} A(u, v)]$$

$$\text{subject to } |S| \leq k$$

where k is a given positive integer. Please prove that profit(S) is neither submodular nor supermodular. Furthermore, give a solution by using the sandwich method.

7. (Positive Influence Maximization) Consider a directed graph $G = (V, E)$. The positive influence process is defined as follows: Every node has two states, active and inactive. Before process, every node is inactive. Initially, select at most k nodes (called seeds) and activate them. In each subsequent step, every inactive node v checks whether the number of active incoming neighbors is not less than the number of inactive incoming neighbors. If yes, then v becomes active. The process will end if no new inactive node becomes active. The influence spread is the number of active nodes at the end of process. Prove that the influence spread is neither submodular or supermodular as a set function with respect to seed set.

8. Algorithm 46 is the modular-modular algorithm for $\min_{A \in 2^X} f(A)$. Show that it ends at a local optimal solution satisfying Condition 1.

9. Based on the sandwich theorem, we can design Algorithm 47 for $\min_{A \in 2^X} f(A)$. Show that this algorithm always ends at a local minimum solution satisfying Condition 1.

Algorithm 46 Modular-Modular Algorithm

Input: a set function $f : 2^X \rightarrow R$ and its DS decomposition $f = g - h$ where g and h are submodular functions.
Output: A subset A of X.
1: choose a set $A \subseteq X$;
2: **while** $f(A^+) < f(A)$ **do**
3:    $A \leftarrow A^+$;
4:    **for** $\sigma \in \Sigma_A$ **do**
5:       compute a lower-bound modular function m_{gl}^σ for g;
6:       compute a lower-bound modular function m_{hl}^σ for h;
7:       compute a minimum solution A_σ^+ for $\min_{Y \in 2^X} [m_{gl}^\sigma(Y) - m_{hl}^\sigma(Y)]$;
8:    **end for**
9:    $\sigma \leftarrow \text{argmin}_{\sigma \in \Sigma_A} f(A_\sigma^+)$;
10:   $A^+ \leftarrow A_\sigma^+$
11: **end while**
12: **return** A.

Algorithm 47 Iterated sandwich method

Input: a set function $f : 2^X \to R$ and its DS decomposition $f = g - h$ where g and h are submodular functions.
Output: A subset A of X.

1: choose a set $A \subseteq X$;
2: **while** $f(A^+) < f(A)$ **do**
3:      $A \leftarrow A^+$;
4:      **for** $\sigma \in \Sigma_A$ **do**
5:          compute upper and lower-bound modular functions m_{gu} and m_{gl}^σ for g, respectively;
6:          compute upper and lower-bound modular functions m_{hu} and m_{hl}^σ for h, respectively;
7:          compute a minimum solution A_u^+ for $\min_{Y \in 2^X}[m_{gu}(Y) - m_{hl}^\sigma(Y)]$;
8:          compute a minimum solution A_l^+ for $\min_{Y \in 2^X}[m_{gl}^\sigma(Y) - m_{hu}(Y)]$;
9:          compute a minimum solution A_0^+ for $\min_{Y \in 2^X}[m_{gl}^\sigma(Y) - m_{hl}^\sigma(Y)]$;
10:          $A_\sigma^+ \leftarrow \min(f(A_u^+), f(A_l^+), f(A_0^+))$;
11:      **end for**
12:      $\sigma \leftarrow \text{argmin}_{\sigma \in \Sigma_A} f(A_\sigma^+)$;
13:      $A^+ \leftarrow A_\sigma^+$
14: **end while**
15: **return** A.

---

10. Please modify Algorithm 43 into one which runs in polynomial-time and a G-L performance ratio not far from $(1 - e^{-1})$.

Historical Notes

As online social networks (OSN) grow rapidly, the influence-driven information technology and influence-based research subjects have been studied extensively in the literature. One of the subjects is the negative influence, rumor. There already exist many research publications on rumor blocking in the literature [57, 130–133, 196, 197, 199, 381, 382, 392, 426, 429, 455]. They are employing various methods. For example, Fan et al. [130, 133] consider community structure of social networks and try to limit the spread of rumor within a community. Tong et al. [381] and Chen et al. [57] formulate the rumor blocking into a noncooperative game model. Above works are built on an assumption that one can get immune from rumor by receiving a positive influence which is sent by protectors. Hence, the placement of protectors is the main task, which becomes an attractive set function optimization problem.

Other methods do not depend on such an assumption. Instead, they may assume that there are monitors which can screen out the rumor from information flow, that is, they can cut the spread of rumor at some nodes or edges [154]. Therefore, the placement of those monitors is the main task, which becomes an interesting set function optimization problem.

For set function optimization, there are many beautiful results on submodular optimizations indicated in Chaps. 10 and 11. However, in recent development of computer technology, many nonsubmodular optimization problems are raised, such as group influence [457], community detection [444], content spread maximization [19], target activation maximization [431, 443], text-mining [430], sentimental analysis [91–93], cloud computing [124], machine learning [414], and viral marketing in social networks [198–200, 327, 342, 456, 460], and misinformation blocking [378]. Therefore, the study of nonsubmodular optimization becomes a hot research direction.

There are four classes of approaches to deal with nonsubmodular optimization problems. The first class consists of efforts along the traditional line. Since the nonsubmodular optimization is, in nature, hard to deal with, one cannot find an efficient algorithm with satisfied guaranteed performance. In this class, the algorithm is often analyzed with some artificial parameter, such as the supermodular degree [136–138] or curvature [15, 108, 225, 389], with which some beautiful results on the approximation performance ratio may be established. However, those parameters are usually hard to be estimated, and in many specific problems, they do not have a significant value.

Due to above, one may give up the approximation performance ratio and establish other standards to evaluate the performance of algorithms. The second class of algorithms are designed based on this point. It consists of data-dependent approximation algorithms, which are evaluated by a new type of performance ratio. Those algorithms are also called sandwich methods [53, 299, 401, 456]. The performance ratio is data-dependent, which does not give clear information about algorithm performance.

The third class consists of algorithms ending at local optimal solutions. They are usually based on the DS decomposition of the set functions [224, 309, 324, 415]. One of the disadvantages of those algorithms is that they are unlikely running in polynomial-time.

The fourth class consists of algorithms which produce approximation solution within a factor of certain bound from a local optimal solution [156–158]. This is a new research direction which has a good potential.

Bibliography

1. P.K. Agarwal, M. van Kreveld, S. Suri: Label placement by maximum independent set in rectangles, *Comput. Geom. Theory Appl.*, 11(118): 209–218 (1998).
2. A.A. Ageev and M. Svirdenko: Pipage rounding: a new method of constructing algorithms with proven performance guarantee, *Journal of Combinatorial Optimization*, 8: 307–328 (2004).
3. C. Ambühl: An optimal bound for the MST algorithm to compute energy efficient broadcast trees in wireless networks, *Proceedings, 32nd International Colloquium on Automata, Languages and Programming*, Springer LNCS 3580: 1139–1150 (2005).
4. C. Ambühl, T. Erlebach, M. Mihalák and M. Nunkesser: Constant-approximation for minimum-weight (connected) dominating sets in unit disk graphs, *Proceedings, 9th International Workshop on Approximation Algorithms for Combinatorial Optimization (APPROX 2006)*, Springer LNCS 4110: 3–14 (2006).
5. E.M. Arkin, J.S.B. Mitchell and G. Narasimhan: Resource-constructed geometric network optimization, *Proceedings, 14th Annual Symposium on Computational Geometry*, Minneapolis, pp.307–316, 1998.
6. S. Arora: Polynomial-time approximation schemes for Euclidean TSP and other geometric problems, *Proceedings, 37th IEEE Symp. on Foundations of Computer Science*, pp. 2–12, 1996.
7. S. Arora: Nearly linear time approximation schemes for Euclidean TSP and other geometric problems, *Proceedings, 38th IEEE Symp. on Foundations of Computer Science*, pp. 554–563, 1997.
8. S. Arora: Polynomial-time approximation schemes for Euclidean TSP and other geometric problems, *Journal of ACM*, 45: 753–782 (1998).
9. S. Arora, M. Grigni, D. Karger, P. Klein and A. Woloszyn: Polynomial time approximation scheme for Weighted Planar Graph TSP, *Proceedings, 9th ACM-SIAM Symposium on Discrete Algorithms*, pp. 33–41, 1998.
10. S. Arora, C. Lund, R. Motwani, M. Sudan and M. Szegedy: Proof verification and hardness of approximation problems, *Proceedings, 33rd IEEE Symposium on Foundations of Computer Science*, pp. 14–23, 1992.
11. S. Arora, C. Lund, R. Motwani, M. Sudan and M. Szegedy: Proof verification and hardness of approximation problems, *Journal of the ACM*, 45: 753–782 (1998).
12. S. Arora, P. Raghavan and S. Rao: Polynomial Time Approximation Schemes for Euclidean k-medians and related problems, *Proceedings, 30th ACM Symposium on Theory of Computing*, pp. 106–113, 1998.

© The Author(s), under exclusive license to Springer Nature Switzerland AG 2022
D.-Z. Du et al., *Introduction to Combinatorial Optimization*, Springer Optimization and Its Applications 196, https://doi.org/10.1007/978-3-031-10596-8

13. S. Arora and S. Safra: Probabilistic checking of proofs: A new characterization of NP, *Proceedings, 33rd IEEE Symposium on Foundations of Computer Science*, pp. 2–13, 1992.

14. S. Arora and S. Safra: Probabilistic checking of proofs: A new characterization of NP, *J. Assoc. Comput. Mach.*, 45: 70–122 (1998).

15. Wenruo Bai, Jeffrey A. Bilmes: Greed is still good: maximizing monotone submodular+supermodular (BP) functions, *Proceedings, ICML*, 314–323, 2018.

16. B.S. Baker: Approximation algorithms for NP-complete problems on planar graphs, *Proceedings, 24th FOCS*, pp. 265–273, 1983.

17. B.S. Baker: Approximation algorithms for NP-complete problems on planar graphs, *Journal of ACM*, 41(1): 153–180 (1994).

18. E. Balkanski, A. Rubinstein and Y. Singer: The limitations of optimization from samples, *Proceedings, 49th Annual ACM SIGACT Symposium on Theory of Computing (STOC)*, Montreal, QC, Canada, June 19–23, pp. 1016–1027, 2017.

19. Anton Barhan and Andrey Shakhomirov: Methods for sentiment analysis of Twitter Messages, *Proceedings, 12th Conference of Fruct Association*, pp. 216–222, 2012.

20. J. Bar-LLan, G. Kortsarz and D. Prleg: Generalized submodular cover problem and applications, *Theoretical Computer Science*, 250: 179–200 (2001).

21. D. Bayer and J.C. Lagarias: The non-linear geometry of linear programming, I. Affine and projective scaling trajectories, II. Legendre transform coordinates, III. Central trajectories, *Preprints*, AT&T Bell Laboratories (Murray Hill, NJ, 1986).

22. E.M. Beale: Cycling in dual simplex algorithm, *Navel Research Logistics Quarterly* 2: 269–276 (1955).

23. M. Bellare, O. Goldreich and M. Sudan: Free bits and nonapproximability, *Proceedings, 36th FOCS*, pp.422–431, 1995.

24. R. Bellman: On a routing problem, *Quarterly of Applied Mathematics*, 16: 87–90 (1958).

25. P. Berman, B. Basgupta, S. Muthukrishnan, S. Ramaswami: Efficient approximation algorithms for tiling and packing problem with rectangles, *J. Algorithms*, 41: 178–189 (2001).

26. P. Berman, G. Calinescu, C. Shah, A. Zelikovsky: Efficient energy management in sensor networks, in *Ad Hoc and Sensor Networks, Wireless Networks and Mobile Computing*, vol. 2, ed. by Y. Xiao, Y. Pan (Nova Science Publishers, Hauppauge, 2005).

27. D.P. Bertsekas: A simple and fast label correcting algorithm for shortest paths, *Networks*, 23(8): 703–709 (1993).

28. Aditya Bhaskara, Moses Charikar, Eden Chlamtac, Uriel Feige: Aravindan vijayaraghavan: detecting high log-densities – an $O(n^{1/4})$ approximation for densest k-subgraph, *Proceedings, 42nd ACM International Symposium on Theory of Computing*, ACM, New York, pp. 201–210, 2010.

29. Arim Blum, Tao Jiang, Ming Li, John Tromp and M. Yannakakis: Linear approximation of shortest superstrings, *Journal of ACM*, 41(4): 630–647 (1994).

30. Otakar Boruvka on Minimum Spanning Tree Problem (translation of both 1926 papers, comments, history) (2000) Jaroslav Nesetril, Eva Milková, Helena Nesetrilová. (Section 7 gives his algorithm, which looks like a cross between Prim's and Kruskal's.)

31. Dimitris Bertsimas, Chung-Piaw Teo, Rakesh Vohra: On dependent randomized rounding algorithms, *Oper. Res. Lett.* 24(3): 105–114 (1999).

32. Robert G. Bland: New finite pivoting rules for the simplex method, *Mathematics of Operations Research* 2 (2): 103–107 (1977).

33. Al Borchers and Ding-Zhu Du: The k-Steiner ratio in graphs, *Proceedings, 27th ACM Symposium on Theory of Computing*, pp. 641–649, 1995.

34. Al Borchers and Ding-Zhu Du: The k-Steiner ratio in graphs, *SIAM J. Comput.*, 26(3): 857–869 (1997).

35. Al Borchers and Prosenjit Gupta: Extending the Quadrangle Inequality to Speed-Up Dynamic Programming. *Inf. Process. Lett.*, 49(6): 287–290 (1994).

36. O. Boruvka: On a minimal problem, *Prace Morask'e Pridovedeké Spolecnosti*, 3: 37–58 (1926).

37. Thomas S. Brylawski: A decomposition for combinatorial geometries, *Transactions of the American Mathematical Society*, 171: 235–282 (1972).

38. Niv Buchbinder, Moran Feldmany, Joseph (Seffi) Naorz, Roy Schwartz: Submodular Maximization with cardinality constraints, *Proceedings, 25th annual ACM-SIAM symposium on Discrete algorithms*, pp. 1433–1452, 2014.

39. R.G. Busacker, P.G. Gowen: A procedure for determining a family of minimum cost network flow patterns, *Operations Research Office Technical Report* 15, John Hopkins University, Baltimore; 1961.

40. J. Byrka, F. Grandoni, T. Rothvoss, L. Sanita: An improved LP-based approximation for Steiner tree, *Proceedings, 42nd ACM Symposium on Theory of Computing*, pp. 583–592, June 5–8, 2010.

41. G. Calinescu, C. Chekuri, M. Pál and J. Vondrák: Maximizing a submodular set function subject to a matroid constraint, *SIAM J. Comp.*, 40(6): 1740–1766 (2011).

42. Gruia Calinescu, Chandra Chekuri, Martin Pai, Jan Vondrak: Maximizing a submodular set function subject to a matroid constraint, *Proceedings, IPCO*, pp. 182–196, 2007.

43. Mihaela Cardei, My T. Thai, Yingshu Li, Weili Wu: Energy-efficient target coverage in wireless sensor networks, *Proceedings, INFOCOM*, pp. 1976–1984, 2005.

44. T.M. Chan: Polynomial-time approximation schemes for picking and piercing fat objects, *J. Algorithms*, 46: 178–189 (2003).

45. T.M. Chan: A note on maximum independent sets in rectangle intersection graphs, *Information Processing Letters*, 89: 19–23 (2004).

46. M. Charikar, C. Chekuri, T. Cheung, Z. Dai, A. Goel, S. Guha and M. Li: Approximation algorithms for directed Steiner problems, *Journal of Algorithms*, 33: 73–91 (1999).

47. A. Charnes: Optimality and degeneracy in linear programming, *Econometrica*, 2: 160–170 (1952).

48. Mmanu Chaturvedi, Ross M. McConnell: A note on finding minimum mean cycle. *Inf. Process. Lett.*, 127: 21–22 (2017).

49. Bernard Chazelle: A minimum spanning tree algorithm with inverse-Ackermann type complexity, *Journal of the Association for Computing Machinery*, 47 (6): 1028–1047 (2000).

50. Bernard Chazelle: The soft heap: an approximate priority queue with optimal error rate, *Journal of the Association for Computing Machinery*, 47 (6): 1012–1027 (2000).

51. Jing-Chao Chen: Iterative rounding for the closest string problem, *CoRR abs/0705.0561: (2007)*.

52. Tiantian Chen, Bin Liu, Wenjing Liu, Qizhi Fang, Jing Yuan, Weili Wu: A random algorithm for profit maximization in online social networks, *Theor. Comput. Sci.*, 803: 36–47 (2020)

53. Wei Chen, Tian Lin, Zihan Tan, Mingfei Zhao, Xuren Zhou: Robus influence maximization, *Proceedings, KDD*, San Francisco, CA, USA, pp. 795–804, 2016,

54. Wei Chen, Xiaoming Sun, Jialin Zhang, Zhijie Zhang: Optimization from structured samples for coverage functions, *Proceedings, ICML*, pp. 1715–1724, 2020.

55. Wei Chen, Xiaoming Sun, Jialin Zhang, Zhijie Zhang: Network Inference and Influence Maximization from Samples, *Proceedings, ICML*, pp. 1707–1716, 2021.

56. W.T. Chen and N.F. Huang: The Strongly connection problem on multihop packet radio networks, *IEEE Transactions on Communications*, 37(3): 293–295 (1989).

57. Xin Chen, Qingqin Nong, Yan Feng, Yongchang Cao, Suning Gong, Qizhi Fang, Ker-I Ko: Centralized and decentralized rumor blocking problems, *J. Comb. Optim.*, 34(1): 314–329 (2017).

58. Maggie Xiaoyan Cheng, Lu Ruan, Weili Wu: Achieving minimum coverage breach under bandwidth constraints in wireless sensor networks, *Proceedings, INFOCOM*, pp. 2638–2645, 2005.

59. Maggie Xiaoyan Cheng, Lu Ruan, Weili Wu: Coverage breach problems in bandwidth-constrained sensor networks, *ACM Trans. Sens. Networks*, 3(2): 12 (2007)

60. X. Cheng, B. DasGupta and B. Lu: A polynomial time approximation scheme for the symmetric rectilinear Steiner arborescence problem, *Journal of Global Optimization*, 21(4): 385–396 (2001).

61. Xiuzhen Cheng, Xiao Huang, Deying Li, Weili Wu, Ding-Zhu Du: A polynomial-time approximation scheme for the minimum-connected dominating set in ad hoc wireless networks, *Networks*, 42(4): 202–208 (2003).
62. X. Cheng, J.-M. Kim, and B. Lu: A polynomial time approximation scheme for the problem of interconnecting highways, *Journal of Combinatorial Optimization*, 5: 327–343, (2001).
63. D. Cheriton and R.E. Tarjan: Finding minimum spanning trees, *SIAM J. Comput.*, 5: 724–742 (1976).
64. J. Cheriy, S. Vempala: A. Vetta: Network design via iterative rounding of setpair relaxations, *Combinatorica*, 26(3): 255–275 (2006).
65. G. Choquet: Etude de certains réseaux de routes, *C R Acad Sci Paris*, 205: 310–313 (1938).
66. N. Christofides: Worst-case analysis of a new heuristic for the travelling salesman problem, *Technical Report*, Graduate School of Industrial Administration, Carnegie-Mellon University, Pittsburgh, PA, 1976.
67. F.R.K. Chung and E.N. Gilbert: Steiner trees for the regular simplex, *Bull. Inst. Math. Acad. Sinica*, 4: 313–325 (1976).
68. F.R.K. Chung and R.L. Graham: A new bound for euclidean Steiner minimum trees, *Ann. N.Y. Acad. Sci.*, 440: 328–346 (1985).
69. F.R.K. Chung and F.K. Hwang: A lower bound for the Steiner tree problem, *SIAM J.Appl.Math.*, 34: 27–36 (1978).
70. V. Chvátal: A greedy heuristic for the set-covering problem, *Mathematics of Operations Research*, 4(3): 233–235 (1979).
71. S.A. Cook: The complexity of theorem-proving procedures, *Proceedings, 3rd ACM Symposium on Theory of Computing*, pp. 151–158, 1971.
72. William J. Cook, William H. Cunningham, William R. Pulleyblank, Alexander Schrijver: *Combinatorial Optimization*, (Wiley, 1997).
73. Thomas H. Cormen, Charles E. Leiserson, Ronald L. Rivest, Clifford Stein: *Introduction to Algorithms, (3rd ed.)*, (MIT Press, 2009).
74. Henry H. Crapo, Gian-Cario Rota: *On the Foundations of Combinatorial Theory: Combinatorial Geometries*, (Cambridge, Mass.: M.I.T. Press, 1970).
75. R. Crourant and H. Robbins, *What Is Mathematics?*, (Oxford Univ. Press, New York, 1941).
76. W. H. Cunningham: Decomposition of submodular functions, *Combinatorica*, 3(1): 53–68 (1983).
77. D. Dai and C. Yu: A $(5 + \varepsilon)$-approximation algorithm for minimum weighted dominating set in unit disk graph, *Theoretical Computer Science*, 410: 756–765 (2009).
78. G.B. Dantzig: Application of the simplex method to a transportation problem, in: *Activity Analysis of Production and Allocation*, (Cowles Commission Monograph 13), T.C. Koopmans (ed.), John-Wiley, New York, pp. 359–373, 1951.
79. G.B. Dantzig: Maximization of a linear function of variables subject to linear inequalities, Chap. XXI of *Activity Analysis of Production and Allocation*, (Cowles Commission Monograph 13), T.C. Koopmans (ed.), John-Wiley, New York, 1951, pp. 339–347.
80. G.B. Dantzig: A. Orden, P. Wolfe: Note on linear programming, *Pacific J. Math*, 5: 183–195 (1955).
81. G.B. Dantzig and P. Wolfe: Decomposition principle for linear programs, *Operations Research*, 8: 101–111 (1960).
82. Robert B. Dial: Algorithm 360: Shortest-Path Forest with Topological Ordering [H], *Communications of the ACM*, 12 (11): 632–633 (1969).
83. E.W. Dijkstra: A note on two problems in connexion with graphs, *Numerische Mathematik*, 1: 269–271 (1959).
84. Ling Ding, Xiaofeng Gao, Weili Wu, Wonjun Lee, Xu Zhu, Ding-Zhu Du: Distributed construction of connected dominating sets with minimum routing cost in wireless networks, *Proceedings, ICDCS*, pp. 448–457, 2010.
85. Ling Ding, Weili Wu, James Willson, Lidong Wu, Zaixin Lu, Wonjun Lee: Constant-approximation for target coverage problem in wireless sensor networks, *Proceedings, INFOCOM*, pp. 1584–1592, 2012.

86. Xingjian Ding, Jianxiong Guo, Deying Li, Weili Wu: Optimal wireless charger placement with individual energy requirement, *Theor. Comput. Sci.*, 857: 16–28 (2021).
87. Xingjian Ding, Jianxiong Guo, Yongcai Wang, Deying Li, Weili Wu: Task-driven charger placement and power allocation for wireless sensor networks, *Ad Hoc Networks*, 119: 102556 (2021).
88. E.A. Dinic: Algorithm for solution of a problem of maximum flow in a network with power estimation, *Soviet Mathematics - Doklady*, 11: 1277–1280 (1970).
89. Yefim Dinitz: Dinitz' algorithm: the original version and Even's version, in Oded Goldreich, Arnold L. Rosenberg, Alan L. Selman (eds.), *Theoretical Computer Science: Essays in Memory of Shimon Even.* (Springer, 2006): pp. 218–240, 2006.
90. I. Dinur, D. Steurer: Analytical approach to parallel repetition, *Proceedings, 46th Annual ACM Symposium on Theory of Computing*, pp. 624–633, 2014.
91. Luobing Dong, Qiumin Guo, Weili Wu: Speech corpora subset selection based on time-continuous utterances features, *J. Comb. Optim.*, 37(4): 1237–1248 (2019).
92. Luobing Dong, Qiumin Guo, Weili Wu, Meghana N. Satpute: A semantic relatedness preserved subset extraction method for language corpora based on pseudo-Boolean optimization, *Theor. Comput. Sci.*, 836: 65–75 (2020).
93. Luobing Dong, Meghana N. Satpute, Weili Wu, Ding-Zhu Du: Two-phase multidocument summarization through content-attention-based subtopic detection, *IEEE Trans. Comput. Soc. Syst.*, 8(6): 1379–1392 (2021).
94. D.E. Drake and S. Hougardy, On approximation algorithms for the terminal Steiner tree problem, *Information Processing Letters*, 89: 15–18 (2004).
95. Stuart Dreyfus: Richard Bellman on the birth of dynamic programming, *Operations Research*, 50(1): 48–51 (2002).
96. Ding-Zhu Du: On heuristics for minimum length rectangular partitions, *Technical Report*, Math. Sci. Res. Inst., Univ. California, Berkeley, 1986.
97. Ding-Zhu Du, R.L. Graham, P.M. Pardalos, Peng-Jun Wan, Weili Wu and W. Zhao: Analysis of greedy approximations with nonsubmodular potential functions, *Proceedings, 19th ACM-SIAM Symposiun on Discrete Algorithms (SODA)*, pp. 167–175, 2008.
98. Ding-Zhu Du, D. Frank Hsu, and K.-J. Xu: Bounds on guillotine ratio, *Congressus Numerantium*, 58: 313–318 (1987).
99. Ding-Zhu Du and Ker-I Ko: *Theory of Computational Complexity (2nd Ed)*, (John Wiley, New York, NY, 2014).
100. Ding-Zhu Du, Ker-I Ko, Xiaodong Hu: *Design and Analysis of Approximation Algorithms*, (Springer, 2012).
101. Ding-Zhu Du and Frank K. Hwang: The Steiner ratio conjecture of Gilbert-Pollak is true, *Proceedings of National Academy of Sciences*, 87: 9464–9466 (1990).
102. Ding-Zhu Du, Frank K. Hwang, M.T. Shing and T. Witbold: Optimal routing trees, *IEEE Transactions on Circuits*, 35: 1335–1337 (1988).
103. Ding-Zhu Du, Zevi Miller: Matroids and subset interconnection design, *SIAM J. Discrete Math.*, 1(4): 416–424 (1988).
104. Ding-Zhu Du, L.Q. Pan, and M.-T. Shing: Minimum edge length guillotine rectangular partition, *Technical Report 0241886*, Math. Sci. Res. Inst., Univ. California, Berkeley, 1986.
105. Ding-Zhu Du, Panos M. Pardalos, Weili Wu: *Mathematical Theory of Optimization*, (Springer, 2010).
106. Ding-Zhu Du, Yan-Jun Zhang: On heuristics for minimum length rectilinear partitions, *Algorithmica*, 5: 111–128 (1990).
107. Ding-Zhu Du, Yanjun Zhang and Qing Feng: On better heuristic for euclidean Steiner minimum trees, *Proceedings, 32nd FOCS*, pp. 431–439, 1991.
108. Hongjie Du, Weili Wu, Wonjun Lee, Qinghai Liu, Zhao Zhang, Ding-Zhu Du: On minimum submodular cover with submodular cost, *J. Global Optimization*, 50(2): 229–234 (2011).
109. Hongjie Du, Weili Wu, Shan Shan, Donghyun Kim, Wonjun Lee: Constructing weakly connected dominating set for secure clustering in distributed sensor network, *J. Comb. Optim.*, 23(2): 301–307 (2012).

110. Hongwei Du, Panos M. Pardalos, Weili Wu, Lidong Wu: Maximum lifetime connected coverage with two active-phase sensors, *J. Glob. Optim.*, 56(2): 559–568 (2013).

111. Hongwei Du, Weili Wu, Qiang Ye, Deying Li, Wonjun Lee, Xuepeng Xu: CDS-based virtual backbone construction with guaranteed routing cost in wireless sensor networks, *IEEE Trans. Parallel Distributed Syst.*, 24(4): 652–661 (2013).

112. Hongwei Du, Qiang Ye, Weili Wu, Wonjun Lee, Deying Li, Ding-Zhu Du, Stephen Howard: Constant approximation for virtual backbone construction with guaranteed routing cost in wireless sensor networks, *Proceedings, INFOCOM*, pp. 1737–1744, 2011.

113. Hongwei Du, Qiang Ye, Jiaofei Zhong, Yuexuan Wang, Wonjun Lee, Haesun Park: Polynomial-time approximation scheme for minimum connected dominating set under routing cost constraint in wireless sensor networks, *Theor. Comput. Sci.*, 447: 38–43 (2012).

114. Xiufeng Du, Weili Wu, Dean F. Kelley: Approximations for subset interconnection designs, *Theoretical Computer Science*, 207(1): 171–180 (1998).

115. Yingfan L. Du, Hongmin W. Du: A new bound on maximum independent set and minimum connected dominating set in unit disk graphs, *J. Comb. Optim.*, 30(4): 1173–1179 (2015).

116. J. Edmonds: Maximum matching and a polyhedron with 0, 1-vertices, *Journal of Research National Bureau of Section B*, 69: 125–130 (1965).

117. J. Edmonds: Minimum partition of a matroid into independent subsets, *Journal of Research National Bureau of Section B*, 69: 67–72 (1965).

118. J. Edmonds: Paths, trees and flowers, *Canadian Journal of Mathematics*, 17: 449–467 (1965).

119. J. Edmonds: Optimum branchings, *Journal of Research National Bureau of Section B*, 71: 233–240 (1967).

120. J. Edmonds: Submodular functions, matroids, and certain polyhedrons, in: *Combinatorial Structure and Their Applications* (R. Guy, H. Hanani, N. Sauer, J. Schönheim, eds.) Gordon and Breach, New York, pp. 69–87, 1970.

121. J. Edmonds: Edge-disjoint branchings, in *Combinatorial Algorithms* (R. Rustin, ed.) Algorithmics Press, New York, pp. 91–96, 1973.

122. J. Edmonds, E.L. Johnson: Matching, Euler Tours, and the Chinese Postman, *Math. Programm.*, 5 : 88–124 (1973).

123. J. Edmonds, R. Karp: Theoretical improvements in algorithmic efficiency for network flow problems, *Journal of the ACM*, 19(2): 248–264 (1972).

124. J. Edmonds, K. Pruhs: Scalably scheduling processes with arbitrary speedup curves, *ACM Trans. Algorithms*, 8(3): 28 (2012).

125. M.A. Engquist: A successive shortest path algorithm for the assignment problem, *Research Report*, Center for Cybernetic Studies (CCS) 375, University of Texas, Austin; 1980.

126. T. Erlebach, T. Grant, F. Kammer: Maximising lifetime for fault tolerant target coverage in sensor networks, *Sustain. Comput. Inform. Syst.*, 1: 213–225 (2011).

127. T. Erlebach, K. Jansen and E. Seidel: Polynomial-time approximation schemes for geometric graphs, *Proceedings, 12th SODA*, pp. 671–679, 2001.

128. T. Erlebach, M. Mihal: A $(4 + \varepsilon)$-approximation for the minimum-weight dominating set problem in unit disk graphs, *Proceedings, WAOA*, pp. 135-1, 2009.

129. Thomas R. Ervolina, S. Thomas McCormick: Two strongly polynomial cut cancelling algorithms for minimum cost network flow, *Discrete Applied Mathematics*, 4: 133–165 (1993).

130. Lidan Fan, Weili Wu: Rumor blocking, *Encyclopedia of Algorithms*, pp. 1887–1892, 2016.

131. Lidan Fan, Weili Wu, Kai Xing, Wonjun Lee: Precautionary rumor containment via trustworthy people in social networks, *Discrete Math., Alg. and Appl.*, 8(1): 1650004:1-1650004:18 (2016).

132. Lidan Fan, Weili Wu, Xuming Zhai, Kai Xing, Wonjun Lee, Ding-Zhu Du: Maximizing rumor containment in social networks with constrained time, *Social Netw. Analys. Mining*, 4(1): 214 (2014).

133. Lidan Fan, Zaixin Lu, Weili Wu, Bhavani M. Thuraisingham, Huan Ma, Yuanjun Bi: Least Cost Rumor Blocking in Social Networks, *Proceedings, ICDCS*, pp. 540–549, 2013.

134. Uriel Feige: A threshold of ln n for approximating set cover, *J. ACM*, 45(4): 634–652 (1998).

135. U. Feige, V. Mirrokni and J. Vondrák: Maximizing nonmonotone submodular functions, *Proceedings, 48th IEEE Foundations of Computer Science*, pp. 461–471, 2007.

136. U. Feige and R. Izsak: Welfare maximization and the supermodular degree, *Proceedings, ACM ITCS*, pp. 247–256, 2013.

137. Moran Feldman, Rani Izsak: Building a good team: Secretary problems and the supermodular degree, *Proceedings, 28th SODA*, pp. 1651–1670, 2017.

138. M. Feldman and R. Izsak: Constrained monotone function maximization and the supermodular degree, *Proceedings, 18th RANDOM / 17th APPROX*, pp. 160–175, 2014.

139. Moran Feldman, Joseph Naor, Roy Schwartz: A unified continuous greedy algorithm for submodular maximization, *Proceedings, 52nd FOCS*, pp. 570–579, 2011.

140. David E. Ferguson: Fibonaccian searching, *Communications of the ACM*, 3 (12): 648 (1960).

141. Yuval Filmus and Justin Ward: A tight combinatorial algorithm for submodular maximization subject to a matroid constraint, *Proceedings, 53rd FOCS*, pp. 659–668, 2012.

142. M. L. Fisher, G. L. Nemhauser and L. A. Wolsey: An analysis of approximations for maximizing submodular set functions – C II. In *Polyhedral Combinatorics*, volume 8 of Mathematical Programming Study, pages 73–87. North-Holland Publishing Company, 1978.

143. Lisa Fleischer, Kamal Jain, David P. Williamson, An iterative rounding 2-approximation algorithm for the element connectivity problem, *Proceedings, 42nd Annual IEEE Symposium on Foundations of Computer Science*, 2001.

144. Robert W. Floyd: Algorithm 97: Shortest Path, *Communications of the ACM*, 5(6): 345 (1962).

145. L.R. Ford, D.R. Fulkerson: Maximal flow through a network, *Canadian Journal of Mathematics*, 8: 399–404 (1956).

146. L.R. Ford, D.R. Fulkerson: Solving the transportation problem, *Manamental Science*, 3: 24–32 (1956-57).

147. L.R. Ford, D.R. Fulkerson: A simple algorithm for finding maximal network flow and an application to Hitchcock problem, *Canadian Journal of Mathematics*, 9: 210–218 (1957).

148. L.R. Foulds and R.L. Graham: The Steiner problem in Phylogeny is NP-complete, *Advanced Applied Mathematics*, 3: 43–49 (1982).

149. M.L. Fredman, R.E. Tarjan: Fibonacci heaps and their uses in improved network optimization algorithms, *Journal of the Association for Computing Machinery*, 34: 596–615 (1987).

150. S. Fujishige: *Submodular Functions and Optimization*, Annals of Discrete Mathematics, volume 58. (Elsevier Science, 2005).

151. D.R. Fulkerson: An out-of-kilter method for minimal cost flow problems, *Journal of the Society for Industrial and Applied Mathematics*, 9(1):18–27 (1961).

152. Harold N. Gabow, Suzanne Gallagher: Iterated rounding algorithms for the smallest k-edge connected spanning subgraph, *Proceedings, 19th SODA*, pp. 550–559, 2008.

153. Harold N. Gabow, Michel X. Goemans, Evá Tardos, David P. Williamson: Approximating the smallest k-edge connected spanning subgraph by LP-rounding, *Networks*, 53(4): 345–357 (2009).

154. Ling Gai, Hongwei Du, Lidong Wu, Junlei Zhu, Yuehua Bu: Blocking rumor by cut, *J. Comb. Optim.*, 36(2): 392–399 (2018).

155. Rajiv Gandhi, Samir Khuller, Srinivasan Parthasarathy, Aravind Srinivasan: Dependent rounding and its applications to approximation algorithms, *J. ACM*, 53(3): 324–360 (2006).

156. Chuangen Gao, Hai Du, Weili Wu, Hua Wang: Viral marketing of online game by DS decomposition in social networks, *Theor. Comput. Sci.*, 803: 10–21 (2020).

157. Chuangen Gao, Shuyang Gu, Ruiqi Yang, Jiguo Yu, Weili Wu, Dachuan Xu: Interaction-aware influence maximization and iterated sandwich method, *Theor. Comput. Sci.*, 821: 23–33 (2020)

158. Chuangen Gao, Shuyang Gu, Ruiqi Yang, Jiguo Yu, Weili Wu, Dachuan Xu: Interaction-aware influence maximization and iterated sandwich method, *Proceedings, AAIM*, pp. 129–141, 2019.

159. Jiawen Gao, Suogang Gao, Wen Liu, Weili Wu, Ding-Zhu Du, Bo Hou: An approximation algorithm for the k-generalized Steiner forest problem, *Optim. Lett.*, 15(4): 1475–1483 (2021).

160. Xiaofeng Gao, Yaochun Huang, Zhao Zhang and Weili Wu: $(6 + \varepsilon)$-approximation for minimum weight dominating set in unit disk graphs, *Proceedings, COCOON*, pp. 551–557. 2008.

161. Xiaofeng Gao, Wei Wang, Zhao Zhang, Shiwei Zhu, Weili Wu: A PTAS for minimum d-hop connected dominating set in growth-bounded graphs, *Optim. Lett.*, 4(3): 321–333 (2010).

162. Xiaofeng Gao, Weili Wu, Xuefei Zhang, Xianyue Li: A constant-factor approximation for d-hop connected dominating sets in unit disk graph, *Int. J. Sens. Networks*, 12(3): 125–136 (2012).

163. M.R. Garey, R.L. Graham and D.S. Johnson, The complexity of computing Steiner minimal trees, *SIAM J. Appl. Math.*, 32: 835–859 (1977).

164. M.R. Garey and D.S. Johnson: The complexity of near-optimal graph coloring, *J. Assoc. Comput. Mach.*, 23: 43–49 (1976).

165. M.R. Garey and D.S. Johnson, The rectilinear Steiner tree is NP-complete, *SIAM J. Appl. Math.*, 32: 826–834 (1977).

166. M.R. Garey and D.S. Johnson: *Computers and Intractability: A Guide to the Theory of NP-Completeness*, (W. H. Freeman and Company, New York, 1979).

167. N. Garg, J. Köemann: Faster and simpler algorithms for multicommodity flows and other fractional packing problems, *Proceedings, 39th Annual Symposium on the Foundations of Computer Science*, pp. 300–309, 1998.

168. N. Garg, G. Konjevod, R. Ravi, A polylogarithmic approximation algorithm for the group Steiner tree problem, *Proceedings, 9th SODA*, vol. 95, p. 253, 1998.

169. Dongdong Ge, Yinyu Ye, Jiawei Zhang: The fixed-hub single allocation problem: a geometric rounding approach, working paper, 2007.

170. Dongdong Ge, Simai Hey, Zizhuo Wang, Yinyu Ye, Shuzhong Zhang: Geometric rounding: a dependent rounding scheme for allocation problems, working paper, 2008.

171. A.M.H. Gerards: A short proof of Tutte's characterization of totally unimodular matrices, *Linear Algebra and Its Applications*, 114/115: 207–212 (1989).

172. E.N. Gilbert and H.O. Pollak: Steiner minimal trees, *SIAM J. Appl. Math.*, 16: 1–29 (1968).

173. M. X. Goemans, A. Goldberg, S. Plotkin, D. Shmoys, E. Tardos and D. P. Williamson: Approximation algorithms for network design problems, *Proceedings, 5th SODA*, pp. 223–232, 1994.

174. M.X. Goemans and D.P. Williamson: New $\frac{3}{4}$-approximation algorithms for the maximum satisfiability problem, *SIAM Journal on Discrete Mathematics*, 7: 656–666 (1994).

175. A.V. Goldberg, S. Rao: Beyond the flow decomposition barrier, *Journal of the ACM*, 45(5): 783 (1998).

176. Andrew V. Goldberg, Robert E. Tarjan: Finding minimum-cost circulations by canceling negative cycles, *Journal of the ACM*, 36 (4): 873–886 (1989).

177. Andrew V. Goldberg, Robert E. Tarjan: Finding minimum-cost circulations by successive approximation. *Math. Oper. Res.*, 15(3): 430–466 (1990).

178. A.V. Goldberg, R.E. Tarjan: A new approach to the maximum-flow problem, *Journal of the ACM*, 35(4): 921 (1988).

179. C. C. Gonzaga: Polynomial affine algorithms for linear programming, *Mathematical Programming*, 49: 7–21 (1990).

180. C. Gonzaga: An algorithm for solving linear programming problems in $O(n3L)$ operations, in: N. Megiddo, ed., *Progress in Mathematical Programming: Interior-Point and Related Methods*, pp. 1–28, Springer, New York, 1988.

181. C. Gonzaga: Conical projection algorithms for linear programming, *Mathematical Programming*, 43: 151–173 (1989).

182. T. Gonzalez, S.Q. Zheng: Bounds for partitioning rectilinear polygons, *Proc. 1st Symp. on Computational Geometry*, pp. 281–287, 1985.

183. T. Gonzalez, S.Q. Zheng: Improved bounds for rectangular and guillotine partitions, *Journal of Symbolic Computation* 7: 591–610 (1989).

184. R.L. Graham: Bounds on multiprocessing timing anomalies, *Bell System Tech. J.*, 45: 1563–1581 (1966).

185. R. L. Graham, Pavol Hell: On the history of the minimum spanning tree problem, *Annals of the History of Computing*, 7(1): 43–57 (1985).

186. R.L. Graham and F.K. Hwang: Remarks on Steiner minimal trees, *Bull. Inst. Math. Acad. Sinica*, 4: 177–182 (1976).

187. M. Grötschel, L. Lovász and A. Schrijver: *Geometric Algorithms and Combinatorial Optimization (2nd edition)*, (Springer-Verlag, 1988).

188. Shuyang Gu, Chuangen Gao, Ruiqi Yang, Weili Wu, Hua Wang, Dachuan Xu: A general method of active friending in different diffusion models in social networks, *Soc. Netw. Anal. Min.*, 10(1): 41 (2020).

189. Shuyang Gu, Ganquan Shi, Weili Wu, Changhong Lu: A fast double greedy algorithm for non-monotone DR-submodular function maximization, *Discret. Math. Algorithms Appl.*, 12(1): 2050007:1-2050007:11 (2020).

190. F. Guerriero, R. Musmanno: Label correcting methods to solve multicriteria shortest path problems, *Journal of Optimization Theory and Applications*, 111(3): 589–613 (2001).

191. S. Guha, S. Khuller: Approximation algorithms for connected dominating sets, *Algorithmca*, 20(4): 374–387 (1998).

192. Leonidas J. Guibas, Jorge Stolfi: On computing all north-east nearest neighbors in the L_1 metric, *Inf. Process. Lett.*, 17(4): 219–223 (1983).

193. Jianxiong Guo, Weili Wu: Adaptive influence maximization: If influential node unwilling to be the seed, *ACM Trans. Knowl. Discov. Data*, 15(5): 84:1-84:23 (2021).

194. Jianxiong Guo, Weili Wu: Influence maximization: Seeding based on community structure, *ACM Trans. Knowl. Discov. Data*, 14(6): 66:1-66:22 (2020)

195. Jianxiong Guo, Weili Wu: Continuous profit maximization: A study of unconstrained Dr-submodular maximization, *IEEE Trans. Comput. Soc. Syst.*, 8(3): 768–779 (2021).

196. Jianxiong Guo, Tiantian Chen, Weili Wu: A multi-feature diffusion model: rumor blocking in social networks, *IEEE/ACM Trans. Netw.*, 29(1): 386–397 (2021).

197. Jianxiong Guo, Yi Li, Weili Wu: Targeted protection maximization in social networks. *IEEE Trans. Netw. Sci. Eng.*, 7(3): 1645–1655 (2020).

198. Jianxiong Guo, Weili Wu: Discount advertisement in social platform: algorithm and robust analysis, *Soc. Netw. Anal. Min.*, 10(1): 57 (2020).

199. Jianxiong Guo, Weili Wu: Viral marketing with complementary products, in *Nonlinear Combinatorial Optimization* (edited by Du, Pardalos, Zhang), Springer, pp. 309–315, 2019.

200. Jianxiong Guo, Weili Wu: A novel scene of viral marketing for complementary products, *IEEE Trans. Comput. Soc. Syst.*, 6(4): 797–808 (2019).

201. Ling Guo, Deying Li, Yongcai Wang, Zhao Zhang, Guangmo Tong, Weili Wu, Ding-Zhu Du: Maximisation of the number of β-view covered targets in visual sensor networks, *Int. J. Sens. Networks*, 29(4): 226–241 (2019)

202. D. Gusfield and L. Pitt, A bounded approximation for the minimum cost 2-sat problem, *Algorithmica*, 8: 103–117 (1992).

203. E. Halperin, R. Krauthgamer: Polylogarithmic inapproximability, *Proceedings, 35th ACM Symposium on Theory of Computing*, pp. 585–594, 2003.

204. T.E. Harris, F.S. Ross: Fundamentals of a Method for Evaluating Rail Net Capacities, *Research Memorandum*, 1955.

205. Refael Hassin: The minimum cost flow problem: A unifying approach to existing algorithms and a new tree search algorithm, *Mathematical Programming*, 25: 228–239 (1983).

206. J. Hastad: Clique is hard to approximate within n to the power $1-\varepsilon$, *Acta Math.*, 182: 105–142 (1999).

207. J. Hastad: Some optimal inapproximability results, *J. Assoc. Comput. Mach.*, 48: 798–859 (2001).

208. D. Hausmann, B. Korte, T.A. Jenkyns: Worst case analysis of greedy type algorithms for independence systems, *Mathematical Programming Study*, 12: 120–131 (1980).

209. M.T. Heideman, D. H. Johnson and C. S. Burrus: Gauss and the history of the fast Fourier transform, *IEEE ASSP Magazine*, 1(4): 14–21 (1984).

210. "Sir Antony Hoare". Computer History Museum. Archived from the original on 3 April 2015. Retrieved 22 April 2015.

211. C. A. R. Hoare: Algorithm 64: Quicksort, *Comm. ACM.*, 4(7): 321 (1961).

212. D.S. Hochbaum: Approximating covering and packing problems: set cover, vertex cover, independent set, and related problems, in D.S. Hochbaum (ed.) *Approximation Algorithms for NP-Hard Problems*, PWS Publishing Company, Boston, pp. 94–143, 1997.

213. D.S. Hochbaum and W. Maass, Approximation schemes for covering and packing problems in image processing and VLSI, *J.ACM*, 32: 130–136 (1985).

214. A.J. Hoffman: Some recent applications of the theory of linear inequalities to extremal combinatorial analysis, in *Combinatorial Analysis* (Yew York, 1958; R. Bellman, M. Hall, Jr, eds.), American Mathematical Society, Providence, Rhode Islands, pp. 113–127, 1960.

215. J.E. Hopcroft, R.M. Karp: An n5/2 algorithm for maximum matchings in bipartite graphs, *SIAM Journal on Computing*, 2 (4): 225–231 (1973).

216. Chenfei Hou, Suogang Gao, Wen Liu, Weili Wu, Ding-Zhu Du, Bo Hou: An approximation algorithm for the submodular multicut problem in trees with linear penalties, *Optim. Lett.*, 15(4): 1105–1112 (2021).

217. S.Y. Hsieh and S.-C. Yang: Approximating the selected-internal Steiner tree, *Theoretical Computer Science*, 38: 288–291 (2007).

218. Luogen Hua: *Exploratory of Optimal Selection*, (Science Publisher, 1971).

219. Yaochun Huang, Xiaofeng Gao, Zhao Zhang, Weili Wu: A better constant-factor approximation for weighted dominating set in unit disk graph, *J. Comb. Optim.*, 18(2): 179–194 (2009).

220. H.B. Hunt III, M.V. Marathe, V. Radhakrishnan, S.S. Ravi, D.J. Rosenkrantz, and R.E. Stearns: Efficient approximations and approximation schemes for geometric problems, *Journal of Algorithms*, 26(2): 238–274 (1998).

221. F.K. Hwang, On Steiner minimal trees with rectilinear distance, *SIAM J. Appl. Math.*, 30: 104–114 (1972).

222. F.K. Hwang, An $O(n \log n)$ algorithm for rectilinear minimal spanning trees, *J. ACM*, 26: 177–182 (1979).

223. O.H. Ibarra and C.E. Kim: Fast approximation algorithms for the knapsack and sum of subset proble, *J. Assoc. Comput. Mach.*, 22: 463–468 (1975).

224. R. Iyer and J. Bilmes: Algorithms for approximate minimization of the difference between submodular functions, *Proceedings, 28th UAI*, pp. 407–417, 2012.

225. R. Iyer and J. Bilmes: Submodular optimization subject to submodular cover and submodular knapsack constraints, *Proceedings, Advances of NIPS*, 2013.

226. Rishabh K. Iyer, Stefanie Jegelka, Jeff A. Bilmes: Fast Semidifferential-based Submodular Function Optimization, *Proceedings, ICML*, (3): 855–863 (2013).

227. K. Jain: A factor 2 approximation algorithm for the generalized Steiner network problem, *Combinatorica*, 21: 39–60 (2001).

228. Thomas A Jenkyns: The efficacy of the "greedy" algorithm, *Congressus Numerantium*, no 17: 341–350 (1976).

229. T. Jiang and L. Wang, An approximation scheme for some Steiner tree problems in the plane, *Lecture Notes in Computer Science*, Vol 834: 414–427 (1994).

230. T. Jiang, E.B. Lawler and L. Wang: Aligning sequences via an evolutionary tree: complexity and algorithms, *Proceedings, 26th STOC*, 1994.

231. D.S. Johnson: Approximation algorithms for combinatorial problems, *Journal of Computer and System Sciences*, 9: 256–278 (1974).

232. R. Jonker, A. Volgenant: A shortest augmenting path algorithm for dense and sparse linear assignment problems, *Computing*, 38(4): 325–340 (1987).

233. L.V. Kantorovich: A new method of solving some classes of extremal problems, *Doklady Akad Sci SSSR*, 28: 211–214 (1940).

234. A. Karczmarz, J. Lacki: Simple label-correcting algorithms for partially dynamic approximate shortest paths in directed graphs, *Proceedings, Symposium on Simplicity in Algorithms*, Society for Industrial and Applied Mathematics, pp. 106–120, 2020.

235. N. Karmakkar: A new polynomial-time algorithm for linear programming, *Proceedings, 16th Annual ACM Symposium on the Theory of Computing*, pp. 302–311, 1984.
236. R.M. Karp: Reducibility among combinatorial problems, in *Complexity of Computer Computations*, (E.E. Miller and J.W. Thatcher eds.), Plenum Press, New York, pp. 85–103, 1972.
237. R.M. Karp: Probabilistic analysis of partitioning algorithms for the traveling salesman problem in the plane, *Mathematics of Operations Research*, 2(3): 209–224 (1977).
238. R. M. Karp: A characterization of the minimum cycle mean in a digraph, *Discrete Mathematics*, 23(3): 309–311 (1978).
239. L. Kou, G. Markowsky and L. Berman, A fast algorithm for Steiner trees, *Acta Informatics*, 15: 141–145 (1981).
240. J.A. Kelner, Y.T. Lee, L. Orecchia, A. Sidford: An almost-linear-time algorithm for approximate max flow in undirected graphs, and its multicommodity generalizations, *Proceedings, 25th Annual ACM-SIAM Symposium on Discrete Algorithms (SODA)*, pp. 217–226, 2014.
241. L.G. Khachiyan: A polynomial algorithm for linear programming, *Doklad. Akad. Nauk. USSR Sec.*, 244: 1093–1096 (1979).
242. S. Khanna, R. Motwani, M. Sudan and U. Vazirani: On syntactic versus computational views of approximability, *SIAM J. Comput.*, 28: 164–191 (1999).
243. S. Khanna, S. Muthukrishnan and M. Paterson: On approximating rectangle tiling and packing, *Proceedings, 9th ACM-SIAM Symp. on Discrete Algorithms*, pp. 384–393, 1998.
244. J. Kiefer: Sequential minimax search for a maximum, *Proceedings of the American Mathematical Society*, 4(3): 502–506 (1953).
245. Donghyun Kim, Baraki H. Abay, R. N. Uma, Weili Wu, Wei Wang, Alade O. Tokuta: Minimizing data collection latency in wireless sensor network with multiple mobile elements, *Proceedings, INFOCOM*, pp. 504–512, 2012.
246. Donghyun Kim, Xianyue Li, Feng Zou, Zhao Zhang, Weili Wu: Recyclable connected dominating set for large scale dynamic wireless networks, *Proceedings, WASA*, pp. 560–569, 2008.
247. Donghyun Kim, Wei Wang, Ling Ding, Jihwan Lim, Heekuck Oh, Weili Wu: Minimum average routing path clustering problem in multi-hop 2-D underwater sensor networks, *Optim. Lett.*, 4(3): 383–392 (2010).
248. Donghyun Kim, Wei Wang, Deying Li, Joonglyul Lee, Weili Wu, Alade O. Tokuta: A joint optimization of data ferry trajectories and communication powers of ground sensors for long-term environmental monitoring, *J. Comb. Optim.*, 31(4): 1550–1568 (2016).
249. Donghyun Kim, Wei Wang, Nassim Sohaee, Changcun Ma, Weili Wu, Wonjun Lee, Ding-Zhu Du: Minimum data-latency-bound k-sink placement problem in wireless sensor networks, *IEEE/ACM Trans. Netw.*, 19(5): 1344–1353 (2011).
250. Donghyun Kim, Wei Wang, Junggab Son, Weili Wu, Wonjun Lee, Alade O. Tokuta: Maximum lifetime combined barrier-coverage of weak static sensors and strong mobile sensors, *IEEE Trans. Mob. Comput.*, 16(7): 1956–1966 (2017).
251. Donghyun Kim, Wei Wang, Weili Wu, Deying Li, Changcun Ma, Nassim Sohaee, Wonjun Lee, Yuexuan Wang, Ding-Zhu Du: On bounding node-to-sink latency in wireless sensor networks with multiple sinks, *Int. J. Sens. Networks*, 13(1): 13–29 (2013).
252. Donghyun Kim, R. N. Uma, Baraki H. Abay, Weili Wu, Wei Wang, Alade O. Tokuta: Minimum latency multiple data MULE trajectory planning in wireless sensor networks, *IEEE Trans. Mob. Comput.*, 13(4): 838–851 (2014).
253. Donghyun Kim, Zhao Zhang, Xianyue Li, Wei Wang, Weili Wu, Ding-Zhu Du: A better approximation algorithm for computing connected dominating sets in unit ball graphs, *IEEE Trans. Mob. Comput.*, 9(8): 1108–1118 (2010).
254. Robert Kingan, Sandra Kingan: A software system for matroids, *Graphs and Discovery*, DIMACS Series in Discrete Mathematics and Theoretical Computer Science, pp. 287–296, 2005.
255. L.M. Kirousis, E. Kranakis, D. Krizanc and A. Pelc: Power consumption in packer radio networks, *Theoretical Computer Science*, 243: 289–305 (2000).

256. L.V. Klee and G.J. Minty: How good is the simplex algorithm, in O. Shisha (ed.) *Inequalities* 3, (Academic, New York, 1972).
257. Morton Klein: A primal method for minimal cost flows with applications to the assignment and transportation problems, *Management Science*, 14 (3): 205–220 (1967).
258. Donald E. Knuth: *The Art of Computer Programming: Volume 3, Sorting and Searching, second edition*, (Addison-Wesley, 1998).
259. Ker-I Ko: *Computational Complexity of Real Functions and Polynomial Time Approximation*, Ph.D. Thesis, Ohio State University, Columbus, Ohio, 1979.
260. Ker-I Ko: *Computational Complexity of Real Functions*, (Birkhauser Boston, Boston, MA, 1991).
261. M. Kojima, S. Mizuno and A. Yoshise: A primal-dual interior point method for linear programming, in: *Progress in Mathematical Programming: Interior-Point and Related Methods* (N. Megiddo, ed.), pp. 29–48, (Springer, New York, 1988).
262. J. Komolos and M.T. Shing: Probabilistic partitioning algorithms for the rectilinear Steiner tree problem, *Networks*, 15: 413–423 (1985).
263. Bernhard Korte, Dirk Hausmann: An analysis of the greedy heuristic for independence systems, Ann. Discrete Math., 2: 65–74 (1978).
264. B. Korte, J. Vygen: *Combinatorial Optimization*, (Springer, 2002).
265. L. Kou, G. Markowsky and L. Berman: A Fast Algorithm for Steiner Trees, *Acta Informatica*, 15: 141–145 (1981).
266. J.B. Kruskal: On the shortest spanning subtree of a graph and the traveling salesman problem, *Proc. Amer. Math. Sot.*, 7: 48–50 (1956).
267. H.W. Kuhn: The Hungarian method for the assignment problem, *Naval Research Logistics Quarterly*, 2: 83–97 (1955).
268. H.W. Kuhn: Variants of the Hungarian method for assignment problems, *Naval Research Logistics Quarterly*, 3: 253–258 (1956).
269. M.K. Kwan: Graphic Programming Using Odd or Even Points, *Chinese Math.*, 1: 273–277 (1962).
270. Lei Lai, Qiufen Ni, Changhong Lu, Chuanhe Huang, Weili Wu: Monotone submodular maximization over the bounded integer lattice with cardinality constraints, *Discret. Math. Algorithms Appl.*, 11(6): 1950075:1-1950075:14 (2019).
271. T. Lappas, E. Terzi, D. Gunopulos and H. Mannila: Finding effectors in social networks, *Proceedings, 16th ACM SIGKDD Int. Conf. Knowl. Discovery Data Mining (KDD)*, pp. 1059–1068, 2010.
272. Eugene Lawler: *Combinatorial Optimization: Networks and Matroids*, (Dover, 2001).
273. D.T. Lee: Two-dimensional Voronoi diagrams in the L_p metric, *J. ACM*, 27: 604–618 (1980).
274. D.T. Lee and C.K. Wang: Voronoi diagrams in L, (L,) metrics with 2-dimensional storage applications, *SIAM J. Comput.*, 9: 200–211 (1980).
275. Jon Lee: *A First Course in Combinatorial Optimization*, (Cambridge University Press, 2004).
276. J. Lee, V. Mirrokni, V. Nagarajan and M. Sviridenko: Nonmonotone submodular maximization under matroid and knapsack constraints, *Proceedings, 41th ACM Symposium on Theory of Computing*, pp. 323–332, 2009.
277. J.K. Lenstra, D.B. Shmoys and E. Tardos: Approximation algorithms for scheduling unrelated parallel machines, *Mathematical Programming*, 46: 259–271 (1990).
278. Jure Leskovec, Andreas Krause, Carlos Guestrin, Christos Faloutsos, Jeanne VanBriesen, and Natalie Glance: Cost-effective outbreak detection in networks, *Proceedings, 13th ACM SIGKDD international conference on Knowledge discovery and data mining (KDD)*, New York, ACM, pp. 420–429, 2007.
279. C. Levcopoulos: Fast heuristics for minimum length rectangular partitions of polygons, *Proceedings, 2nd Symp. on Computational Geometry*, pp. 100–108, 1986.
280. Anany V. Levitin: *Introduction to the Design and Analysis of Algorithms*, (Addison Wesley, 2002).
281. Deying Li, Hongwei Du, Peng-Jun Wan, Xiaofeng Gao, Zhao Zhang, Weili Wu: Construction of strongly connected dominating sets in asymmetric multihop wireless networks, *Theor. Comput. Sci.*, 410(8-10): 661–669 (2009).

282. Deying Li, Hongwei Du, Peng-Jun Wan, Xiaofeng Gao, Zhao Zhang, Weili Wu: Minimum power strongly connected dominating sets in wireless networks, *Proceedings, ICWN*, pp. 447–451, 2008.

283. Deying Li, Donghyun Kim, Qinghua Zhu, Lin Liu, Weili Wu: Minimum total communication power connected dominating set in wireless networks, *Proceedings, WASA*, pp. 132–141, 2012.

284. Deying Li, Qinghua Zhu, Hongwei Du, Weili Wu, Hong Chen, Wenping Chen: Conflict-free many-to-one data aggregation scheduling in multi-channel multi-hop wireless sensor networks, *Proceedings, ICC*, pp. 1–5, 2011.

285. Guanfeng Li, Hui Ling, Taieb Znati, Weili Wu: A Robust on-Demand Path-Key Establishment Framework via Random Key Predistribution for Wireless Sensor Networks, *EURASIP J. Wirel. Commun. Netw.*, 2006: 091304 (2006).

286. J. Li, Y. Jin, A PTAS for the weighted unit disk cover problem, in Automata, Languages, and Programming, *Proceedings, ICALP*, pp. 898–909, 2015.

287. Xianyue Li, Xiaofeng Gao, Weili Wu: A Better Theoretical Bound to Approximate Connected Dominating Set in Unit Disk Graph. WASA 2008: 162–175.

288. G.-H. Lin and G. Xue: Steiner tree problem with minimum number of Steiner points and bounded edge-length, *Information Processing Letters*, 69: 53–57 (1999).

289. G.-H. Lin and G. Xue: On the terminal Steiner tree problem, *Information Processing Letters*, 84: 103–107 (2002).

290. H. Lin and J. Bilmes: Optimal selection of limited vocabulary speech corpora, In *Interspeech*, 2011.

291. A. Lingas, R.Y. Pinter, R.L. Rivest and A. Shamir: Minimum edge length partitioning of rectilinear polygons, *Proceedings, 20th Allerton Conf. on Comm. Control and Compt.*, pp. 53–63, Illinos, 1982.

292. A. Lingas: Heuristics for minimum edge length rectangular partitions of rectilinear figures, *Proceedings, 6th GI-Conference*, pp. 199–210, Dortmund, Springer-Verlag, 1983.

293. Bin Liu, Xiao Li, Huijuan Wang, Qizhi Fang, Junyu Dong, Weili Wu: Profit Maximization problem with coupons in social networks, *Theor. Comput. Sci.*, 803: 22–35 (2020).

294. Bin Liu, Xiao Li, Huijuan Wang, Qizhi Fang, Junyu Dong, Weili Wu: Profit maximization problem with coupons in social networks, *Proceedings, AAIM*, pp. 49–61, 2018.

295. Bin Liu, Yuxia Yan, Qizhi Fang, Junyu Dong, Weili Wu, Huijuan Wang: Maximizing profit of multiple adoptions in social networks with a martingale approach, *J. Comb. Optim.*, 38(1): 1–20. (2019).

296. Siwen Liu, Hongmin W. Du: Constant-approximation for minimum weight partial sensor cover, *Discret. Math. Algorithms Appl.*, 13(4): 2150047:1-2150047:8 (2021).

297. L. Lovász: On the ratio of optimal integral and fractional covers, *Discrete Mathematics*, vol 13 (1975) 383–390.

298. B. Lu, L. Ruan: Polynomial time approximation scheme for the rectilinear Steiner arborescence problem, *Journal of Combinatorial Optimization*, 4: 357–363 (2000).

299. Wei Lu, Wei Chen, Laks V.S. Lakshmanan: From competition to complementarity: comparative influence diffusion and maximization, *Proceedings, the VLDB Endowsment*, 9(2): 60–71 (2015).

300. Zaixin Lu, Travis Pitchford, Wei Li, Weili Wu: On the maximum directional target coverage problem in wireless sensor networks, *Proceedings, MSN*, pp. 74–79, 2014.

301. Zaixin Lu, Weili Wu, Wei Wayne Li: Target coverage maximisation for directional sensor networks, *Int. J. Sens. Networks*, 24(4): 253–263 (2017).

302. Zaixin Lu, Wei Zhang, Weili Wu, Joonmo Kim, Bin Fu: The complexity of influence maximization problem in the deterministic linear threshold model, *J. Comb. Optim.*, 24(3): 374–378 (2012).

303. Zaixin Lu, Wei Zhang, Weili Wu, Bin Fu, Ding-Zhu Du: Approximation and inapproximation for the influence maximization problem in social networks under deterministic linear threshold model, *ICDCS Workshops*, pp. 160–165, 2011.

304. Zaixin Lu, Zhao Zhang, Weili Wu: Solution of Bharathi-Kempe-Salek conjecture for influence maximization on arborescence, *J. Comb. Optim.*, 33(2): 803–808 (2017).

305. C. Lund, M. Yanakakis: On the hardness of approximating minimization problems, *J. ACM*, 41(5): 960–981 (1994).

306. Chuanwen Luo, Wenping Chen, Deying Li, Yongcai Wang, Hongwei Du, Lidong Wu, Weili Wu: Optimizing flight trajectory of UAV for efficient data collection in wireless sensor networks, *Theor. Comput. Sci.*, 853: 25–42 (2021).

307. Chuanwen Luo, Lidong Wu, Wenping Chen, Yongcai Wang, Deying Li, Weili Wu: Trajectory optimization of UAV for efficient data collection from wireless sensor networks, *Proceedings, AAIM*, pp. 223–235, 2019.

308. Saunders Mac Lane: Some interpretations of abstract linear dependence in terms of projective geometry, *American Journal of Mathematics*, 58 (1): 236–240 (1936).

309. Takanori Maehara, Kazuo Murota: A framework of discrete DC programming by discrete convex analysis, *Math. Program.*, 152(1-2): 435–466 (2015).

310. I. Mandoiu and A. Zelikovsky: A note on the MST heuristic for bounded edge-length Steiner trees with minimum number of Steiner points, *Information Processing Letters*, 75(4): 165–167 (2000).

311. N. Megiddo and M. Shub: Boundary behaviour of interior point algorithms in linear programming, *Research Report RJ 5319*, IBM Thomas J. Watson Research Center (Yorktown Heights, NY, 1986).

312. V. Melkonian and E. Tardos: Algorithms for a network design problem with crossing supermodular demands, *Networks*, 43: 256–265 (2004).

313. S. Micali, V.V. Vazirani: An $O(\sqrt{|V|} \cdot |E|)$ algorithm for finding maximum matching in general graphs, *Proc. 21st IEEE Symp. Foundations of Computer Science*, pp. 17–27 (1980).

314. Manki Min, Hongwei Du, Xiaohua Jia, Christina Xiao Huang, Scott C.-H. Huang, Weili Wu: Improving Construction for Connected Dominating Set with Steiner Tree in Wireless Sensor Networks, *J. Glob. Optim.*, 35(1): 111–119 (2006).

315. M. Min, S.C.-H. Huang, J. Liu, E. Shragowitz, W. Wu, Y. Zhao and Y. Zhao, An approximation scheme for the rectilinear Steiner minimum tree in presence of obstructions, *Novel Approaches to Hard Discrete Optimization*, Fields Institute Communications Series, American Math. Society, vol 37: 155–163 (2003).

316. George J. Minty: On the axiomatic foundations of the theories of directed linear graphs, electrical networks and network-programming, *Journal of Mathematics and Mechanics*, 15: 485–520 (1966).

317. J.S.B. Mitchell: Guillotine subdivisions approximate polygonal subdivisions: A simple new method for the geometric k-MST problem. *Proceedings, 7th ACM-SIAM Symposium on Discrete Algorithms*, pp. 402–408, 1996.

318. J.S.B. Mitchell: Guillotine subdivisions approximate polygonal subdivisions: Part II - A simple polynomial-time approximation scheme for geometric k-MST, TSP, and related problem, *SIAM J. Comput.*, 28: 1298–1307 (1999).

319. J.S.B. Mitchell: Guillotine subdivisions approximate polygonal subdivisions: Part III - Faster polynomial-time approximation scheme for geometric network optimization, *Proceedings, 9th Canadian Conference on Computational Geometry*, pp. 229–232, 1997.

320. J.S.B. Mitchell, A. Blum, P. Chalasani, S. Vempala: A constant-factor approximation algorithm for the geometric k-MST problem in the plane, *SIAM J. Comput.*, 28: 771–781 (1999).

321. R.C. Monteiro and I. Adler: An $O(n^3 L)$ primal-dual interior point algorithm for linear programming, *Manuscript*, Department of Industrial Engineering and Operations Research, University of California (Berkeley, CA, 1987).

322. J. Munkres: Algorithms for the assignment and transportation problems, *Journal of the Society for Industrial and Applied Mathematics*, 5(1): 32–38 (1957).

323. K. Nagano, Y. Kawahara and K. Aihara: Size-constrained submodular minimization through minimum norm base, *Proceedings, 28th International Conference on Machine Learning*, Bellevue, WA, USA, 2011.

324. M. Narasimhan and J. Bilmes: A submodular-supermodular procedure with applications to discriminative structure learning, *Proceedings, UAI*, 2005.
325. George L. Nemhauser and L.E. Trotter: Vertex packings: structural properties and algorithms, *Math. Program.*, 8: 232 (1975).
326. George L. Nemhauser, Laurence A. Wolsey and Marshall L. Fisher: An analysis of approximations for maximizing submodular set functions - I, *Mathematical Programming*, 14(1): 265–294 (1978).
327. Qiufen Ni, Smita Ghosh, Chuanhe Huang, Weili Wu, Rong Jin: Discount allocation for cost minimization in online social networks, *J. Comb. Optim.*, 41(1): 213–233 (2021).
328. F. Nielsen, Fast stabbing of boxes in high dimensions, *Theoret. Comput. Sci.*, 246: 53–72 (2000).
329. Hirokazu Nishimura, Susumu Kuroda (eds.): *A lost mathematician, Takeo Nakasawa. The forgotten father of matroid theory*, (Basel: Birkhäuser Verlag, 2009).
330. J. B. Orlin: A faster strongly polynomial time algorithm for submodular function minimization, *Mathematical Programming*, 118: 237–251 (2009).
331. J.B. Orlin: Max flows in O(nm) time, or better, *Proceedings, 45th annual ACM symposium on Symposium on theory of computing (STOC '13)*, pp. 765–774, 2013.
332. James B. Orlin: A polynomial time primal network simplex algorithm for minimum cost flows, *Mathematical Programming*, 78(2): 109–129 (1997).
333. K. J. Overholt: Efficiency of the Fibonacci search method, *BIT Numerical Mathematics*, 13(1): 92–96 (1973).
334. James Oxley: *Matroid Theory*, (Oxford: Oxford University Press, 1992).
335. Christos H. Papadimitriou, Kenneth Steiglitz: *Combinatorial Optimization : Algorithms and Complexity*, (Dover, July, 1998).
336. C. Papadimitriou and M. Yannakakis: Optimization, approximations, and complexity classes, *Proceedings, 20th ACM Symposium on Theory of Computing*, pp. 229–234, 1988.
337. Panos M. Pardalos, Ding-Zhu Du, Ronald L. Graham (ed.): *Handbook of Combinatorial Optimization*, (Springer, 2013).
338. C-M. Pintea: *Advances in Bio-inspired Computing for Combinatorial Optimization Problem*, Intelligent Systems Reference Library, (Springer, 2014).
339. R. C. Prim: Shortest connecting networks and some generahzattons, *Bell Syst Tech J*, 36: 1389–1401 (1957).
340. Erich Prisner: Two algorithms for the subset interconnection design problem, *Networks*, 22(4): 385–395 (1992).
341. Kirk Pruhs: Speed Scaling, *Encyclopedia of Algorithms*, pp. 2045–2047, 2016.
342. Guoyao Rao, Yongcai Wang, Wenping Chen, Deying Li, Weili Wu: Maximize the probability of union-influenced in social networks, *Proceedings, COCOA*, pp. 288–301, 2021.
343. Guoyao Rao, Yongcai Wang, Wenping Chen, Deying Li, Weili Wu: Matching influence maximization in social networks, *Theor. Comput. Sci.*, 857: 71–86 (2021).
344. S.B. Rao, W.D. Smith: Approximating geometrical graphs via "spanners" and "banyans", *Proceedings, ACM STOC'98*, pp. 540–550, 1998.
345. S K. Rao, P. Sadayappan, F.K. Hwang and P.W. Shor: The rectilinear Steiner arborescence problem, *Algorithmica*, 7(2-3): 277–288 (1992).
346. R. Ravi and J. D. Kececioglu, Approximation methods for sequence alignment under a fixed evolutionary tree, *Proceedings, 6th Symp. on Combinatorial Parrern Matching. Springer LNCS*, 937: 330–339 (1995).
347. R. Raz and S. Safra: A sub-constant error-probability low-degree test, and a subconstant error-probability PCP characterization of NP, *Proceedings, 28th ACM Symposium on Theory of Computing*, pp. 474–484, 1997.
348. András Recski: *Matroid Theory and its Applications in Electric Network Theory and in Statics*, Algorithms and Combinatorics, vol. 6, (Berlin and Budapest: Springer-Verlag and Akademiai Kiado, 1989).
349. J. Renegar: A polynomial-time algorithm based on Newton's method for linear programming, *Mathematical Programming*, 40: 59–94 (1988).

350. G. Robin and A. Zelikovsky, Improved Steiner trees approximation in graphs, *Proceedings, 11th SIAM-ACM Symposium on Discrete Algorithms (SODA)*, San Francisco, CA, pp. 770–779, January 2000.

351. Lu Ruan, Hongwei Du, Xiaohua Jia, Weili Wu, Yingshu Li, Ker-I Ko: A greedy approximation for minimum connected dominating sets, *Theor. Comput. Sci.*, 329(1-3): 325–330 (2004).

352. J.H. Rubinstein and D.A. Thomas, The Steiner ratio conjecture for six points, *J. Combinatoria Theory, Ser.A*, 58: 54–77 (1991).

353. S. Sahni: Approximate algorithms for the 0/1 knapsack problem, *J. Assoc. Comput. Mach.*, 22: 115–124 (1975).

354. S. Sahni and T. Gonzalez: P-complete approximation algorithms, *J. Assoc. Comput. Mach.*, 23: 555–565 (1976).

355. D. Sankoff: Minimal mutation trees of sequences, *SIAM J. Appl. Math.*, 28: 35–42 (1975).

356. P. Schreiber: On the history of the so-called Steiner weber problem, *Wiss. Z. Ernst-Moritz-Arndt-Univ. Greifswald, Math.-nat.wiss. Reihe*, 35(3): (1986).

357. A. Schrijver: *Theory of Linear and Integer Programming*, (Wiley, Chichester, 1986).

358. Alexander Schrijver: *Combinatorial Optimization: Polyhedra and Efficiency*, Algorithms and Combinatorics. 24. (Springer, 2003).

359. A. Schrijver: A combinatorial algorithm minimizing submodular func- tions in strong polynomial time, *J. Combinatorial Theory (B)*, 80: 346–355 (2000).

360. A. Schrijver: On the history of the transportation and maximum flow problems, *Mathematical Programming*, 91(3): 437–445 (2002).

361. A. Schrijver: On the history of the shortest path problem, *Documenta Math*, Extra Volume ISMP: 155–167 (2012).

362. H. H. Seward: "Internal Sorting by Floating Digital Sort", Information sorting in the application of electronic digital computers to business operations (PDF), *Master's thesis, Report R-232*, Massachusetts Institute of Technology, Digital Computer Laboratory, pp. 25–28, 1954.

363. M. I. Shamos and D. Hoey: Closest point problems, *Proceedings, 16th Annual Symp Foundations of Computer Science*, pp 151–162, 1975.

364. Shan Shan, Weili Wu, Wei Wang, Hongjie Du, Xiaofeng Gao, Ailian Jiang: Constructing minimum interference connected dominating set for multi-channel multi-radio multi-hop wireless network, *Int. J. Sens. Networks*, 11(2): 100–108 (2012).

365. J. Sherman: Nearly maximum flows in nearly linear time, *Proceedings, 54th Annual IEEE Symposium on Foundations of Computer Science (FOCS)*, pp. 263–269, 2013.

366. Alfonso Shimbel: Structural parameters of communication networks, *Bulletin of Mathematical Biophysics*, 15(4): 501–507 (1953).

367. Gerard Sierksma, Yori Zwols: *Linear and Integer Optimization: Theory and Practice*, (CRC Press, 2015).

368. Gerard Sierksma, Diptesh Ghosh: *Networks in Action; Text and Computer Exercises in Network Optimization*, (Springer, 2010).

369. A.J. Skriver, K.A. Andersen: A label correcting approach for solving bicriterion shortest-path problems, *Computers & Operations Research*, 27(6): 507–524 (2000).

370. Petr Slavik: A tight analysis of the greedy algorithm for set cover, *Journal of Algorithms*, 25(2): 237–254 (1997).

371. M. Sviridenko: A note on maximizing a submodular set function subject to knapsack constraint, *Operations Research Letters*, 32: 41–43 (2004).

372. Z. Svitkina and L. Fleischer: Submodular approximation: Sampling-based algorithms and lower bounds, *SIAM Journal on Computing*, 40(6): 1715–1737 (2011).

373. E. Tardos: A strongly polynomial minimum cost circulation algorithm, *Combinatorica*, 5(3): 247–255 (1985).

374. J. Tarhio, E. Ukkonen: A greedy approximation algorithm for constructing shortest common superstrings, *Theoretical Computer Science*, 57(1): 131–145 (1988).

375. M. Todd and B. Burrell: An extension of Karmarkar's algorithm for linear programming using dual variables, *Algorithmica*, 1: 409–424 (1986).

376. M.J. Todd and Y. Ye: A centered projective algorithm for linear programming, *Technical Report 763*, School of Operations Research and Industrial Engineering, Cornell University (Ithaca, NY, 1987).

377. N. Tomizawa: On some techniques useful for solution of transportation network problems, *Networks*, 1(2): 173–194 (1971).

378. Guangmo Amo Tong, Ding-Zhu Du, Weili Wu: On misinformation containment in online social networks, *Proceedings, NeurIPS*, pp. 339–349, 2018.

379. Guangmo Amo Tong, Shasha Li, Weili Wu, Ding-Zhu Du: Effector detection in social networks, *IEEE Trans. Comput. Soc. Syst.*, 3(4): 151–163 (2016).

380. Guangmo Tong, Ruiqi Wang, Xiang Li, Weili Wu, Ding-Zhu Du: An approximation algorithm for active friending in online social networks, *Proceedings, ICDCS*, pp. 1264–1274, 2019

381. Guangmo Amo Tong, Weili Wu, Ding-Zhu Du: Distributed Rumor Blocking in Social Networks: A Game Theoretical Analysis, *IEEE Transactions on Computational Social Systems*, 5(2): 468–480 (2018).

382. Guangmo Amo Tong, Weili Wu, Ling Guo, Deying Li, Cong Liu, Bin Liu, Ding-Zhu Du: An efficient randomized algorithm for rumor blocking in online social networks, *Proceedings, INFOCOM*, pp. 1–9, 2017.

383. J.S. Turner: Approximation algorithms for the shortest common superstring problem, *Information and Computation*, 83(1): 1–20 (1989).

384. W.T. Tutte: *Introduction to the theory of matroids*, Modern Analytic and Computational Methods in Science and Mathematics, vol. 37, (New York: American Elsevier Publishing Company, 1971).

385. Pravin M. Vaidya: An algorithm for linear programming which requires $O(((m+n)n^2 + (m+n)^{1.5}n)L)$ arithmetic operations, *Mathematical Programming*, 47: 175–201 (1990).

386. S.A. Vavasis: Automatic domain partitioning in tree dimensions, *SIAM J. Sci. Stat. Comput.*, 12(4): 950–970 (1991).

387. Vijay V. Vazirani: *Approximation Algorithms*, (Berlin: Springer, 2003).

388. Jan Vondrák: Optimal approximation for the submodular welfare problem in the value oracle model, *Proceedings, STOC*, pp. 67–74, 2008.

389. Peng-Jun Wan, Ding-Zhu Du, Panos M. Pardalos, Weili Wu: Greedy approximations for minimum submodular cover with submodular cost. *Comp. Opt. and Appl.*, 45(2): 463–474 (2010).

390. Chen Wang, My T. Thai, Yingshu Li, Feng Wang, Weili Wu: Minimum coverage breach and maximum network lifetime in wireless sensor networks, *Proceedings, GLOBECOM*, pp. 1118–1123, 2007.

391. Chen Wang, My T. Thai, Yingshu Li, Feng Wang, Weili Wu: Optimization scheme for sensor coverage scheduling with bandwidth constraints, *Optim. Lett.*, 3(1): 63–75 (2009)

392. Ailian Wang, Weili Wu, Junjie Chen: Social network rumors spread model based on cellular automata, *Proceedings, MSN*, pp. 236–242, 2014.

393. Ailian Wang, Weili Wu, Lei Cui: On Bharathi-Kempe-Salek conjecture for influence maximization on arborescence, *J. Comb. Optim.*, 31(4): 1678–1684 (2016).

394. L. Wang and D.-Z. Du: Approximations for bottleneck Steiner trees, *Algorithmica*, 32: 554–561 (2002).

395. L. Wang and D. Gusfield: Improved approximation algorithms for tree alignment, *Proceedings, 7th Symp. on Combinatorial Parrern Matching. Springer LNCS*, 1075: 220–233 (1996).

396. L. Wang and T. Jiang: An approximation scheme for some Steiner tree problems in the plane, *Networks*, 28: 187–193 (1996).

397. L. Wang, T. Jiang and D. Gusfield: A more efficient approximation scheme for tree alignment, *Proceedings, 1st annual international conference on computational biology*, pp. 310–319, 1997.

398. L. Wang, T. Jiang and E.L. Lawler: Approximation algorithms for tree alignment with a given phylogeny, *Algorithmica*, 16: 302–315 (1996).

399. Wei Wang, Donghyun Kim, Nassim Sohaee, Changcun Ma, Weili Wu: A PTAS for minimum d-hop underwater sink placement problem in 2-d underwater sensor networks, *Discret. Math. Algorithms Appl.*, 1(2): 283–290 (2009).

400. Wei Wang, Donghyun Kim, James Willson, Bhavani M. Thuraisingham, Weili Wu: A better approximation for minimum average routing path clustering problem in 2-*d* underwater sensor networks, *Discret. Math. Algorithms Appl.*, 1(2): 175–192 (2009).

401. Zhefeng Wang, Yu Yang, Jian Pei and Enhong Chen, Activity maximization by effective information diffusion in social networks, *IEEE Transactions on Knowledge and Data Engineering*, 29(11): 2374–2387 (2017).

402. Stephen Warshall: A theorem on Boolean matrices, *Journal of the ACM*, 9(1): 11–12 (1962).

403. D.J.A. Welsh: *Matroid Theory*, L.M.S. Monographs, vol. 8, (Academic Press, 1976).

404. Neil White (ed.): *Theory of Matroids*, Encyclopedia of Mathematics and its Applications, vol. 26, (Cambridge: Cambridge University Press, 1986).

405. Neil White (ed.): *Combinatorial geometries*, Encyclopedia of Mathematics and its Applications, vol. 29, (Cambridge: Cambridge University Press, 1987).

406. Hassler Whitney: On the abstract properties of linear dependence, *American Journal of Mathematics*, 57(3): 509–533 (1935).

407. Chr. Wiener, Ueber eine Aufgabe aus der Geometria situs, *Mathematik Annalen*, 6: 29–30 (1873).

408. David P. Williamson, David B. Shmoys: *The Design of Approximation Algorithms*, (Cambridge University Press, 2011).

409. James Willson, Weili Wu, Lidong Wu, Ling Ding, Ding-Zhu Du: New approximation for maximum lifetime coverage, *Optimization*, 63(6): 839–847 (2014).

410. James Willson, Zhao Zhang, Weili Wu, Ding-Zhu Du: Fault-tolerant coverage with maximum lifetime in wireless sensor networks, *Proceedings, INFOCOM*, pp. 1364–1372, 2015.

411. Laurence A. Wolsey: Heuristic analysis, linear programming and branch and bound, *Mathematical Programming Study* 13: 121–134 (1980).

412. Laurence A. Wolsey: Maximizing real-valued submodular function: primal and dual heuristics for location problems, *Math. of Operations Research* 7: 410–425 (1982).

413. Laurence A. Wolsey: An analysis of the greedy algorithm for the submodular set covering problem, *Combinatorica*, 2(4): 385–393 (1982).

414. Baoyuan Wu, Siwei Lyu, Bernard Ghanem: Constrained submodular minimization for missing labels and class imbalance in multi-label learning, *Proceedings, AAAI*, pp. 2229–2236, 2016.

415. Chenchen Wu, Yishui Wang, Zaixin Lu, P.M. Pardalos, Dachuan Xu, Zhao Zhang, Ding-Zhu Du: Solving the degree-concentrated fault-tolerant spanning subgraph problem by DC programming, *Math. Program.*, 169(1): 255–275 (2018).

416. Lidong Wu, Hongwei Du, Weili Wu, Deying Li, Jing Lv, Wonjun Lee: Approximations for minimum connected sensor cover, *Proceedings, INFOCOM*, pp. 1187–1194, 2013.

417. Lidong Wu, Hongwei Du, Weili Wu, Yuqing Zhu, Ailian Wang, Wonjun Lee: PTAS for routing-cost constrained minimum connected dominating set in growth bounded graphs, *J. Comb. Optim.*, 30(1): 18–26 (2015).

418. Lidong Wu, Huijuan Wang, Weili Wu: Connected set-cover and group Steiner tree, *Encyclopedia of Algorithms*, pp. 430–432, 2016.

419. Weili Wu, Xiuzhen Cheng, Min Ding, Kai Xing, Fang Liu, Ping Deng: Localized outlying and boundary data detection in sensor networks, *IEEE Trans. Knowl. Data Eng.*, 19(8): 1145–1157 (2007).

420. Weili Wu, Hongwei Du, Xiaohua Jia, Yingshu Li, Scott C.-H. Huang: Minimum connected dominating sets and maximal independent sets in unit disk graphs, *Theor. Comput. Sci.*, 352(1-3): 1–7 (2006).

421. Weili Wu, Zhao Zhang, Chuangen Gao, Hai Du, Hua Wang, Ding-Zhu Du: Quality of barrier cover with wireless sensors, *Int. J. Sens. Networks*, 29(4): 242–251 (2019).

422. Weili Wu, Zha Zhang, Wonjun Lee, Ding-Zhu Du: *Optimal Coverage in Wireless Sensor Networks*, (Springer, 2020).

423. Biaofei Xu, Yuqing Zhu, Deying Li, Donghyun Kim, Weili Wu: Minimum (k, ω)-angle barrier coverage in wireless camera sensor networks, *Int. J. Sens. Networks*, 21(3): 179–188 (2016).

424. Wen Xu, Zaixin Lu, Weili Wu, Zhiming Chen: A novel approach to online social influence maximization, *Soc. Netw. Anal. Min.*, 4(1): 153 (2014).

425. Wen Xu, Weili Wu: *Optimal Social Influence*, (Springer, 2020).

426. Ruidong Yan, Deying Li, Weili Wu, Ding-Zhu Du, Yongcai Wang: Minimizing influence of rumors by blockers on social networks: algorithms and analysis, *IEEE Trans. Netw. Sci. Eng.*, 7(3): 1067–1078 (2020).

427. D-N Yang, H-J Hung, W-C Lee, W Chen: Maximizing acceptance probability for active friending in online social networks, *Proceedings, 19th ACM SIGKDD international conference on Knowledge discovery and data mining*, pp. 713–721, 2013.

428. Ruiqi Yang, Shuyang Gu, Chuangen Gao, Weili Wu, Hua Wang, Dachuan Xu: A constrained two-stage submodular maximization, *Theor. Comput. Sci.*, 853: 57–64 (2021).

429. Ruidong Yan, Yi Li, Weili Wu, Deying Li, Yongcai Wang: Rumor blocking through online link deletion on social networks, *ACM Trans. Knowl. Discov. Data*, 13(2): 16:1-16:26 (2019).

430. Wenguo Yang, Jianmin Ma, Yi Li, Ruidong Yan, Jing Yuan, Weili Wu, Deying Li: Marginal gains to maximize content spread in social networks, *IEEE Trans. Comput. Soc. Syst.*, 6(3): 479–490 (2019).

431. Wenguo Yang, Jing Yuan, Weili Wu, Jianmin Ma, Ding-Zhu Du: Maximizing Activity Profit in Social Networks, *IEEE Trans. Comput. Soc. Syst.*, 6(1): 117–126 (2019).

432. M. Yannakakis: On the approximation of maximum satisfiability, *Journal of Algorithms*, 3: 475–502 (1994).

433. A.C. Yao: On constructing minimum spanning trees in k-dimensional spaces and related problems, *SIAM J. Comput.*, 11: 721–736 (1982).

434. F.F. Yao: Efficient dynamic programming using quadrangle inequalities, *Proceedings, 12th Ann. ACM Symp. on Theory of Computing*, pp. 429–435, 1980.

435. Jing Yuan, Weili Wu, Yi Li, Ding-Zhu Du: Active friending in online social networks, *Proceedings, BDCAT*, pp. 139–148, 2017.

436. Jing Yuan, Weili Wu, Wen Xu: Approximation for influence maximization, *Handbook of Approximation Algorithms and Metaheuristics*, (2) 2018.

437. A. Zelikovsky, The 11/6-approximation algorithm for the Steiner problem on networks, *Algorithmica*, 9: 463–470 (1993).

438. A. Zelikovsky, A series of approximation algorithms for the acyclic airected Steiner tree Problem, *Algorithmica*, 18: 99–110 (1997).

439. F.B. Zhan, C.E. Noon: A comparison between label-setting and label-correcting algorithms for computing one-to-one shortest paths, *Journal of Geographic information and decision analysis*, 4(2): 1–11 (2000).

440. Jianzhong Zhang, Shaoji Xu: *Linear Programming*, (Schiece Press, 1987).

441. Ning Zhang, Incheol Shin, Feng Zou, Weili Wu, My T. Thai: Trade-off scheme for fault tolerant connected dominating sets on size and diameter, *Proceedings, FOWANC*, pp. 1–8, 2008.

442. Wei Zhang, Weili Wu, Wonjun Lee, Ding-Zhu Du: Complexity and approximation of the connected set-cover problem, *J. Glob. Optim.*, 53(3): 563–572 (2012).

443. Yapu Zhang, Jianxiong Guo, Wenguo Yang, Weili Wu: Targeted Activation Probability Maximization Problem in Online Social Networks, *IEEE Trans. Netw. Sci. Eng.*, 8(1): 294–304 (2021).

444. Yapu Zhang, Jianxiong Guo, Wenguo Yang, Weili Wu: Mixed-case community detection problem in social networks: Algorithms and analysis, *Theor. Comput. Sci.*, 854: 94–104 (2021)

445. Yapu Zhang, Wenguo Yang, Weili Wu, Yi Li: Effector detection problem in social networks, *IEEE Trans. Comput. Soc. Syst.*, 7(5): 1200–1209 (2020).

446. Zhao Zhang, Xiaofeng Gao, Weili Wu: Algorithms for connected set cover problem and fault-tolerant connected set cover problem, *Theor. Comput. Sci.*, 410(8-10): 812–817 (2009).

447. Zhao Zhang, Xiaofeng Gao, Weili Wu, Ding-Zhu Du: PTAS for minimum connected dominating set in unit ball graph, *Proceedings, WASA*, pp. 154–161, 2008.

448. Zhao Zhang, Xiaofeng Gao, Weili Wu, Ding-Zhu Du: A PTAS for minimum connected dominating set in 3-dimensional Wireless sensor networks. *J. Glob. Optim.*, 45(3): 451–458 (2009).

449. Zhao Zhang, Xiaofeng Gao, Xuefei Zhang, Weili Wu, Hui Xiong: Three approximation algorithms for energy-efficient query dissemination in sensor database system, *Proceedings, DEXA*, pp. 807–821, 2009.

450. Zhao Zhang, Joonglyul Lee, Weili Wu, Ding-Zhu Du: Approximation for minimum strongly connected dominating and absorbing set with routing-cost constraint in disk digraphs, *Optim. Lett.*, 10(7): 1393–1401 (2016).

451. Zhao Zhang, James Willson, Zaixin Lu, Weili Wu, Xuding Zhu, Ding-Zhu Du: Approximating maximum lifetime k-coverage through minimizing weighted k-cover in homogeneous wireless sensor networks, *IEEE/ACM Trans. Netw.*, 24(6): 3620–3633 (2016).

452. Zhao Zhang, Weili Wu, Jing Yuan, Ding-Zhu Du: Breach-free sleep-wakeup scheduling for barrier coverage with heterogeneous wireless sensors, *IEEE/ACM Trans. Netw.*, 26(5): 2404–2413 (2018).

453. Zhao Zhang, Weili Wu, Lidong Wu, Yanjie Li, Zongqing Chen: Strongly connected dominating and absorbing set in directed disk graph, *Int. J. Sens. Networks*, 19(2): 69–77 (2015).

454. Jiao Zhou, Zhao Zhang, Weili Wu, Kai Xing: A greedy algorithm for the fault-tolerant connected dominating set in a general graph, *J. Comb. Optim.*, 28(1): 310–319 (2014).

455. Jianming Zhu, Smita Ghosh, Weili Wu: Robust rumor blocking problem with uncertain rumor sources in social networks, *World Wide Web*, 24(1): 229–247 (2021)

456. Jianming Zhu, Smita Ghosh, Weili Wu: Group influence maximization problem in social networks, *IEEE Trans. Comput. Soc. Syst.*, 6(6): 1156–1164 (2019).

457. Jianming Zhu, Junlei Zhu, Smita Ghosh, Weili Wu and Jing Yuan: Social influence maximization in hypergraph in social networks, *IEEE Transactions on Network Science and Engineering*, 6(4): 801–811 (2019).

458. Yuqing Zhu, Deying Li, Ruidong Yan, Weili Wu, Yuanjun Bi: Maximizing the influence and profit in social networks, *IEEE Trans. Comput. Soc. Syst.*, 4(3): 54–64 (2017).

459. Yuqing Zhu, Weili Wu, Yuanjun Bi, Lidong Wu, Yiwei Jiang, Wen Xu: Better approximation algorithms for influence maximization in online social networks, *J. Comb. Optim.*, 30(1): 97–108 (2015).

460. Yuqing Zhu, Zaixin Lu, Yuanjun Bi, Weili Wu, Yiwei Jiang, Deying Li: Influence and profit: Two sides of the coin, *Proceedings, ICDM*, pp. 1301–1306, 2013.

461. Feng Zou, Xianyue Li, Donghyun Kim, Weili Wu: Construction of minimum cnnected dominating set in 3-dimensional wireless network, *Proceedings, WASA*, pp. 134–140, 2008.

462. Feng Zou, Xianyue Li, Donghyun Kim and Weil Wu: Two constant approximation algorithms for node-weighted Steiner tree in unit disk graphs, *Proceedings, COCOA*, pp. 278–285, 2008.

463. Feng Zou, Yuexuan Wang, XiaoHua Xu, Xianyue Li, Hongwei Du, Peng-Jun Wan, Weili Wu: New approximations for minimum-weighted dominating sets and minimum-weighted connected dominating sets on unit disk graphs, *Theor. Comput. Sci.*, 412(3): 198–208 (2011).

464. D. Zuckerman: Linear degree extractors and the inapproximability of max clique and chromatic number, *Proceedings, 38th ACM Symposium on Theory of Computing*, pp. 681–690, 2006.

465. D. Zuckerman: Linear degree extractors and the inapproximability of Max Clique and Chromatic Number, *Theory Comput.*, 3: 103–128 (2007).

Printed in the United States
by Baker & Taylor Publisher Services